SELENIUM RESEARCH FOR ENVIRONMENT AND HUMAN HEALTH: PERSPECTIVES, TECHNOLOGIES AND ADVANCEMENTS

PROCEEDINGS OF THE 6TH INTERNATIONAL CONFERENCE ON SELENIUM IN THE ENVIRONMENT AND HUMAN HEALTH, YANGLING/XI'AN, SHAANXI, CHINA, 27–30 OCTOBER 2019

Selenium Research for Environment and Human Health: Perspectives, Technologies and Advancements

Editors

Gary Bañuelos
USDA-ARS, Water Management Research Unit, USA

Zhi-Qing Lin
Department of Environmental Sciences & Department of Biological Sciences, Southern Illinois University – Edwardsville, USA

Dongli Liang
Department of Environmental Science and Engineering, College of Natural Resources and Environment, Northwest A&F University, China

Xue-bin Yin
Department of Environmental Sciences, Advanced Lab for Functional Agriculture, University of Science and Technology of China, China

CRC Press is an imprint of the
Taylor & Francis Group, an **informa** business
A BALKEMA BOOK

CRC Press/Balkema is an imprint of the Taylor & Francis Group, an informa business

Typeset by MPS Limited, Chennai, India

Library of Congress Cataloging-in-Publication Data

Applied for

Published by: CRC Press/Balkema
Schipholweg 107C, 2316 XC Leiden, The Netherlands

First issued in paperback 2023

ISBN: 978-1-03-257061-7 (pbk)
ISBN: 978-1-138-39014-0 (hbk)
ISBN: 978-0-429-42348-2 (ebk)

DOI: https://doi.org/10.1201/9780429423482

Publisher's Note
The publisher has gone to great lengths to ensure the quality of this reprint but points out that some imperfections in the original copies may be apparent.

Selenium Research for Environment and Human Health: Perspectives, Technologies and Advancements – Bañuelos, Lin, Liang & Yin (eds)
© 2020 Taylor and Francis Group, London, ISBN 978-1-138-39014-0

Table of Contents

Uptake and accumulation of Se

Biofortification

Effects of Se on plant, animal, and fish health

Interactions with heavy metals

Selenoproteins

Selenium's effect on epidemiology, health, injury, and disease

Selenium Research for Environment and Human Health: Perspectives, Technologies and Advancements – Bañuelos, Lin, Liang & Yin (eds)
© 2020 Taylor and Francis Group, London, ISBN 978-1-138-39014-0

Preface

The selenium (Se) content in crops and human daily dietary intake has become a topic of great interest in public health around the world, since Se is an essential and impactful trace element for animals and humans. There are estimations that more than a billion people are affected to some degree by Se deficiency, which compromises human health including the immune system, thyroid, metabolism, and spermatogenesis. Selenium containing amino acids such as selenocysteine and selenomethionine form various selenoproteins that have distinctive and essential functions for normal biochemical and physiological processes. Worldwide, plant-based products are the major source of Se for human dietary intake, and consequently Se deficiencies can arise if dietary Se supply and intake is abnormal. Most nutritional Se is provided in the form of two amino acids – selenocysteine or selenomethionine – and their derivatives, mainly in the form of ingested proteins, as well as low molecular weight selenocompounds.

Among many environmental and biological factors, plant Se concentrations will, however, primarily depend upon soil Se content. The metalloid Se is ubiquitous in the environment with total concentrations in most soils ranging from 0.01 to 2.0 mg/kg; however, high concentrations can naturally occur in seleniferous areas. Soil Se concentrations and Se bioavailability will, however, vary with environmental and soil physio-chemical conditions and its distribution in soils is usually heterogeneous or site specific. Sources of naturally occurring Se can be of geogenic or anthropogenic origin or result from volcanic emission and atmospheric deposition. Its bioavailability is controlled by many chemical and biochemical processes within the soil profile that will also determine the speciation of Se. The speciation of Se in the soil is one of the most important factors that determine Se bioavailability and accumulation in crops and feed materials. Although Se is not essential for higher plants, plants can readily take up Se due to its similarity to sulfur (S). Its uptake is largely determined by the phyto-availability of Se in the soil, as well as significant influences on its accumulation promoted via soil microbial communities, conventional breeding, and molecular and genetic engineering techniques. Indeed, better understanding these processes could be helpful for improving the new agronomy-based strategy "biofortification" that can result in the production of Se-enriched food and feed products. Food-chain-based approaches like Se biofortification are designed to increase Se intake through the dietary materials and the strategy may provide a practical and cost-effective way to increase Se intake for people living in low Se areas around the world.

The importance of Se on human and animal health has been acknowledged and increasingly recognized by the public worldwide. The Se research community is excitedly and fervently working on the multi-faceted functionality of Se, and its influence on so many parts of human, animal, and even plant life. There is so much site-specific detail to learn about Se, and its impact on the different biological environments. Only through continued interactions with fellow Se researchers from around the world, can our overall knowledge expand on selenium's impact throughout all facets of biological, chemical and physical sciences. To discover the various roles that Se plays in today's complex world, the 6th International Conference on Selenium in the Environment and Human Health will be held on 27–30 October 2019 in Yangling/Xi'an, China, as a continuation of our biannual conference series after Suzhou, China (2009, 2011), Hefei, China (2013), Sao Paulo, Brazil (2015), and Stockholm, Sweden (2017).

This conference has attracted some of the finest and diverse minds of researchers from the international Se research community. The format of the conference strongly promotes exchange of knowledge across different disciplines representing more than 18 countries. This conference provides a venue where we all interact together in the same house. Surrounded by the ancient beauty of Xi'an and having the excellent conference facilities at Northwest A & F University, a very Se-enriching experience is waiting for all of us. There is a lot happening in the rapidly growing research world of Se, as evident from 103 extended abstracts that have been peer reviewed and included in this proceeding. The publication of this book is attributed to the international participants representing a multitude of disciplines from the worldwide Se research community. Importantly, it is refreshing to see that the interest to pursue Se topics is alive and continually growing, as we see by the large number of abstracts coauthored by many graduate students and junior scientists from around the world. Ultimately, it is our goal to utilize this 6th International Conference on Selenium to listen, learn, share, and grow wiser on understanding the many powers of Se.

Gary S. Bañuelos, *California, USA*
Zhi-Qing Lin, *Illinois, USA*
Dongli Liang, *Shaanxi, China*
Xue-bin Yin, *Jiangsu, China*

Selenium Research for Environment and Human Health: Perspectives, Technologies and Advancements – Bañuelos, Lin, Liang & Yin (eds)
© 2020 Taylor and Francis Group, London, ISBN 978-1-138-38014-0

Preface

The selenium (Se) content in crops and human daily dietary intake has become a topic of great interest [illegible] health around the world, since Se is an essential and important trace element for animals and humans. [illegible] estimations [illegible] more than a billion people [illegible] affected [illegible] by Se deficiency [illegible] human health [illegible] and [illegible] that have [illegible] essential functions for normal [illegible] Worldwide, plant-based products [illegible] Se [illegible] dietary intake [illegible] Se supply and intake is abnormal. Most nutritional Se [illegible] selenomethionine [illegible] and their derivatives, [illegible] as well as low molecular weight selenocompounds.

[illegible] environmental and biological [illegible] will, however, primarily depend upon [illegible]. The [illegible] Se [illegible] concentrations [illegible] soil [illegible] from [illegible] naturally occur in seleniferous [illegible] environmental and soil [illegible] sources of anthropogenic [illegible] Se [illegible] and atmospheric [illegible] within the soil [illegible] that will also determine the [illegible] of Se [illegible] is one of the most important factors that determine Se [illegible] crops and food materials. Although Se is not essential for higher plants, plants can [illegible] take up Se [illegible] similar to sulfur (S) [illegible] uptake is largely determined by [illegible] accumulation [illegible] molecular and [illegible] techniques, [illegible] better understanding [illegible] strategies [illegible] food and feed products [illegible] based approaches [illegible]

[illegible] human and animal health [illegible] and increasing [illegible] Se [illegible] functions [illegible] animal and even plant life. There is so much [illegible] about Se, and its impact on the different biological [illegible] selenium [illegible] that Se plays [illegible] Selenium in the Environment and Human Health [illegible] Sao Paulo, Brazil (2015) and Stockholm, Sweden (2017).

This [illegible] international Se [illegible] of knowledge across different disciplines [illegible] provides a venue where [illegible] in the same [illegible] Surrounded by the [illegible] excellent [illegible] University [illegible] There is [illegible] happening in the rapidly growing research world of Se [illegible] extended abstracts that have been peer reviewed and included in this proceeding. The [illegible] of this book is [illegible] participants [illegible] Se research [illegible] that the [illegible] to pursue Se [illegible] the large number of abstracts [illegible] around the world [illegible] this [illegible] international conference [illegible]

Selenium Research for Environment and Human Health: Perspectives, Technologies and Advancements – Bañuelos, Lin, Liang & Yin (eds)
© 2020 Taylor and Francis Group, London, ISBN 978-1-138-39014-0

About the editors

Gary Bañuelos

Dr. Gary Bañuelos has more than two decades of research experience, most of which has focused on developing selenium phytoremediation and biofortification strategies for saline soils and waters. Dr. Bañuelos has produced over 200 publications, of which 111 are senior authorships, including 16 book chapters and 5 patents. In addition, Dr. Bañuelos has co-edited 6 books (5 as senior editor with Professor Z.Q. Lin). Furthermore, his book with Professor Lin *Development and Uses of Biofortified Agricultural Products* was translated into Chinese and sold on Amazon.com in China. Dr. Bañuelos is nationally and internationally recognized for his expertise and insights into salt-, boron-, and selenium-tolerant crops and Se phytoremediation and biofortification strategies, as indicated by over 200 invitations at the national and international level to speak on these selenium related topics. Moreover, Dr. Bañuelos was recognized as the USDA-ARS early Career Scientist of the Year in 1992, USDA-ARS Pacific West Senior Scientist of the Year in 2017, and recently presides as President of the International Society for Selenium Research. Dr. Bañuelos' research has been highly regarded throughout their career, as evidenced by extramural support totaling more than 3.5 million dollars. The results of his research on the phytoremediation of Se have the potential to be executed upon selenium-impacted soils of the western United States, predominately located in Colorado and the west side of central California. Moreover, the plant-based technology, phytoremediation, is being extensively practiced in selenium-rich areas of the world, while biofortification is especially important in China with over 500 million selenium-deficient people.

Zhi-Qing Lin

Dr. Zhi-Qing Lin is Professor in the Department of Environmental Sciences and Department of Biological Sciences at Southern Illinois University – Edwardsville in the U.S. Dr. Lin received his BS degree in Environmental Biology from Liaoning University in 1983, MS degree in Pollution Ecology from Chinese Academy of Sciences in 1986, and Ph.D. degree in Renewable Resources from McGill University in Canada in 1996. He was Post-doctoral Research Fellow and Post-graduate Researcher on Environmental Biogeochemistry at University of California – Berkeley from 1996 to 2002. His research is related to the biogeochemistry of environmentally important trace elements, with special focus on biological accumulation, transformation, and volatilization processes of selenium in the water-soil-plant system. He has authored or coauthored over 94 peer-reviewed articles published in scientific journals, proceedings, and monographs.

Dongli Liang

Dr. Dongli Liang is Professor in the College of Natural Resources and Environment at Northwest Agriculture & Forestry University, China. She currently serves as Chair of Department of Environmental Science and Engineering. Dr. Liang received her B.S. degree in Agriculture from China Agriculture University and Ph.D. degree from Northwest A&F University. Her research interest is in the field of Environmental Soil Chemistry. Her research program at Northwest A&F University primarily focuses on environmental behaviors and fates of heavy metals in soil, harmful effects of metal pollutants on human health through food chains, and also on environmental remediation of metal-polluted soil and water. In 2006, Dr. Liang started her research on selenium transport and transformation in soil-plant systems, particularly on soil selenium bioavailability and has published over 100 peer-reviewed research papers.

Xue-bin Yin

Dr. Xue-bin Yin is an Associate Professor in the Department of Environmental Sciences and the Director of Advanced Lab for Functional Agriculture, at University of Science and Technology of China (USTC) in Suzhou. He also serves as Dean of Shanxi Institute of Functional Agriculture (SIFA at Shanxi). He currently serves as Secretary of International Society for Selenium Research. Dr. Yin received his Bachelor of Science degree (2000) in Geochemistry, Ph.D. degree of Environmental Science (2005) from USTC. Dr. Yin started his research on selenium biogeochemistry, phytoremediation and biofortification at USTC in 1998, and continued at the Chinese Academy of Sciences in 2005. His current research is related to the biofortification of trace/macro elements (such as selenium, zinc, calcium, iron, and iodine), with a special emphasis on their biological transformation process, health issue and utilization in the agriculture and functional foods. He is the author or coauthor of over 50 refereed journal and review articles, proceedings, and book chapters. He has also chaired or co-chaired the International Selenium Conference for five times since 2008.

Selenium Research for Environment and Human Health: Perspectives, Technologies and Advancements – Bañuelos, Lin, Liang & Yin (eds)
© 2020 Taylor and Francis Group, London, ISBN 978-1-138-39014-0

An Agenda for Selenium Research – from Yesterday to Tomorrow

Gerald F. Combs, Jr., Ph.D.[1]

I was introduced to selenium (Se) by Milt Scott and have thought about it for the better part of five decades. In the late 1950s, Milt collaborated with Klaus Schwarz at National Institutes of Health and found Se to "spare" vitamin E in livestock and experimental animals. As Milt's graduate student in 1971, I was interested in the metabolic basis of that effect, and wanted to know whether Se was a nutrient *per se* or simply an effector of vitamin E function. This question was clarified in 1972 when John Rotruck (in Bill Hoekstra's lab at the University of Wisconsin) found Se to be an essential constituent of the antioxidant enzyme glutathione peroxidase. That finding led to the mushrooming of research in this field. Within the next two decades other selenoenzymes were identified[2]; each was found to contain Se in a previously unidentified amino acid, selenocysteine, which was found to be incorporated *via* a novel co-translational mechanism; and dietary levels of ca. 0.1-0.3 mg/kg were found to be required for maximal expression of the major selenoenzymes.

At the same time, others found that Se reduced tumorigenesis in virtually all animal and cell models studied. Those studies showed both organic and inorganic forms of Se effective at non-toxic intakes greater than those needed to prevent vitamin E deficiency signs or maximize selenoenzyme expression. In the 1960s through 1990s, results from epidemiological studies suggested that cancer risk may be inversely related to Se status. Yet, only a few randomized controlled trials were conducted. The Nutritional Prevention of Cancer (NPC) Trial, which Larry Clark, Bruce Turnbull and I conducted, found a two-third reduction in prostate and colorectal cancer risks with a daily 200 μg Se supplement (as Se-enriched yeast) in a cohort of some 1300 Se-adequate Americans, with no protection among subjects with baseline plasma Se levels <121 ng/ml. The much larger Selenium and Vitamin E Cancer Prevention Trial (SELECT) found no such effects of a similar Se supplement (as selenomethionine) in cohort of Americans most of whom had baseline plasma Se levels >121 ng/ml. These results were consistent with a U-shaped relationship of Se intake and cancer risk.

By the 1980s, it had become clear that Se was an essential nutrient, at least for animals, with the property of also being antitumorigenic at supranutritional exposures. The question arose: "Is Se essential for humans?" This question would be answered by researchers in China.

In the spring of 1978, I received a manuscript from an individual whom I didn't know at the Institute of Food Hygiene in Beijing. It purported to show that a heart disease occurred in children with levels of plasma Se of only 10–20 ng/ml[3]. Several months later, I received a call from the same individual, Dr. Chen Junshi, asking to spend a year in my lab at Cornell. I arranged for him to work with me and a colleague, Colin Campbell. In a productive year, the three of us learned a lot from each other and became good friends. This led to Colin and me being invited in 1980 to go to China with two European scientists to conduct a "selenium" seminar at Chen's Institute[4,5]. From Beijing we traveled by overnight train to Xichang City, an area of endemic Se deficiency in southwest Sichuan Province. There we met with 'barefoot' doctors and their pediatric patients with the multifocal myocarditis called Keshan Disease. Chen's colleagues and researchers in Xi'an had found in large-scale trials that Keshan Disease was preventable (but not treatable) with supplemental Se (sodium selenite, 0.5–1 mg Se/child/wk; or selenite-fortified table salt containing 10–15 μg Se/g). Subsequent research raised questions as to whether Se deficiency was a direct cause of Keshan Disease, which may also involve cardiotrophic RNA viruses. Indeed, severe Se deficiency is now seen as, at least, a major predisposing factor.

On my return to Beijing, I had dinner with Dr. Liu Jinxu, of the Chinese Academy of Agricultural Sciences and a 1952 Cornell Ph.D. in Animal Nutrition. He described the rebuilding of his Institute after the Cultural Revolution, and I asked what I could do to help. Dr. Liu responded: "*We are going to get foreign experts here to talk. But our young people need to learn how foreign scientists think about science – scientific reasoning!*" Within the year, Dr. Liu had arranged support from the Ministry of Agriculture for me to spend 13 weeks working in the

[1] Senior Scientist, Jean Mayer USDA Human Nutrition Research Center on Aging, Tufts University, Boston, MA; and Professor of Nutrition Emeritus, Division of Nutritional Sciences, Cornell University, Ithaca, NY.

[2] Ultimately, some 25 selenocysteine-containing proteins would be identified.

[3] This compares to plasma Se levels of 80–200 ng/ml in the US and 35–100 ng/ml in Europe.

[4] This meeting was made possible by the efforts of Dr. Curtis Hames, Claxton, GA.

[5] The Institute staff included Dr. Chen Chungming, who would be influential in Chinese national public health programming, and Dr. Jin Daxun, who to everyone's surprise, had been a fellow graduate student with my father at Cornell in the late 1940s.

Institute of Animal Science in Beijing. Over the next several years I made four such visits[6], and brought Chinese colleagues to spend comparable times in my lab at Cornell[7]. In 1984, I chaired the first international scientific meeting held in modern China – the Third International Symposium on Selenium in Biology and Medicine[8]. This attracted some 250 participants, half being Chinese scientists.[9] We used the two-volume proceedings[10] to get Chinese Se research into the English language literature.

By the 1990s, Se was established as an essential nutrient for humans, and a Recommended Daily Allowance (RDA) was established. It was based on a single study from China that had found a daily intake of 41 μg Se supported the maximal expression of plasma glutathione peroxidase (GPX3) in a small group of Se-deficient Chinese men. The RDA was revised in 2000 to reflect the amount of Se necessary to prevent both Keshan Disease and support maximal expression of two selenoenzymes (GPX3 and selenoprotein P [SePP1]). Various national advisory panels subsequently produced a range of recommended reference intakes (25–125 μg/d). When I compared the reported Se contents of different national food supplies, I found that sub-clinical Se deficiency (Se intakes <50 μg/d) likely affected 10–50% of residents in most countries for which data were available – only Canada, Japan, Norway, and the US appeared not to show prevalent low Se status.

Thus, over five decades I have seen Se added to the Nutrition agenda as an essential nutrient with potential to reduce cancer risk. I have also noted that consequences of Se deprivation can be sub-clinical in nature, requiring other precipitating factors (e.g. vitamin E deficiency, viruses, carcinogens, iodine deficiency, protein insufficiency) to reveal manifestations of sub-optimal expression of selenoenzymes and/or insufficient amounts of active Se-metabolites. It has also became clear that Se can have both healthful and adverse effects at supranutritional intakes, raising questions about its window of safety. This has left me with three questions which I consider pressing:

1. *Can Se be part of a useful practical strategy for reducing cancer risk in humans?*
The clinical significance of Se in cancer prevention remains a subject of debate. The (only) nine randomized clinical trials conducted to date have yielded inconsistent results. Systematic reviews have found that Se may be effective in preventing cancer in individuals of low to adequate, but not high Se status, and a U-shaped dose-response relationship has been proposed. Yet, these conclusions are weakened by the fact that several trials have not followed robust protocols, and none has been conducted in recent years or in cohorts with individuals of both marginal and adequate Se status.

2. *Is Se safe?*
Excess type 2 diabetes (T2D) was noted among subjects in the upper quintile of plasma Se in two US NHANES (National Health and Nutrition Examination Survey) cohorts, as well as among Se-supplemented subjects in the NPC trial who achieved plasma Se levels averaging ca. 190 ng/ml. Other trials have not found excess T2D, and studies in animal models have revealed no negative effects of Se on glycemic control. Resolving the question of whether supranutritional intakes of Se can increase T2D risk will demand well controlled trials with subjects randomized by T2D risk factors (high BMI, elevated fasting glucose) and followed with unequivocal diagnostic indicators (fasting glucose, HbA1c, oral glucose tolerance). Until then, there would appear to be no justification for any healthy adult, regardless of his/her baseline Se status, to consume more than 100 μg Se/day.[11]

3. *Who can benefit from Se?*
This may be the ultimate question about the health value of Se. Se-supplementation of adults with Se intakes <50 μg/d will increase selenoenzyme expression; those increases will depend on the extent to which baseline plasma Se levels were below approximately 90 ng/ml. Whether individuals with plasma Se levels >90 ng/ml may also benefit from increased Se intakes is suggested from results of the NPC Trial in which supplemental Se reduced cancer risk among adults with plasma Se levels in the range of 80–121 ng/ml, a range that includes some 10% of Americans. Thus, individuals with plasma Se levels >90 ng/ml may benefit from cancer protection unrelated to selenoenzyme expression[12]. If this can be confirmed, it would mean that large numbers of adults are

---

[6] And my taking a Cornell summer total immersion course in Mandarin.
[7] Their visits were facilitated through generous support from American agribusinesses.
[8] This would not have been possible without the help of Julan Spallholz and S.P. Yang, Texas Tech Univ.; Jin Daxun and Niu Shiru, China Nat. Center for Prev. Med.; Orville Levander, USDA; and Jim Oldfield, Oregon State Univ.
[9] To everyone's surprise and delight, I opened the symposium in Mandarin.
[10] Combs, G.F., Jr., J.E. Spallholz, O.A. Levender and J.E. Oldfield (eds). 1987. *Proceedings of the Third International Symposium on Selenium in Biology and Medicine.* AVI Publishing Co., Westport, CT, vols A and B, 1138 pp.
[11] The toxicity of high doses of Se has been described in case reports of accidental exposures to various seleniferous products, and from naturally occurring chronic selenosis in Enshi County, Hubei Province, China. The latter involved very high Se levels in soils, water and throughout the local food system, with some residents consuming >1.5 mg Se/p/day (almost 4 times the UL set by WHO and IOM) showing hair, nail and skin lesions.
[12] It should be remembered that *all* individuals, regardless of Se status, will show increases in plasma total Se when supplemented with the dominant food form, SeMet. In fact, that response will be nearly linearly over the entire range of practical doses, reflecting to increases in Se nonspecifically bound in albumin and other plasma proteins.

likely to benefit from increased Se intakes – some to reduce cancer risks, others also to maximize selenoprotein expression.

Today's Selenium Research Agenda must address these key questions with robust research that produces unequivocal and, thus, useful results. Pursuing this agenda is the challenge for those who will be thinking about Se over the *next* decades.

Selenium Research for Environment and Human Health: Perspectives, Technologies and Advancements – Bañuelos, Lin, Liang & Yin (eds)
© 2020 Taylor and Francis Group, London, ISBN 978-1-138-39014-0

List of Corresponding Authors

Corresponding authors (some annotated with * in chapters) are listed alphabetically.

Ahmad, S.: *School of Biosciences, University of Nottingham, UK;* plxsa13@nottingham.ac.uk

Arnér, Elias S.J.: *Division of Biochemistry, Department of Medical Biochemistry and Biophysics, Karolinska Institutet, SE-171 77 Stockholm, Sweden;* clias.arncr@ki.se

Bañuelos, G.S.: *US Department of Agriculture, ARS, Parlier, California, USA;* gary.banuelos@usda.gov

Bao, Z.Y.: *Engineering Research Center of Nano-Geo Materials of Ministry of Education, Faculty of Materials Science and Chemistry, China University of Geosciences, Wuhan, Hubei, China; Zhejiang Institute, China University of Geosciences (Wuhan), Hangzhou, 311305, China;* zybao@cug.edu.cn

Berntssen, M.H.G.: *Institute for Marine Research (IMR), Bergen, Norway;* marc.berntssen@hi.no

Caton, J.S.: *Department of Animal Sciences, North Dakota State University, ND 58108, USA;* joel.caton@ndsu.edu

Chen, Q.Q.: *Environmental Science, College of Life Science and Resources and Environment, Yichun University, Yichun, Jiangxi, China;* chenqq@nju.edu.cn

Chen, S.Z.: *Beijing Center for Diseases Prevention and Control, Beijing 100013, China;* csz1987buct@163.com

Chi, F.Q.: *Soil Fertilizer and Environment Resources Institute, Heilongjiang Academy of Agriculture Sciences, Key Laboratory of Soil Environment and Plant Nutrition of Heilongjiang Province, Harbin, 150086, China;* fqchi2013@163.com

Chilimba, A.D.C.: *AGRISO Consultants and Ngolojere Investments, P.O. Box 399, Zomba, Malawi;* achilimba@gmail.com

Dai, X.X.: *School of Public health, Xi'an Jiaotong University Health Science Center, No.76 Yanta West Road, Xi'an, Shaanxi 710061, China;* xxiadai@mail.xjtu.edu.cn

Du Laing, G.: *Ghent University, Ghent, Belgium;* gijs.dulaing@ugent.be

Feinberg, A.: *Swiss Federal Institute of Technology (ETH), Institute of Biogeochemistry and Pollutant Dynamics & Institute for Atmospheric and Climate Science, Zürich, Switzerland; Eawag: Swiss Federal Institute of Aquatic Science and Technology, Duebendorf, Switzerland;* aryeh.feinberg@env.ethz.ch

Feldmann, J.: *TESLA-Trace Element Speciation Laboratory, University of Aberdeen, Aberdeen, Scotland, UK;* j.feldmann@abdn.ac.uk

Gu, M.: *Cultivation Base of Guangxi Key Laboratory for Agro-Environment and Agro-Products Safety, College of Agriculture, Guangxi University, Nanning 530004, China;* gumh@gxu.edu.cn

Guilherme, L.R.G.: *Department of Soil Science, Federal University of Lavras, Lavras, Minas Gerais, Brazil;* guilherm@ufla.br

Guo, Y.B.: *College of Resources and Environmental Sciences, China Agricultural University, Beijing, China;* guoyb@cau.edu.cn

Hackler, J.: *Institut für Experimentelle Endokrinologie, Charité-Universitätsmedizin Berlin, D-13353 Berlin, Germany;* julian.hackler@charite.de

Heller, R.A.: *University Hospital Heidelberg, Heidelberg Trauma Research Group, Department of Trauma and Reconstructive Surgery, Center for Orthopedics, Trauma Surgery and Spinal Cord Injury, Heidelberg, Germany;* raban.heller@outlook.com

Hoffmann, P.R.: *John A. Burns School of Medicine, University of Hawaii, Honolulu, Hawaii, USA;* peterrh@hawaii.edu

Huang, D.: *College of Natural Resources and Environment, Northwest A&F University, Yangling, China;* dlynnhuang@nwsuaf.edu.cn

Huang, Z.: *SeNA Research Institute, Life Science College, Sichuan University, Chengdu, China; Department of Chemistry, Georgia State University, Atlanta, Georgia, USA;* huang@senaresearch.org
Huang, Z.: *College of Life Science and Technology, Jinan University, Guangzhou, Guangdong, China;* 1547461148@qq.com
Hughes, D.J.: *Cancer Biology and Therapeutics Group, Conway Institute, University College Dublin, Dublin, Ireland;* david.hughes@ucd.ie
Jia, W.X.: *Jiangsu Bio-Engineering Research Center for Selenium, Suzhou, Jiangsu, China; School of Resources and Environment, Shanxi Agricultural University, Taigu, Shanxi, China;* 1085875885@qq.com
Jiang, L.: *School of Biosciences, Sutton Bonington Campus, University of Nottingham, Leicestershire, UK;* linxi.jiang@nottingham.ac.uk
Lei, X.G.: *Department of Animal Science, Cornell University, Ithaca, NY 14853, USA;* xl20@cornell.edu
Li, H.F.: *Upland Crops Research Institute, Guangxi Academy of Agricultural Sciences, Nanning, China;* lihuifeng2010@126.com
Li, H.R.: *Key Laboratory of Land Surface Pattern and Simulation, Institute of Geographical Sciences and Natural Resources Research, Chinese Academy of Sciences, Beijing, China; College of Resources and Environment, University of Chinese Academy of Sciences, Beijing, China;* lihr@igsnrr.ac.cn
Li, J.: *Laboratory of Analytical Chemistry and Applied Ecochemistry, Faculty of Bioscience Engineering, Ghent University, Ghent, Belgium;* jun.li@ugent.be
Li, L.: *Robert W. Holley Center for Agriculture and Health, USDA-ARS, Cornell University, Ithaca, NY 14853, USA*
Li, M.: *Key Laboratory of Agri-Food Safety of Anhui Province, School of Plant Protection, Anhui Agriculture University, Hefei, Anhui, China;* plantprotection2006@126.com
Li, T.: *Jiangsu Key Laboratory of Crop Genetics and Physiology/ Key Laboratory of Plant Functional Genomics of the Ministry of Education/ Jiangsu Key Laboratory of Crop Genomics and Molecular Breeding/Jiangsu Co-Innovation Center for Modern Production Technology of Grain Crops, Yangzhou University, Yangzhou 225009, China;* taoli@yzu.edu.cn
Liang, D.L.: *College of Natural Resources and Environment, Northwest A&F University, Yangling, Shaanxi 712100, China; Key Laboratory of Plant Nutrition and the Agri-environment in Northwest China, Ministry of Agriculture, Yangling, Shaanxi 712100, China;* dlliang@nwafu.edu.cn
Lin, Z.-Q.: *Department of Environmental Sciences, Southern Illinois University, Edwardsville, Illinois 62026, USA;* zhlin@siue.edu
Liu, Li: *Pomology Institute, Shanxi Academy of Agricultural Sciences, Shanxi Key Laboratory of Germplasm Improvement and Utilization in Pomology, Taiyuan, Shanxi, China;* hlyl0210@yeah.net
Liu, X.W.: *Microelement Research Center, Huazhong Agricultural University, Wuhan 430070, China;* lxw2016@mail.hzau.edu.cn
Luo, K.L.: *Institute of Geographic Sciences and Natural Resources Research, Chinese Academy of Sciences, Beijing, China;* luokl@igsnrr.ac.cn
Lyons, G.H.: *School of Agriculture, Food and Wine, The University of Adelaide, Waite Campus, Urrbrae, South Australia 5064, Australia;* graham.lyons@adelaide.edu.au
Ming, J.J.: *Enshi Tujia & Miao Autonomous Prefecture Academy of Agricultural Sciences, Enshi, Hubei, China; Hubei Selenium Industry Technology Research Institute, Enshi, Hubei, China;* jiajiaming77@163.com
Moraes, M.F.: *Federal University of Mato Grosso, Cuiaba, Brazil;* moraesmf@yahoo.com.br
Otieno, S.B.: *Department of Community Health, Nairobi Outreach Centre, Great Lakes University, Westlands, Nairobi, Kenya;* samwelbotieno@yahoo.com
Qiao, Y.H.: *College of Resources and Environmental Sciences, China Agricultural University, Beijing, China;* qiaoyh@cau.edu.cn
Qin, H.-B.: *State Key Laboratory of Environmental Geochemistry, Institute of Geochemistry Chinese Academy of Sciences, Guiyang 550081, China;* qinhaibo@vip.gyig.ac.cn
Ralston, N.V.C.: *Earth System Science and Policy, University of North Dakota, Grand Forks, ND, USA; Sage Green NRG, Grand Forks, ND, USA;* nick.ralston@und.edu

Ramkissoon, C.: *University of Adelaide, South Australia; University of Nottingham, United Kingdom;* stxcr5@nottingham.ac.uk

Reis, A.R.: *UNESP – Univ Estadual Paulista, Postal Code 17602-496, Tupã-SP, Brazil; UNESP – Univ Estadual Paulista, Postal Code 14884-900, Jaboticabal-SP, Brazil;* andre.reis@unesp.br

Schiavon, M.: *Department of Agricultural Biotechnology, University of Padova, 35020 Legnaro (PD), Italy;* michela.schiavon@unipd.it

Schomburg, L.: *Institute for Experimental Endocrinology, Charité-Medical School Berlin, Germany;* lutz.schomburg@charite.de

Schweizer, U.: *Rheinische Friedrich-Wilhelms-Universität Bonn, Germany;* uschweiz@uni-bonn.de

Seelig, J.: *Institute for Experimental Endocrinology, Charité – Universitätsmedizin Berlin, Berlin, Germany;* julian.seelig@charite.de

Song, J.P.: *School of Earth and Space Sciences, University of Science and Technology of China, Hefei, Anhui, China; Jiangsu Bio-Engineering Research Center for Selenium, Suzhou, Jiangsu, China;* 1655297314@qq.com

Sun, Q.: *Charité -Universitätsmedizin Berlin, Institut für Experimentelle Endokrinologie, Germany;* qian.sun@charite.de

Sunde, R.A.: *Department of Nutritional Sciences, University of Wisconsin, Madison, WI 53706, USA;* sunde@nutrisci.wisc.edu

Tang, D.J.: *Ankang Se-enriched Product Research and Development Center, Ankang, China; Key Laboratory of Se-enriched Products Development and Quality Control, Ministry of Agriculture and Rural Affairs, Ankang, China;* 724833026@qq.com

Teju Prakash, N.: *School of Energy and Environment, Thapar Institute of Engineering and Technology, Patiala, India;* tejoprakash@gmail.com

Tolu, J.: *Swiss Federal Institute of Technology (ETH), Institute of Biogeochemistry and Pollutant Dynamics & Institute for Atmospheric and Climate Science, Zürich, Switzerland; Eawag: Swiss Federal Institute of Aquatic Science and Technology, Duebendorf, Switzerland;* julie.tolu@eawag.ch

Tsuji, P.: *Towson University, Towson, Maryland, USA;* ptsuji@towson.edu

Tu, S.X.: *College of Resources and Environment, Huazhong Agricultural University, Wuhan, China; Microelement Research Center, Huazhong Agricultural University, Wuhan, China;* stu@mail.hzau.edu.cn

Wang, CH.Y.: *Crop Science Research Institute, Shanxi Academy of Agricultural Sciences, Taiyuan, Shanxi 030000, China;* wrwcy@139.com

Waters, D.J.: *Center for Exceptional Longevity Studies, Gerald P. Murphy Cancer Foundation, West Lafayette, IN, USA;* dwaters@gpmcf.org

Wei, C.Y.: *Institute of Agricultural Quality Standards and Testing Technology, Jilin Academy of Agricultural Sciences, Changchun, 130033, China;* weichy@yeah.net

White, P.J.: *The James Hutton Institute, Invergowrie, Dundee DD2 5DA, United Kingdom;* philip.white@hutton.ac.uk

Winkel, L.H.E.: *Swiss Federal Institute of Technology (ETH), Institute of Biogeochemistry and Pollutant Dynamics, Zurich, Switzerland; Eawag: Swiss Federal Institute of Aquatic Science and Technology, Duebendorf, Switzerland;* lenny.winkel@eawag.ch

Wu, F.Y.: *College of Natural Resources and Environment, Northwest A&F University, Yangling, Shaanxi, 712100, China;* fuyongwu@nwsuaf.edu.cn

Wu, W.L.: *College of Resources and Environmental Sciences, China Agricultural University, Beijing, China;* wuwenl@cau.edu.cn

Xiong, Y.M.: *School of Public Health, Xi'an Jiaotong University Health Science Center, Key Laboratory of Trace Elements and Endemic Diseases of National Health Commission of the People's Republic of China, Xi'an, China;* xiongym@mail.xjtu.edu.cn

Yang, L.S.: *Key Laboratory of Land Surface Pattern and Simulation, Institute of Geographical Sciences and Natural Resources Research, Chinese Academy of Sciences, Beijing, China; College of Resources and Environment, University of Chinese Academy of Sciences, Beijing, China;* yangls@igsnrr.ac.cn

Yang, X.Y.: *Xi'an Jiaotong University, Xi'an, China;* 675396406@qq.com

Yang, ZH.P.: *College of Agricultural, Shanxi Agricultural University, Taigu, Shanxi 030801, China;* yangzp.2@163.com

Yin, X.B.: *Key Lab for Functional Agriculture, USTC (Suzhou), Jiangsu, China; School of Earth and Space Sciences, University of Science and Technology of China, Hefei, China; Engineering Center for Functional Agriculture, Nanjing University (Suzhou), Jiangsu, China; Shanxi Institute for Functional Agriculture, Taigu, Shanxi, China; Key Laboratory of Se-enriched Products Development and Quality Control, Ministry of Agriculture and Rural Affairs and National-local Joint Engineering Laboratory of Se-enriched Food Development, Ankang, Shaanxi, China; Guangxi Selenium-rich Agricultural Research Center, Nanning, Guangxi, China;* xbyin@ustc.edu.cn

Yin, Y.L.: *Institute of Subtropical Agriculture, the Chinese Academy of Sciences, Changsha, China;* yinyulong@isa.ac.cn

Yu, T.: *School of Science, China University of Geosciences, Beijing, China;* yutao@cugb.edu.cn

Yuan, F.: *Xing Zhi College, Xi'an University of Finance and Economics, Xi'an, China;* yuanfangyf2005@163.com

Yuan, L.X.: *Jiangsu Bio-Engineering Research Center for Selenium, Suzhou, Jiangsu, China;* yuanlinxi001@gmail.com

Yuan, Y.Q.: *School of Environmental Science and Engineering, Sun Yat-sen University, Guangzhou, China;*

Zang, H.W.: *School of Plant Protection, Anhui Agricultural University, Hefei, Anhui, China; Jiangsu Bio-Engineering Research Center for Selenium, Suzhou, Jiangsu, China;* zanghw01@163.com

Zhang, L.H.: *Agricultural Faculty, Henan University of Science and Technology, Luoyang, Henan, China;* lhzhang2007@126.com

Zhang, R.Q.: *School of Public Health, Shaanxi University of Chinese Medicine, Xianyang, Shaanxi, China; Institute of Endemic Diseases, School of Public Health, Health Science Center, Xi'an Jiaotong University, Xi'an, Shaanxi, China;* Zhangrqxianyang@163.com

Zhao, X.H.: *College of Resources and Environment, Huazhong Agricultural University, Wuhan, China;* xhzhao@mail.hzau.edu.cn

Zhao, Z.G.: *Yichun University, Yichun, Jiangxi 336000, China;* zhaozg_77@163.com

Zou, Y.-P.: *Selenium Technology Innovation Center, College of Life Sciences and Resource Environment, Yichun University, Yichun, China;* zyping66@163.com

Origin of Se from geological, biochemical, microbial and global cycling

Selenium Research for Environment and Human Health: Perspectives, Technologies and Advancements – Bañuelos, Lin, Liang & Yin (eds)
© 2020 Taylor and Francis Group, London, ISBN 978-1-138-39014-0

Safe utilization and zoning on selenium-enriched land resources: A case study in Enshi, China

T. Yu
School of Science, China University of Geosciences, Beijing, China

Z.F. Yang, Q.Y. Hou, X.Q. Xia, W.L. Hou & Y.T. Li
School of Earth Science and Resources, China University of Geosciences, Beijing, China

W.J. Ma
School of Public Health, Peking University Health Science Center, Beijing, China

B.Z. Yan
Lamont-Doherty Earth Observatory, Columbia University, New York, USA

1 INTRODUCTION

Selenium (Se) is known as an important trace element. Enshi in Central China has abundant Se resources and is recognized as the "World Capital of Selenium," a title given at the 14th International Symposium on Trace Elements in Man and Animals in 2011.

The distribution of Se-enriched soil in the survey area is related to the stratum, particularly in the black strata of the Permian (Wang et al. 2018). Se in soil is often enriched with heavy metals, such as cadmium (Cd). Thus far, Se poisoning has occurred in the area, but no Cd poisoning has been reported yet. Cd can accumulate in the human body to potentially toxic levels; as such, Cd contamination in Se-rich soils cannot be ignored. Previous studies showed that Se-enriched plants contain less heavy metals, the amount of which exceeds the crop safety standard because of the antagonism of Se and Cd (Rayman 2000). How to safely develop Se-enriched land resources and reasonably use the high-level distribution of land resources of heavy metals in soil remain unclear.

In this research, methods of land resource safety zoning were established based on element concentrations of the crop as a bridge and considering the crop–human dose effect and the relationship between the concentrations of crops and topsoil.

2 MATERIALS AND METHODS

2.1 *Study area*

The study area is a 3067-ha arable land located in Shadi Town, Enshi City, which has subtropical monsoon humid climate. The annual precipitation, annual average sunshine, relative humidity, and annual average temperature are 1100–1400 mm, 1350–1860 h, 70–80%, and 15°C, respectively. The terrain of the study area is mountainous and undulating, with an elevation of 273–1428 m. The main soil types in the area are yellow brown soil and yellow soil. The main land use types are dry land, paddy field, and forest land. The economic development of the area is dominated by agriculture. Various crops, such as corn, rice, tobacco, konjac, and vegetables, are grown in the area.

2.2 *Sample collection and preparation*

Grid points were used in combination with land use map, and one to two sampling points were set per 0.25 km^2. At each sampling point, a wooden shovel was used to vertically collect 1.5 kg of 0-20 cm soil columns. Weeds and gravels were removed. A total of 839 soil samples was collected. At the same time, 110 homologous crop samples were also obtained from the soil sampling sites. The soil samples were air dried at room temperature. After drying in the shade, the samples were ground with a wooden hammer. The sample was passed through a nylon sieve with a pore size of 2 mm. About 300 g of evenly mixed sample was packed and sent to the laboratory for further processing and testing. The corn samples were threshed, rinsed three times with deionized water, and stirred with a clean glass rod. After draining, the crop samples were transferred to an enamel pan and placed in an oven. Drying was conducted for about 24 h at 60°C until a constant weight was achieved. The corn seeds were ground to a specified particle size (about 60 mesh) by using a grain mill prior to analysis.

Biological samples such as blood, urine, and hair of local residents were collected by collaborators from Peking University who also conducted interviews.

2.3 *Chemical analysis*

Soil pH was measured using a 1:25 soil to water suspension (w / v) through the ion-selective electrode method. Ca, K, and S contents of the soil samples were determined through X-ray fluorescence spectrometry (XRF, PW2440, Philips Co., Dutch). The soil samples were digested by HNO_3-HCl-HF-$HClO_4$ to determine Cd concentrations by inductively coupled plasma mass spectrometry (ICP-MS, X-SERIES II, Thermo Electron, USA) with a detection limit of 0.03 mg/kg. Concentrations of Se were determined by atomic fluorescence spectrometry (AFS-230E; Kechuang Haiguang Instrument, China), with a detection limit of 0.01 mg/kg. The analysis was performed using national standard soil samples to monitor the accuracy of the test. Coded samples were used to monitor the precision of the test analysis. The accuracy and precision of the analysis were within the monitoring limit and were satisfactory. The corn samples were acid digested in HNO_3 and H_2O_2 for measurement of Cd, Cu, Hg, Pb, Zn, and Se levels through ICP-MS (Agilent 7700x, Santa Clara, USA).

The blood, urine, and hair samples were acid digested using a high-pressure microwave vessel system (Ultra WAVE, Milestone, Italy). Cd and Se concentrations were measured using ICP-MS.

3 RESULTS AND DISCUSSION

3.1 *Cd and Se concentrations in soil and corn*

The average concentration of Se in the topsoil is 0.84 ± 1.39 mg/kg, and that of Cd is 0.93 ± 1.63 mg/kg. The coefficients of variation of Se and Cd are as high as 1.65 and 1.79, respectively, which indicates that the concentrations of Se and Cd in the region vary greatly and are highly dispersed among different geological bodies. The average concentrations of Se and Cd in corn are 0.22 ± 0.96 and 0.15 ± 0.32 mg/kg, respectively, which are mainly related to their high concentrations in soil.

3.2 *Biomarker measurements in the exposed and control groups*

The concentrations of Cd and Se in the selected biomarkers, namely, blood, urine, and hair samples, were significantly higher in the treatment group than those in the control group ($p < 0.05$). In the exposed group, 28.13% of the population had urine Cd levels beyond the standards of FAO/WHO. No subject had urine Cd levels beyond the standards in the control group.

3.3 *Corn absorption models for Se and Cd*

A semi-empirical model with a theoretical process and an empirical component was established. Correlation analysis showed that the factors affecting Se content in corn are the concentrations of Se and K in soil and soil pH. Meanwhile, Cd content in corn Cd is totally connected with soil Cd.

The models are listed below:

$$\log(Se_{corn}) = 1.485\log(Se_{soil}) - 0.0642\log(K_{soil}) + 0.071pH_{soil} - 1.793 \quad (1)$$

where Se_{corn} is Se concentration in corn (mg/kg), Se_{soil} is Se concentration in soil (mg/kg), K_{soil} is the K concentration in soil (%), and pH_{soil} is the soil pH.

$$\log(Cd_{corn}) = 0.924\log(Cd_{soil}) - 0.983 \quad (2)$$

where Cd_{corn} is Cd concentration in corn (mg/kg) and Cd_{soil} is Cd concentration in soil (mg/kg).

The accuracy and precision of the models were judged by normalized mean error (NME) and normalized root mean square difference (NRMSE). In addition, the models were tested and verified by parallel samples.

3.4 *Safe utilization and zoning of selenium-enriched land*

Combined with previous dose-response data and models, a Se-enriched safe utilization area was planned based on the provisions of the Action Plan for Soil Pollution Control (or the Soil Ten Articles) (MEE, 2016). The proportions of priority utilization, safe utilization, planting adjustment, and strict management of agricultural land in the study area were 58.8%, 22.9%, 17.0%, and 1.2%, respectively.

4 CONCLUSIONS

Soil Se and Cd were highly associated with each other in the study area. The ecological effect and the dose limit model were used to calculate the elemental limit value. The safe use area of the Se-enriched land was then zoned, which is greatly enlarged than the original Soil Ten Articles referred.

The authors gratefully acknowledge the support of China Geological Survey and China Scholarship Council.

REFERENCES

Ministry of Ecology and Environment (MEE). 2016. *Announcement of the Chinese State Council on Issuing the Action Plan for Soil Pollution Control.* (in Chinese) http://zfs.mee.gov.cn/fg/gwyw/201605/t20160531_352665.shtml (accessed on June 12, 2019)

Rayman, M.P. 2000. The importance of selenium to human health. *Lancet* 356: 233–241.

Wang, R., Yu, T., Yang Z.F. et al. 2018. Bioavailability of soil selenium and its influencing factors in selenium-enriched soil. *Resources and Environment in the Yangtze Basin* 27(7): 1647–1654. (in Chinese).

Selenium Research for Environment and Human Health: Perspectives, Technologies and Advancements – Bañuelos, Lin, Liang & Yin (eds)
© 2020 Taylor and Francis Group, London, ISBN 978-1-138-39014-0

Comparison of selenium level in different time periods in typical selenium-deficient areas of Shaanxi

X.Y. Wang, H.R. Li & L.S. Yang
Key Laboratory of Land Surface Pattern and Simulation, Institute of Geographical Sciences and Natural Resources Research, Chinese Academy of Sciences, Beijing, China
College of Resources and Environment, University of Chinese Academy of Sciences, Beijing, China

1 INTRODUCTION

1.1 *Change of selenium content*

Selenium (Se) is an essential trace element for humans. The relationship between Se nutrition and health has always been a hot research field. There is a low-Se belt in the natural environment of China, which extends from northeast to southwest of the country. Since the 1980s, with the socioeconomic development, the Se nutrition level of the population within this belt has been obviously improved. The time-space difference of the corresponding soil-grain-human hair Se level changes is worth a systemic study.

1.2 *Selection of study area*

We selected the typical low-Se area in the middle of Shaanxi Province as the study area, analyzed the temporal variation characteristics of Se content throughout different periods.

2 MATERIALS AND METHODS

We collected Se content data of soil, grain, and human hair (both adults and children) in the 1980s, 1990s and 2010s of 26 counties in the middle of Shaanxi Province from historical studies. Mathematical statistics and geostatistical methods were used to analyze the temporal variation characteristics of Se content in different periods of Shaanxi Province. The statistical software program used was SPSS version 19.0 and ArcMap version 10.2.

3 RESULTS AND DISCUSSION

3.1 *Selenium content in soil*

The average content of Se in soil of Shaanxi Province in the 1980s and the 2010s was 0.085 ± 0.024 mg/kg and 0.172 ± 0.058 mg/kg, respectively (Huang 2018, Li et al. 1982). The latter is significantly higher than the former ($p < 0.01$). The Se content of the soil increased twofold, which changed from the Se deficiency state

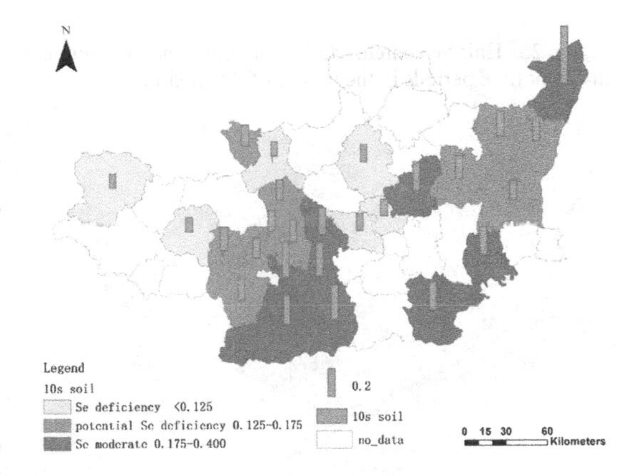

Figure 1. Soil Se content in the middle of Shaanxi Province in the 2010s (mg/kg).

in the 1980s to the potential Se deficiency state in the 2010s according to Tan's classification of soil Se content (Tan 1989). The soil Se content in some areas has reached an appropriate, sufficient level in the 2010s, but some areas (about 19.2%) are still in the state of soil Se deficiency (Fig. 1).

3.2 *Selenium content in hair of adults and children*

The average content of Se in hair of adults in Shaanxi Province in the three time periods was 0.100 ± 0.023 mg/kg, 0.236 ± 0.056 mg/kg, and 0.351 ± 0.092 mg/kg, respectively (Zhu et al. 2018, Yang et al. 2018, Sun et al. 1992, Zhang et al. 2001). The level increased from the Se deficiency level in the 1980s, to the marginal Se deficiency level in the 1990s, and to the medium Se level in the 2010s. The content of Se in hair of children was generally higher than that of adults. The average content of Se in hair of children in the different periods was 0.140 ± 0.043 mg/kg, 0.272 ± 0.051 mg/kg, and 0.425 ± 0.096 mg/kg, respectively (Zhu et al. 2018, Yang et al. 2018, Sun et al. 1992, Du et al. 2018, Cao et al. 2003). The Se content in hair of children had been in the middle Se level since the 1990s (Fig. 2). There are significant differences in hair

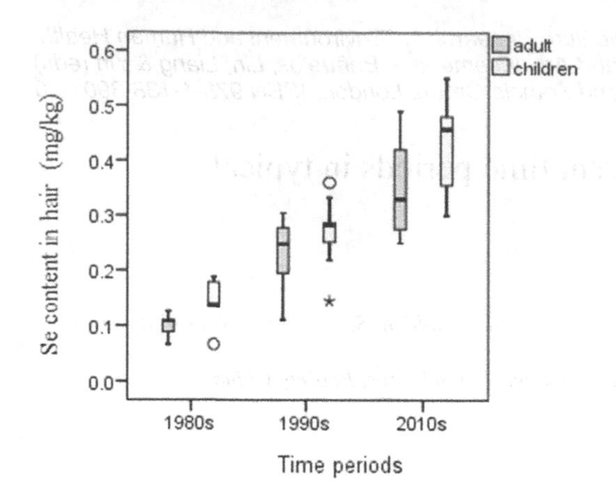

Figure 2. Hair Se content (mg/kg) of adults and children in different time periods in the middle of Shaanxi Province.

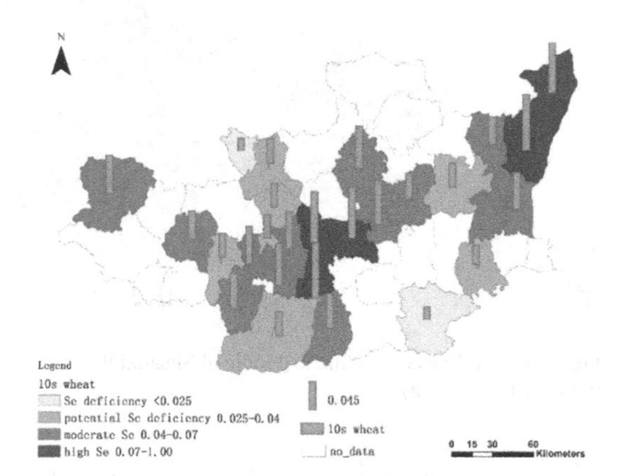

Figure 3. Wheat Se content (mg/kg) in the middle of Shaanxi Province in the 2010s.

Se content among adults and children in the different time periods ($p < 0.01$).

3.3 *Selenium content in wheat*

The average content of Se in wheat in Shaanxi Province in the different time periods was 0.012 ± 0.004 mg/kg, 0.033 ± 0.021 mg/kg, and 0.056 ± 0.022 mg/kg, respectively (Sun et al. 1992, Huang 2018, Li et al. 1982). The Se content has increased from the Se deficiency level in the 1980s, to the critical Se deficiency level in the 1990s, and to the moderate Se level in the 2010s (Fig. 3). There is a significant difference in Se content among the different time periods ($p < 0.01$). The Se content of wheat in some areas has reached high Se level in the 2010s. However, some other areas were still in the Se deficiency level, accounting for 7.7% (Fig. 4).

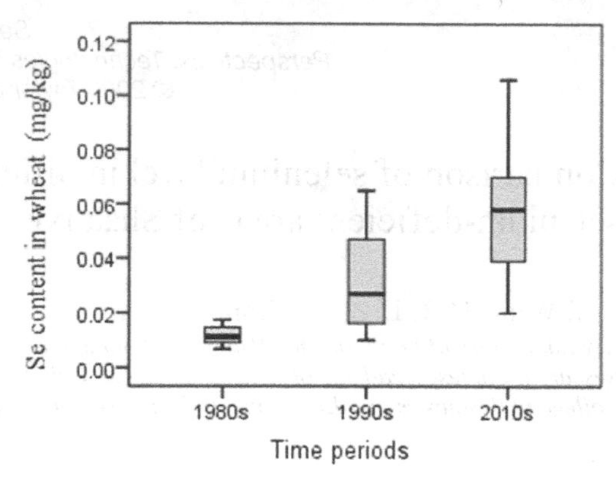

Figure 4. Wheat Se content (mg/kg) of different periods in the middle of Shaanxi Province.

4 CONCLUSIONS

Since the reform and opening up, with the socioeconomic development in China, the living conditions and standards of people have continuously improved. Sources of food have been generally more Se-enriched, and the Se level of the inhabitants has been significantly improved. There are obvious differences in the improvement of Se level. Although soil Se has been separated from the state of Se deficiency as a whole in Shaanxi, some areas are still on the verge of Se deficiency or potential Se deficiency. Future research needs to focus on those low Se regions.

REFERENCES

Cao, X.G., Lv, X.Y., Xu, G.Y., Zhang, B.D., Liu, H.L. & Deng, J.X. 2003. Investigation on Kashin-Beck disease in Shaanxi Province in 2002. *Disease Prevention and Control Circular* 18(3): 27–29. (in Chinese).

Du, B., Zhou, J. & Zhou, J. 2018. Selenium status of children in Kashin–beck disease endemic areas in Shaanxi, China: Assessment with mercury. *Environ Geochem Health* 40(2): 903–913.

Li J.Y., Chen, D.Z. & Ren, S.X. 1982. Study on the relationship between selenium and Kashin-Beck disease in Shaanxi Province. *Journal of Environmental Science* 2(2): 91–100. (in Chinese).

Sun, J. & Xu, J.F. 1992. Investigation on the decline of Kashin-Beck disease and selenium level in internal and external environment in Baoji area. *Disease Prevention and Control Circular* (4): 16–18. (in Chinese).

Tan, J.A. 1989. *The Atlas of Endemic Diseases and Their Environment.* Beijing: Science Press.

Yang, X.D., Dai, H.X., Ren, Y.F. & Du, Y. 2018. Monitoring of hair selenium content in Kashin-Beck disease area of Shaanxi Province. *Chinese Journal of Local Medicine* 37(4): 330. (in Chinese).

Zhang, W., Neve, J., Xu, J., Vanderpas, J. & Wang, Z. 2001. Selenium, iodine and fungal contamination in Yulin district (People's Republic of China) endemic for Kashin-beck disease. *Intern Orthopaedics* 25(3): 188–190.

Selenium Research for Environment and Human Health: Perspectives, Technologies and Advancements – Bañuelos, Lin, Liang & Yin (eds)
© 2020 Taylor and Francis Group, London, ISBN 978-1-138-39014-0

Reference materials of selenium-enriched rocks and soils

Z.Y. Bao & L.Y. Yao
Zhejiang Institute, China University of Geosciences at Wuhan, China

F. Tian, Z.Z. Ma & B.L. Fan
Faculty of Materials Science and Chemistry, China University of Geosciences (Wuhan), Wuhan, China

1 INTRODUCTION

Selenium (Se) is an essential element to animals and humans. Intake of Se originates primarily from consumption of Se-enriched crops and vegetation growing in Se-rich soils, which originated, in most cases, from Se-rich parent materials (rocks). An accurate measurement of the concentration and speciation of Se in Se-rich rocks and soils is therefore necessary and depends strongly on the type and origin of Se-rich rocks and soils. In this paper, we present results from preparation and certification of four reference materials for Se-rich rocks, and five for Se-rich soils.

2 MATERIALS AND METHODS

2.1 *Samples preparation*

Two Se-rich rocks were collected from Enshi city including Yutangba Village (YTB) and Baiyanping Village (BYP) in Hubei Province, and other two samples collected from Ziyang County including Gaoqiao Town (GQ) and Haoping Town (HP) in Shanxi Province. These four locations are the most famous Se-rich areas in China, known as "Selenium Capital of the World" and "Selenium Valley of China", respectively. Mixed rock samples were prepared from the four Se-rich rocks as a candidate of reference material for Se fractionation.

Five Se-rich soil samples as candidate of reference materials were taken from five Se-rich regions in China.

2.2 *Chemical analysis*

Concentrations of Se and other eight elements Ag, As, Cd, Cu, Mo, Pb, V, and Zn were analyzed in Se-rich rocks, while in Se-rich soils concentrations of Se and other six elements Cd, Cr, Cu, Ni, Pb, and Zn were determined. In addition, Se fractionation analysis in the rock and soil samples, including the fractions of water soluble, exchangeable, organic matter binding, sulfide/selenide, and residual Se, was also carried out by a partial extraction procedure modified from Martens & Suarez (1996), Kulp & Pratt (2004), and Zhu et al. (2006). The certified Se fractionation result was reported elsewhere.

3 RESULTS AND DISCUSSION

Concentrations of Se and other elements in Se-rich rock and soil samples were independently analyzed by ten qualified analytical laboratories. After a strict statistical analysis of the data, the certified concentrations and standard deviation values of all elements are shown in Tables 1 and 2 for each reference material.

Table 1. Certified concentration of Se and other elements in Se-rich rock reference materials (mg/kg).*

	GBW07397	GBW07398	GBW07399	GBW07400
	YTB	BYP	GQ	HP
Se	960 ± 20	1030 ± 50	49 ± 4	39 ± 5
Cu	140 ± 20	82 ± 20	520 ± 40	75 ± 5
Pb	16 ± 4	7.8 ± 0.7	37 ± 3	22 ± 2
Zn	27 ± 5	60 ± 4	1500 ± 200	150 ± 20
Cd	3.1 ± 0.2	1.7 ± 0.1	19 ± 2	3.7 ± 0.3
As	8.1 ± 0.5	11 ± 1	250 ± 30	45 ± 5
Mo	204 ± 20	120 ± 30	410 ± 20	90 ± 5
V	3000 ± 200	2180 ± 90	3900 ± 200	1510 ± 90
Ag	(0.7)	(0.5)	(4)	(5)

*YTB: Yutangba Village; BYP: Baiyanping Village; GO: Gaoqiao Town; HP: Haoping Town.

Table 2. Certified concentration of Se and other elements in Se-rich soil reference materials (mg/kg).

	Soils Collected from Five Regions in China				
	ES	ZY	HN	FC	ZHJ
Se	19.5 ± 2.4	9.4 ± 1.3	2.7 ± 0.4	0.91 ± 0.11	0.83 ± 0.09
Cr	175 ± 14	151 ± 12	488 ± 31	147 ± 14	92 ± 8
Ni	101 ± 10	140 ± 17	157 ± 12	23 ± 3	24 ± 2
Cu	79 ± 7	157 ± 13	82 ± 7	26 ± 3	29 ± 3
Zn	160 ± 18	248 ± 19	118 ± 10	74 ± 7	85 ± 7
Cd	11.8 ± 1.1	5.8 ± 0.7	(0.04)	0.20 ± 0.05	0.42 ± 0.03
Pb	30 ± 4	33 ± 4	13.7 ± 1.5	36 ± 3	34 ± 3

REFERENCES

Kulp, T.R. & Pratt, L.M. 2004. Speciation and weathering of selenium in Upper Cretaceous chalk and shale from South Dakota and Wyoming, USA. *Geochim Cosmochim Acta* 68(18): 3687–3701.

Martens, D.A. & Suarez, D.L. 1996. Selenium speciation of soil/sediment determined with sequential extractions and hydride generation atomic absorption spectrophotometry. *Environ Sci Tech* 31(1): 133–139.

Zhu, J.M., Han, W.L., Lei, L. & Zhao, Y.Z. 2006. Selenium speciation of Se-rich rocks from Yutangba of Enshi, China. *Geochim Cosmochim Acta (Suppl)* 70: A754–A754.

Selenium Research for Environment and Human Health: Perspectives, Technologies and Advancements – Bañuelos, Lin, Liang & Yin (eds)
© 2020 Taylor and Francis Group, London, ISBN 978-1-138-39014-0

Trace elements in pyrite of Se-enriched rocks from Ziyang, Central China

H. Tian & C.H. Wei
Faculty of Materials Science and Chemistry, China University of Geosciences, Wuhan, China

L.Y. Yao & Z.Y. Bao*
Zhejiang Institute, China University of Geosciences (Wuhan), Hangzhou, China

S.Y. Xie & H.Y. Zhang
State Key Laboratory of Geological Processes and Mineral Resources (GPMR), Faculty of Earth Sciences, China University of Geosciences, Wuhan, China

1 INTRODUCTION

The Naore village, Shuang'an Town in the Ziyang County, southern Shaanxi, is one of the few known selenium (Se) enriched areas in China, and in these locations, the early cases of human Se poisoning were recorded in the 1980's (Cheng & Mei 1980). However, the Se-rich soils stimulated agriculture production in the Naore area, which resulted in the name "Chinese Selenium Valley" (Tian 2017). The Se source in soil is attributed to pyrite-bearing and Se-enriched tuffs and carbonaceous rocks of Lower Cambrian Lujiaping Formation (Luo et al. 2004). In Naore, pyrite commonly occurs in these Se-enriched rocks that have been identified as the main source of Se (Tian et al. 2016a, b), similar to the western San Joaquin Valley of California, USA (Wu et al. 2000). For geochemical exploration, the accumulation and distribution of trace elements in the pyrite is controlled by its sedimentary process and formation environment, which are significant for understanding the possible origin and forming mechanism of the Se enrichment in Lujiaping Formation.

2 MATERIALS AND METHODS

2.1 *Samples*

Seven samples (labeled as NR-1, NR-2, NR-3-1, NR-3-2, NR-4-1, NR-4-2, and NR-4-3), mainly as black pyritic carbonaceous slate and volcanic tuff of Lujiaping Formation of Early Cambrian period (K.L. Luo, pers. comm.), were collected from Naore Village, Ziyang, China. NR-3-1 is a relative fresh rock, while NR-3-2 is the weathered rock. A total of 89 pyrite grains were hand-selected for microanalysis by EPMA and LA-ICPMS in Center of Geosciences, Universität Göttingen in Germany.

2.2 *Methods*

Major and trace elements of the samples were determined by ICP-OES (PE800) and ICP-MS (PE300D). Selenium and arsenic (As) contents were analyzed by HG-AFS (Beijing Titan Instruments).

For EPMA point analysis of pyrite, an accelerating voltage 20 kV, a beam current 80 nA, and a beam diameter 15 micron were used. For EPMA element mapping, the scanning was performed at 20 kV and 300 nA, with a counting time of 60 ms/step.

Trace element analyses by LA-ICP-MS were carried out with a 193-nm ArF excimer laser ablation system (RESOlution M-50) coupled to an Element2 sector field ICP-MS. Sulfide MASS-1 is served as the external reference material and NIST SRM 610 (glass) was used for quality control. The ^{57}Fe of pyrite was used as the internal standard element, which has been quantified with EPMA.

3 RESULTS AND DISCUSSION

3.1 *Bulk composition*

The SiO_2 contents of the pyrite-bearing rocks ranged from 42.96 to 69.15% with an average of 55.62%. The contents of FeO_T and S in rocks NR-3-1, NR-3-2 and NR-4-2 were higher than those in the other samples, with the lowest values (0.95% FeO and 0.007% S) for sample NR-4-1. The enrichment degrees of Se, As, Mo, Cd, and Cr were 1-2 orders of magnitude above the continental crust abundance. Sample NR-3 was the most enriched in Se, As, and Cd with the highest concentrations being 73.1 μg/g (NR-3-2), 195 μg/g, and 2.94 μg/g (NR-3-1), respectively. However, other samples had Se contents <3.5 μg/g, with the lowest Se concentration (0.56 μg/g) in sample NR-4-1, which did not contain S and Fe.

3.2 *Trace elements concentrations in pyrite*

The average concentrations of As, Se, Cu, Ni, Co, and Zn were 670 ± 530 μg/g (n = 154), 280 ± 340 μg/g (n = 280), 140 ± 70 μg/g (n = 25), 240 ± 230 μg/g (n = 43), 530 ± 590 μg/g (n = 46), and 140 ± 20 μg/g (n = 14), respectively. Also, with the

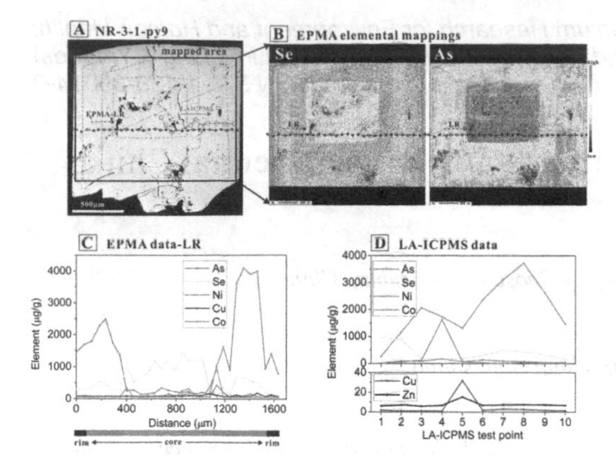

Figure 1. Photomicrograph (A), element analyses (C, D) and EPMA elemental mappings (B) of fresh pyrite grain NR-3-1-py9 (type-I) from Ziyang, China. Note: Figure 1B red to blue colors indicate high to low concentrations.

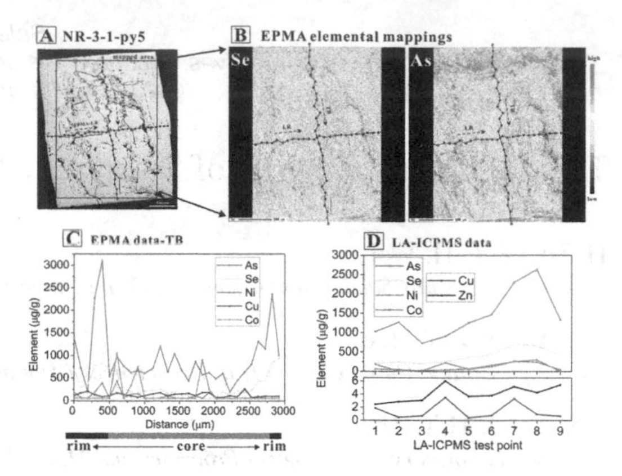

Figure 2. Photomicrograph (A), element analyses (C, D) and EPMA elemental mappings (B) of fresh pyrite grain NR-3-1-py5 (type-II) from Ziyang, China. Figure 2B red to blue colors indicate high to low concentrations.

decreasing pyrite size, the maximum and average value of Se increased, showing a negative relationship between Se content and pyrite size. Similarly, Se and As concentrations in pyrite samples NR-3-1 (520 ± 370 μg/g, n = 56 and 1030 ± 650 μg/g, n = 52, respectively) and NR-3-2 (740 ± 360 μg/g, n = 44, and 720 ± 400 μg/g, n = 41, respectively) were also higher than those of other samples (100 μg/g and 300 μg/g, respectively). Therefore, samples NR-3-1 and NR-3-2 were chosen for further EPMA elemental profiles and mapping.

3.3 *Distribution of trace-elements in pyrite*

Grain-scaled elemental mappings of Se-enriched pyrite revealed two types (I and II) of Se and As distribution patterns with alternating Se or As contents during the growth of zoned pyrite (Figs 1, 2). Type-I pyrite is characterized by high Se (120–1870 μg/g) and low As (<150–1580 μg/g) concentrations in the core and moderate Se and high As concentrations near the pyrite crystal margins. Type-II pyrite presents a similar distribution pattern of Se and As, being both poor (Se 60–680 μg/g and As < 150–1310 μg/g) at the core and Se-enriched along the edge. Meanwhile, these results were confirmed by the measurements by LA-ICP-MS. We also found that these pyrites are overgrown by Co- and Ni-enriched zones at the edges of the crystals (Figs 1C, D, 2C, D). Additionally, high-Se (19.3 wt%) and high-As (4.53 wt%) "veins" and patchy areas were found in pyrite (Fig. 1B), which are distributed mostly in high-Se cores/zones and can be divided into two kinds of categories. These new findings still require further study for understanding their possible secondary enrichment processes.

4 CONCLUSIONS

Pyrite is the major carrier for Se of Lujiaping Formation, and Se was incorporated mainly by isomorphic replacement. The texture of pyrite elemental zonations reflects multi-stage hydrothermal pulses with changing trace element concentrations. Type-II pyrite formed later than type-I pyrite. This study extends our understanding of Se geochemical formation processes and highlights the mechanistic aspects of Se during its cycle.

REFERENCES

Cheng, Q. & Mei, Z. 1980. A preliminary investigation report of selenium poisoning in ziyang county, Shaanxi Province. *Shaanxi Journal of Agricultural Sciences* (6): 19–31.

Tian, H. 2017. *The Occurrence State and Speciation of Selenium and Its Environmental Behaviors in Rock-Soil-Plant from Typical High-Se Areas*. Wuhan: China University of Geosciecnes.

Tian, H., Bao, Z., Wei, C. et al. 2016a. Improved selenium bioavailability of selenium-enriched slate via calcination with a Ca-based sorbent. *J Geochem Explor* 169:

Tian, H., Ma, Z., Chen, X. et al. 2016b. Geochemical characteristics of selenium and its correlation to other elements and minerals in selenium-enriched rocks in Ziyang County, Shaanxi Province, China. *J Earth Sci* 27(5): 763–776.

Luo, K., Xu, L., Tan, J., Wang, D. & Xiang, L. 2004. Selenium source in the selenosis area of the Daba region, South Qinling Mountain, China. *Environ Geol* 45(3): 426–432.

Wu, L., Banuelos, G. & Guo, X. 2000. Changes of soil and plant tissue selenium status in an upland grassland contaminated by selenium-rich agricultural drainage sediment after ten years transformed from a wetland habitat. *Ecotoxicol Environ Saf* 47(2): 201–209.

Selenium Research for Environment and Human Health: Perspectives, Technologies and Advancements – Bañuelos, Lin, Liang & Yin (eds)
© 2020 Taylor and Francis Group, London, ISBN 978-1-138-39014-0

Soil selenium distributions in dry arable land in Zhongwei, China

M.R. Farooq, X.B. Yin & Z.D. Long
School of Earth and Space Sciences, University of Science and Technology of China, Hefei, Anhui, China
Jiangsu Bio-Engineering Research Center for Selenium, Suzhou, Jiangsu, China

L.X. Yuan*
Jiangsu Bio-Engineering Research Center for Selenium, Suzhou, Jiangsu, China

Z.Z. Zhang
State Key Laboratory of Biogeology and Environmental Geology, China University of Geosciences, Wuhan, China

X.D. Liu
School of Earth and Space Sciences, University of Science and Technology of China, Hefei, Anhui, China

1 INTRODUCTION

Geology exerts a fundamental control on concentrations of soil selenium (Se), which influenced the human dietary Se intake (Fordyce 2013). Researchers have tried to solve the uncertainty about nutrient content in soil and its connection to bacterial strains, competitive ions, organic matter (OM), pH, and redox status, as well as dry and harsh climate, evapotranspiration, precipitation, temperature, land use type, and topography (Xu et al. 2018). The influential topographic factors and their relationships with nutrients in soil have been widely studied (Li et al. 2017). However, there are few studies on dry topographic effect on Se distribution in soils. Hence, the present study was based on studying the relationship between dry soil and Se concentrations in the Northwest region of China.

2 MATERIALS AND METHODS

2.1 *Study area*

Zhongwei city is located at west bank of Yellow river near Tengger desert (Fig. 1). The terrain is high in the southwest and low in the northeast. The study area is characterized by scant precipitation, strong winds in the winter, and high temperatures with intense evaporation in the summer. The multi-year average temperature is 9.37°C. The highest and lowest temperatures are usually observed in July and January, respectively. The multi-year mean precipitation is 185.18 mm, with the majority (72%) occurring between June and September. The annual mean evaporation of the area is 1774.25 mm, and over 73% of the annual evaporation occurs between April and September (Li et al. 2014). There are four types of landform classes: Desert, Yellow River alluvial plains, platform mountain, and large basin geomorphic unit, as well as seven types of soil: Sierozem, loessal, aeolian sandy, anthropogenic-alluvial, alluvial, saline, and red clay (Wang et al. 2016).

Figure 1. Total Se concentration in top soils growing watermelons in agricultural areas of Zhongwei, Ningxia Province, China.

2.2 *Sampling and preparation*

A total of 695 soil samples were collected in watermelon cultivated areas from three different counties (Shapotou, Zhongning, and Haiyuan) of Zhongwei city in years 2017 to 2018 (Fig. 1). Systematic grid sampling was applied to measure the distribution of Se in randomly selected sampling sites (Xi et al. 2014). Topsoil from 0 to 20 cm was collected at a density of one sample per 0.5 to 1 km^2. Each site was located in a different soil type of each respective area within the selective grids. GPS coordinates marked spatial distribution within sampling area at study sites.

Table 1. Parameters of Zhongwei dry soil (mg/kg).

	Mean	GM*	Min.	Max.	S.D	CV (%)
T-Se	0.20	0.18	0.01	0.34	0.07	32.82
B-Se	0.03	0.03	0.00	0.06	0.01	39.52
B/T-Se%	14.42	13.71	5.88	39.16	5.15	35.69
pH	8.70	8.70	8.04	9.48	0.32	3.68

*GM: geometric mean

2.3 *Chemical analysis and statistical methods*

Concentrations of Se in soil samples (free of plant roots and detritus) were measured after acid digestion. The digest material was cooled to room temperature and 5 ml HCl (12 M) were added to reduce Se(VI) to Se(IV) for 3–4 h. The detailed procedure was described by Gao et al. (2011). The recovery of the standard reference materials ranged from 85.5% to 117.8%, and the relative standard deviation (RSD) of reference materials was calculated as 0.76% (Yuan et al. 2013). All statistical analyses were carried out by Origin, SPSS 23 (IBM SPSS Statistics). Spatial interpolation kriging method (ArcGIS 10.5 Esri Inc.) was used to determine Se spatial distribution in study area.

3 RESULTS AND DISCUSSION

3.1 *Selenium distribution in topsoil*

The geometric and arithmetic means of Se concentrations for all top soils were 0.18 and 0.20 mg/kg, respectively. The lowest value is 0.01 mg/kg, and the highest value is 0.324 mg/kg (Table 1). The arithmetic means of Se concentration are higher than the geometric means because of skewed distribution of the data from different areas. The average topsoil Se concentrations are as follows: Shapotou County> Haiyuan County> Zhongnig County. The maximum average concentration of Se was 0.25 ± 0.07 mg/kg in Shapotou County. This value is slightly higher than those in the other two counties. The organic matter in Zhongwei soil was <1.0 %, which is low (Wang et al. 2016).

3.2 *Influencing factors for Se distributions*

Uneven distribution of Se in soil has coefficient of variation (CV) with average value of 32.82%. Different chemical processes (pH, redox potential, organic matter, and geochemical competitive ions) can affect Se content in the soils of Zhongwei. High bioavailable/total Se (B/T-Se) percentages (B/Se%) (mean 14.42%) were high in the present study compared with other areas in China (Wang et al., 2017). The high pH (8.7 ± 0.32) could highly improve the bioavailable Se percentages (Table 1) (Dinh et al. 2018). However, further studies will be required to identify other factors that influence soil Se distribution in Zhongwei.

4 CONCLUSIONS

The average total Se content and the bioavailable Se content of top soils in Zhongwei are 0.2 mg/kg and 0.03 mg/kg, respectively. The ratios of average B/T-Se are 14.42%, which is higher than those in other regions of China. The variable Se distribution in Zhongwei arable land is controlled and influenced by different geochemical factors like high pH and low organic matter.

This study was supported by Selenium Industry Project in Zhongwei, Ningxia (Grant No: 2018-FXCY-001; 2018-FXCY-002).

REFERENCES

Dinh, Q.T., Wang, M., Tran, T.A.T. et al. 2018. Bioavailability of selenium in soil-plant system and a regulatory approach. *Crit Rev Env Sci Tec* 49(6): 443–517.

Fordyce, F.M. 2013. Selenium deficiency and toxicity in the environment. In O. Selinus (ed), *Essentials of Medical Geology*: 375–416. Dordrecht: Springer.

Li, P., Wu, J. & Qian, H. 2014. Hydrogeochemistry and quality assessment of shallow groundwater in the southern part of the Yellow River Alluvial Plain (Zhongwei Section), Northwest China. *Earth Sci Res J* 18(1): 27–38.

Li, Z., Liang, D., Peng, Q., Cui, Z., Huang, J. & Lin, Z. 2017. Interaction between selenium and soil organic matter and its impact on soil selenium bioavailability: A review. *Geoderma* 295: 69–79.

Wang, C., Wang, J. & Wang, F. 2016. Spatial variability of soil organic matter in sands of Zhongwei City. *Soil Bulletin* 47(2): 287–293.

Wang, D., Zhou, F., Yang, W., Peng, Q., Man, N. & Liang, D.L. 2017. Selenate redistribution during aging in different Chinese soils and the dominant influential factors. *Chemosphere* 182: 284–292.

Xi, X., Xiao, G., Zhou, G., Ye, J. & Li, Z. 2014. National multi-purpose regional geochemical survey in China. *J Geochem Explor* 139: 21–30.

Xu, Y., Li, Y., Li, H. et al. 2018. Effects of topography and soil properties on soil selenium distribution and bioavailability (phosphate extraction): A case study in Yongjia County, China. *Sci Total Environ* 633: 240–248.

Yuan, L., Zhu, Y., Lin, Z.-Q., Banuelos, G., Li, W., & Yin, X. 2013. A novel selenocystine accumulating plant in selenium-mine drainage area in Enshi, China. *PLoS ONE* 8(6): e65615.

Selenium Research for Environment and Human Health: Perspectives, Technologies and Advancements – Bañuelos, Lin, Liang & Yin (eds)
© 2020 Taylor and Francis Group, London, ISBN 978-1-138-39014-0

Selenium fractionation in soil from typical selenium-rich areas in China

B.L. Fan & Z.Y. Bao
Faculty of Materials Science and Chemistry, China University of Geosciences, Wuhan, Hubei, China

L.Y. Yao
Zhejiang Institute, China University of Geosciences, Hangzhou, Zhejiang, China

1 INTRODUCTION

Selenium (Se) is both an essential micronutrient and a toxic environmental pollutant, with a narrow range between dietary adequacy and toxic level. The WHO recommends a daily intake of 50 to 200 μg of Se for adults. The human intake of Se is mainly from food consumption and dependent on the level of Se in the soil in which the plant grows (Rayman 2000). Selenium in plants is controlled by soil bioavailable Se, which is determined by soil total Se and Se speciation.

Many researchers have applied different sequential extraction methods to evaluate the fractionation and speciation of soil Se, and several common trends have been gradually identified (Kulp et al. 2004). For example, water-soluble and exchangeable Se is generally considered to be bioavailable, while some Se associated with organic matter is potentially available to plants. The rest of Se (selenide-mineral and residue) is not considered available for plants (Tian et al. 2016).

As a trace element, Se is heterogeneously distributed in soils of China. Enshi Prefecture and Ziyang County are the most typical areas with high Se content in China. Previous research mainly focused on the the spatial distribution of soil total Se and the transformation of Se fractionation and speciation. Few studies have investigated the distribution of soil Se fractionation in different types of Se-rich soils.

This study explored the characteristics of soil selenium fractionation in typical selenium-rich areas in China. The aims of this study are (1) to obtain insight into soil Se fractionation in different types of Se-rich areas, (2) to determine the reasons for different Se fractionation distribution in each soil type, and (3) to provide information for national utilization of typical selenium-rich areas in China.

2 MATERIALS AND METHODS

2.1 *Soil collection*

Soil samples were collected from three typical selenium-rich areas: Ziyang County of southern Shaanxi Province (ZY), Enshi Prefecture of Hubei Province (ES) and Wenchang County of Hainan Province (HN) in China. Thirteen of surface topsoil samples were collected from each respective region. The soils were classified as yellow brown soil, gray black soil, and latosol, respectively.

2.2 *Analytical methods*

Concentrations of Al_2O_3, K_2O, Na_2O, CaO and P_2O_5 were measured by XRF. Total Se was analyzed by HG-AFS. The accuracy of determinations was verified using the Chinese standardized reference materials (GBW07405, GBW07406). The recoveries of reference elements were within 3% of the actual values.

2.3 *Sequential extraction*

For extraction of soil Se fractionation, 1 g of homogenized soil was weighed into 50-ml tubes and mixed with 10 ml of solutions to extract Se of five operationally defined soil fractionations in sequence. The method was modified by Kulp et al. (2004) and Tang et al. (2018) as follows: Water-soluble Se (F1) by extracting with Milli-Q water for 2 h, ligand-exchangeable Se (F2) by extracting with 0.1 M P-buffer for 2 h, organic matter-associated Se (F3) by extracting with 0.1 M NaOH for 2 h at 90°C, acid soluble Se (F4) was mixed with 3 M HCl for 2 h at 90°C, and residual Se (F5) digested by mixed acid of HNO_3, HF, and $HClO_4$. All the extraction experiments were performed in duplicate.

3 RESULTS AND DISCUSSION

3.1 *Total soil Se*

The Se content of soil in three selenium-rich areas was quite different from each other. The soil Se concentrations from Enshi were the highest (15.29 ± 4.53 μg/g; n = 4), followed by Ziyang (13.69 ± 6.73 μg/g; n = 5) and Wenchang (2.70 ± 0.19 μg/g; n = 4) (Table 1). Soils from Enshi and Ziyang showed a high Se content with a wide range. Ranking the lowest among the three regions, soil Se content of Wenchang was still more than six times of the world average (about 0.4 μg/g).

Table 1. Selenium contents and fractionations in soil samples.

Sample ID	F1	F2	F3	F4	F5	Se	CIA*
	%	%	%	%	%	μg/g	
ES01	3.10	5.56	42.3	25.7	23.4	4.27	83.1
ES02	2.12	9.05	66.7	8.37	13.7	23.8	80.2
ES03	1.81	7.42	70.5	6.49	13.7	14.1	80.0
ES04	2.35	9.52	65.8	6.86	15.4	18.9	79.5
ZY01	1.38	10.5	14.4	40.1	33.6	9.70	67.1
ZY02	1.96	8.11	14.0	42.5	33.4	5.64	71.3
ZY03	0.99	3.39	24.5	44.4	26.8	14.0	66.1
ZY04	0.82	3.48	17.8	55.3	11.1	23.5	47.2
ZY05	1.06	3.60	10.9	75.0	19.5	15.6	67.4
HN01	1.77	5.86	69.0	4.66	18.7	2.85	99.2
HN02	1.87	6.02	63.2	6.71	22.4	2.88	98.8
HN03	1.78	4.48	68.5	7.98	17.2	2.50	99.4
HN04	2.02	5.61	69.9	7.12	15.29	2.58	98.8

*CIA =100 × n (Al_2O_3) / [n (Al_2O_3)+n (CaO) + n (Na_2O) + n (K_2O)] where CaO is the amount of CaO incorporated in silicate fraction (Nesbit & Young 1982).

3.2 *Fractionations of Se in Se-rich soils*

Selenium fractionations were obtained by the sequential extraction procedure. Organic matter associated Se accounted for the largest proportion in soil from Enshi and Wenchang, followed by residual Se. The percentage of organic matter-associated Se and residual Se were 61.34% and 16.58% for soil of Enshi, respectively, and 67.7% and 18.4% for soil of Wenchang, respectively. Meanwhile, Se mainly existed as acid soluble Se, residual Se and organic matter associated Se for soil of Ziyang, where the acid soluble Se accounted for more than half of the total Se (51.4%), while organic matter associated Se and residual Se accounted for 16.32% and 24.88% of the total Se, respectively. Water-soluble Se and ligand-exchangeable Se were generally less than 10% in all the three soil types.

Soils in Enshi were derived from weathered carbonaceous shale, carbonaceous slate and fertilized with stone coal for a long period of farming (Zhu et al. 2008). Selenium in Ziyang soil originated from carbonaceous slate and tuff, where the main fractionation of acid soluble Se was formed by weathering of pyrite of the parent rocks (Tian et al. 2016). After strong chemical weathering and long-term cultivation in Wenchang, soils remained as a small proportion of acid soluble Se and a large proportion of organic Se. Result shows that there was a strong negative correlation between soil total Se and CIA ($r = -0.764$, $p < 0.01$), indicating total soil Se decreased with stronger weathering intensity. After long-term weathering and farming, the existing form of Se in soil gradually altered from acid-soluble to organic matter associated Se.

4 CONCLUSIONS

Soil Se concentrations of Enshi, Ziyang and Wenchang in this study were 15.29 ± 4.53 μg/g, 13.69 ± 6.73 μg/g, and 2.70 ± 0.19 μg/g, respectively. The Se concentrations and fractionation distribution in soils from several selenium-rich areas was different. The main Se fractionation of Enshi and Wenchang soil was organic matter associated Se, followed by residual Se. Acid soluble Se and residual Se accounted for a large proportion of soil Se in Ziyang. Generally, the parent rock has an important effect on the total Se in soil. Chemical weathering and long-term tillage reduced the proportion of acid-soluble Se and increased the proportion of organic-matter-associated Se.

REFERENCES

Kulp, T.R. & Pratt, L.M. 2004. Speciation and weathering of selenium in Upper Cretaceous chalk and shale from South Dakota and Wyoming, USA. *Geochim Cosmochim Acta* 68(18): 3687–3701.

Nesbitt, H.W. & Young, G.M. 1982. Early Proterozoic climates and plate motions inferred from major element chemistry of lutites. *Nat* 299(5885): 715–717.

Rayman, M.P. 2000. The importance of selenium to human health. *Lancet* 356(9225): 233–241.

Tang, M., Bao, Z., Fan, B. et al. 2018. HG-AFS speciation analysis for 5 species of selenium in Se-rich soil with separation by sequential extraction. *Physical Testing and Chemical Analysis Part B: Chemical Analysis* 54(4): 408–412 (in Chinese).

Tian, H., Ma, Z., Chen, X. et al. 2016. Geochemical characteristics of selenium and its correlation to other elements and minerals in selenium-enriched rocks in Ziyang county, Shaanxi province, China. *J Earth Sci* 27(5): 763–776.

Zhu, J., Wang, N., Li, S. et al. 2008. Distribution and transport of selenium in Yutangba, China: Impact of human activities. *Sci Total Environ* 392(2–3): 252–261.

Selenium Research for Environment and Human Health: Perspectives, Technologies and Advancements – Bañuelos, Lin, Liang & Yin (eds)
© 2020 Taylor and Francis Group, London, ISBN 978-1-138-39014-0

Influential factors of soil selenium content and distribution in a Se-rich area of Yichun

N. He, Y.D. Ding, C.Y. Han & Z.G. Zhao
Yichun University, Yichun, Jiangxi, China

D. Yu
School of Public Administration, China University of Geosciences (Wuhan), Wuhan, China

1 INTRODUCTION

Selenium (Se) is an essential trace element for humans and animals, and is primarily supplied via forages and food crops. Selenium in forages and food crops is dependent on Se content of the soil in areas where there is no other supplementation (Hawkesford et al. 2007, Winkel et al. 2015). The distribution of Se on the earth's surface varies widely, forming Se-deficient and Se-excessive ecosystems, which affect human and animal health. Yuanzhou District is one of the typical Se-rich areas in China. It is located in the middle of the northern foot of the Luoxiao Mountains, the northern foot of the Wugongshan Mountains in the western part of Jiangxi Province. Its south, west, and north borders are surrounded by mountains. The central and eastern areas are wide and hilly, and the terrain is low. Red earth generally distributes in mountains, accounting for 40% of the study area; lime soil accounts for about 30%; paddy soil accounts for about 17% and fluvo-aquic soil mainly distributed on both sides of Yuanhe River. Information regarding Se background levels in soils is important to prevent Se deficiency, optimize human nutrition, and to prevent toxicological effects (Nakamaru et al. 2005, Gabos et al. 2014). This paper discussed the effects of land use patterns, soil types, and elevations on soil Se content, and explores the relationship between soil Se content and selected ecological factors in the study areas.

2 MATERIALS AND METHODS

2.1 *Sample preparation*

According to the soil distribution and field exploration in Yuanzhou District, the sampling site and reasonable sample density are determined by the township boundary. We mainly selected thicker soil types such as farmland, gardenland, woodland, grassland and mountainous hills. We avoided sampling localized areas, exposed areas of rocks, dumping areas, ditches, and transportation areas. A total of 4000 soil samples were collected with a stainless-steel shovel at a depth of 0–20 cm.

2.2 *Sample analysis*

All samples were analyzed for total Se by hydride generation-atomic fluorescence spectrometry (He et al. 2002). The detection limit was 0.002 μg/ml, and the precision of the method was 3.6 CV% (n = 6). The national standard method (NY/T 1104-2006) for soil Se measurements was followed.

2.3 *Data analysis*

Data processing and chart production were done using ArcGIS9.3, SPSS16.0, and MS Excel 2003.

3 RESULTS AND DISCUSSION

3.1 *Distribution of soil Se*

The concentration of Se in most of the soil samples ranged from 0.001 to 1.03 mg/kg with an average of 0.18 mg/kg (Fig. 1).

Most of areas are Se-medium (0.16–0.40 mg/kg), while Se-rich (> 4.0 mg/kg) areas are mainly distributed in the central and southern regions, and selenium-deficient (<0.10 mg/kg) areas are mainly distributed in the north of Yuanzhou district.

3.2 *Effect of soil type on Se content*

There are some differences in Se content in soil types of different parent materials. From the statistics, the Se content of fluvo-aquic soil is relatively stable, with the highest average value of 0.34 mg/kg, and the coefficient of variation is the smallest, indicating that the spatial difference of Se content is low in fluvo-aquic soil. The average Se content of red, paddy, and brown lime soils differed little, and the coefficient of variation exceeded 60%, indicating that the Se content of these three soil types varied greatly (Table 1).

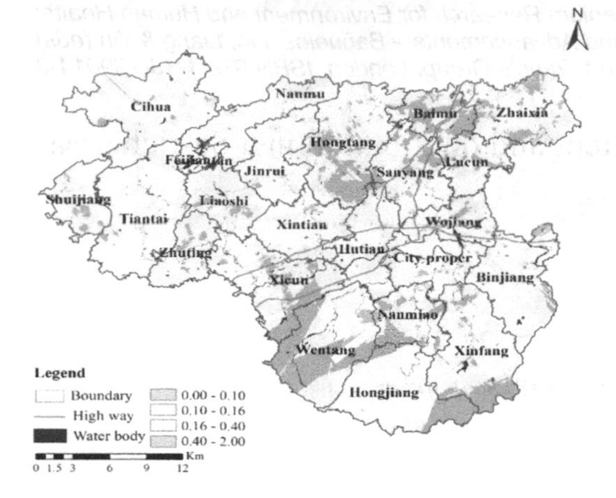

Figure 1. Distribution of soil Se concentrations in Yuanzhou District, Yinchun, China.

Table 1. The soil Se concentrations of different soil types in Yuanzhou County, Yichun, China.

Soil Type	Sample Number	Mean ± STD (mg/kg)	CV (%)
Fluvo-aquic soil	26	0.34 ± 0.13	38.24
Red earth	1153	0.18 ± 0.12	66.67
Paddy soil	2246	0.18 ± 0.11	61.11
Lime soil	534	0.17 ± 0.12	70.59

Table 2. The soil Se concentrations of different land use types (mg/kg).

Land Use	Sample Number	Mean ± STD	CV (%)
Paddy	2100	0.18 ± 0.11	61.11
Dryland	1859	0.2 ± 0.12	60.00

3.3 *Effect of land use on Se content*

Selenium content is significantly different under different land use patterns. From the comparison of average values, dry soil was 0.20 mg/kg, which was slightly higher than that of water. The minimum and maximum values were higher than that of dry soil. We observed that the soil water permeability and leaching have a certain influence on the soil Se content, and the loss of water will cause the loss of Se (Table 2).

3.4 *Effect of elevation on Se content*

There is no significant difference in the average value of soil Se content of each elevation level. The range of Se is the highest in the range of 250-300 m, and the lowest in the range of 100-150 m, with the values at 0.20 and 0.17 mg/kg, respectively. From the comparison of coefficient of variation, with an increase in elevation, the variation of soil Se content is greater.

Table 3. The soil Se concentrations of different elevation (mg/kg).

Elevation (m)	Sample Number	Mean ± STD	CV (%)
65-100	385	0.19 ± 0.11	57.89
100-150	2255	0.17 ± 0.11	64.71
150-200	847	0.18 ± 0.11	61.11
200-250	258	0.18 ± 0.11	55.56
250-300	103	0.20 ± 0.14	70.00
300-350	40	0.19 ± 0.14	73.68
350-404	71	0.17 ± 0.16	94.12

We observed that with the increase of elevation, the difference of soil Se content is greater (Table 3).

4 CONCLUSIONS

The Se content of soil showed certain spatial distribution characteristics. The Se-deficient areas are mainly located in the north of Yuanzhou District, and the Se-rich areas are mainly located in the central and southern part. The Se content of different soil types is significantly different. The content and spatial variability of fluvo-aquic soil is stable and higher than that of red, paddy, and brown lime soils. The Se content is greatly affected by land use, which is higher in dryland soils than that of paddy soil. The average value of soil Se content has no obvious relationship with the change of elevation, but with the increase of elevation, the coefficient of variation of soil Se content increases, indicating that the higher the elevation, the more unstable the soil Se content.

ACKNOWLEDGEMENT

This study received financial support from the Social Science Twelfth Five-Year Project of Jiangxi (No. 15YJ13) and the Local Development Research Center Project of Yichun University (No. 18YB222).

REFERENCES

Gabos, M.B., Alleoni, L.R. & Abreu, C.A. 2014. Background levels of selenium in some selected Brazilian tropical soils. *J Geochem Explor* 145: 35–39.

Hawkesford, M.J. & Zhao, F.J. 2007. Strategies for increasing the selenium content of wheat. *J Cereal Sci* 46 (3): 282–292.

He, B., Liang, L. & Jiang, G.B. 2002. Distributions of arsenic and selenium inselected Chinese coal mines. *Sci Total Environ* 296:19–26.

Nakamaru, Y., Tagami, K. & Uchida, S. 2005. Distribution coefficient of selenium in Japanese agricultural soils. *Chemosphere* 58 (10): 1347–1354.

Winkel, L.H., Vriens, B., Jones, G.D. et al. 2015. Selenium cycling across soil-plant-atmosphere interfaces: a critical review. *Nutrients* 7(6):4199-4239.

Selenium Research for Environment and Human Health: Perspectives, Technologies and Advancements – Bañuelos, Lin, Liang & Yin (eds)
 ISBN 978-1-138-39014-0

Modelling atmospheric selenium transport and deposition on a global scale

A. Feinberg
Swiss Federal Institute of Technology, Institute of Biogeochemistry and Pollutant Dynamics, Zürich, Switzerland
Institute for Atmospheric and Climate Science, Zürich, Switzerland
Eawag: Swiss Federal Institute of Aquatic Science and Technology, Duebendorf, Switzerland

A. Stenke & T. Peter
Swiss Federal Institute of Technology, Institute for Atmospheric and Climate Science, Zürich, Switzerland

L.H.E. Winkel
Swiss Federal Institute of Technology, Institute of Biogeochemistry and Pollutant Dynamics, Zürich, Switzerland
Eawag: Swiss Federal Institute of Aquatic Science and Technology, Duebendorf, Switzerland

1 INTRODUCTION

Selenium (Se) is an essential dietary element for humans and animals. Insufficient or excessive Se intakes can cause health issues, with Se deficiency being a more widespread problem globally and affecting an estimated 0.5 - 1 billion people (Fairweather-Tait et al. 2011). The amount of Se in food crops depends on the amount of bioavailable Se in the soil where the food is grown. The concentrations of Se in soils depends not only on the local bedrock concentration, but also on inputs from the atmosphere. Atmospheric deposition of Se is an important source of Se in soils, especially through wet deposition. Several studies have linked precipitation levels with soil Se concentrations in field studies (e.g. Låg & Steinnes 1978, Blazina et al. 2014). It is therefore important to understand how Se is transported in the atmosphere and, ultimately, where it is removed back to the Earth surface.

An estimated 13,000–19,000 tons of Se is emitted annually into the atmosphere (Wen & Carignan 2007). Anthropogenic sources are responsible for around 40% of total Se emissions and natural sources contribute 60% (Mosher & Duce 1987). Anthropogenic sources include coal combustion, metal smelting, and biomass burning, whereas the major natural sources are volcanoes, the marine biosphere, and the continental biosphere. Volatile species of Se are oxidized in the atmosphere and are expected to condense on available particle surfaces. The atmospheric cycle is closed when soluble gas phase Se species and particle-bound Se are washed out of the atmosphere (wet deposition) or collide with the Earth surface (dry deposition).

Until now, research into the atmospheric Se cycle has been limited. For example, it is unknown which atmospheric chemical species of Se are important, how high Se can travel in the atmosphere, and how far Se can be transported in the atmosphere before being removed by wet or dry deposition. For these reasons, we built the first global atmospheric Se model. Using this model, we can trace where Se deposits at the Earth surface and predict how Se deposition changes with variations in emissions or climate.

2 METHODS

We implemented Se in the aerosol-chemistry-climate model SOCOL-AER. This model has successfully reproduced many aspects of the atmospheric sulfur (S) cycle, showing good agreement with satellite observations of stratospheric sulfate aerosol properties and with measurement networks of S wet and dry deposition (Sheng et al. 2015, Feinberg et al. 2019). Sulfur and Se are in the same group in the periodic table and, therefore, they undergo similar chemical transformations. Since the atmospheric cycles of Se and S are hypothesized to be quite similar, SOCOL-AER is an appropriate base model for including the Se cycle. SOCOL-AER comprises the climate model ECHAM5, which calculates atmospheric dynamical behavior, and the atmospheric chemistry model MEZON that includes a comprehensive set of 89 chemical species and 315 chemical reactions. SOCOL-AER also includes an aerosol microphysics module which tracks the sulfate particle size distribution in 40 size bins between 0.39 nm and 3.2 μm.

We implemented in SOCOL-AER seven gas phase species of Se (DMSe, OCSe, CSe_2, CSSe, H_2Se, oxidized organic Se, oxidized inorganic Se) and 40 tracers of condensed Se in each aerosol size bin. Chemical reaction rates of Se were either taken from the literature or estimated from the analogous S reaction rates. Emission maps of Se, which serve as input for the model, are scaled from existing S emission inventories. Wet and dry deposition of gas phase oxidized Se and aerosol Se are calculated based on gas species solubility, particle radius, and grid cell meteorology.

To test the effect of changing anthropogenic emissions and climate variability on Se deposition, we ran several types of simulations. The spatial distribution of anthropogenic Se emissions was varied to represent different years: 1985, 2000, and 2014. We also ran simulations with emission conditions fixed to the year 2000 and applying the sea surface temperatures from El Niño and La Niña years. These simulations test the impact of climatic variability on the Se deposition patterns.

To evaluate our Se model, we have compiled databases of past measurements of Se in precipitation (~100 stations) and aerosol (~400 stations). Although the measurement data are highly heterogeneous, with different sampling and analysis techniques, measurement frequencies, and time periods, they provide a general picture of atmospheric Se levels in different environments (urban, marine, and remote sites).

3 RESULTS AND DISCUSSION

Our model predicts that Se species are quickly oxidized and condense on available particles. Therefore, Se is mainly transported in the aerosol phase. The predicted global residence time of Se is on the order of only 4-5 days, which is similar to the residence time of submicrometer aerosol particles. Still, this residence time is long enough that Se undergoes far-range transport and travels over thousands of kilometers. For example, we find that deposition over Australia is mainly driven by marine biogenic emissions of methylated Se species that are later oxidized and transported to the land. Transport of Se through the free troposphere (>500 hPa, ~5 km) also plays an important role in the atmospheric cycling of Se. In certain regions, up to 70% of the Se that deposits over land was transported through the free troposphere.

We compare our model with available Se measurements spanning almost 70 years of data from the 1950s to the present. There are very few long-term measurements of atmospheric Se, meaning that long-term Se trends are difficult to identify. The modelled Se concentrations and fluxes generally agree with observations made by previous field studies within an order of magnitude. The model can also be used to identify unique environments where it would be interesting to conduct future measurement campaigns.

Inferred from changes in SO_2 emissions, there have been strong regional shifts in Se emissions between 1985 and 2014 (decreases in North America and Europe, increases in East and South Asia), which affect Se deposition patterns in these regions. Interannual variability in climate, for example, episodic El Niño and La Niña events, shifts precipitation patterns, which also impact the distribution of Se deposition. Further work could investigate whether predicted climate change could also alter the delivery of Se to agricultural soils.

4 CONCLUSIONS

In this work, we presented the first global atmospheric Se model of its kind. The atmospheric Se model can be used to evaluate hypotheses related to Se cycling, for example, whether Se deposition has changed in the past due to variations in anthropogenic emissions and climate. The model estimates of Se deposition fluxes can be further coupled to biogeochemical soil models, to predict which regions may be at risk for Se deficiency.

REFERENCES

Blazina, T., Sun, Y., Voegelin, A. Lenz, M., Berg, M. & Winkel, L.H. 2014. Terrestrial selenium distribution in China is potentially linked to monsoonal climate. *Nat Commun* 5: 4717.

Fairweather-Tait, S.J., Bao, Y., Broadley, M.R. et al. 2011. Selenium in human health and disease. *Antioxid. Redox Signal* 14(7): 1337–1383.

Feinberg, A., Sukhodolov, T., Luo, B.P. et al. 2019. Improved tropospheric and stratospheric sulfur cycle in the aerosol-chemistry-climate model SOCOL-AERv2. *Geosci Model Dev Discussions* doi:10.5194/gmd-2019-138.

Låg, J. & Steinnes, E. 1978. Regional distribution of selenium and arsenic in humus layers of Norwegian forest soils. *Geoderma* 20(1): 3–14.

Mosher, B.W. & Duce, R.A. 1987. A global atmospheric selenium budget. *J Geophys Res* 92(D11): 13289–13298.

Sheng, J.X., Weisenstein, D.K., Luo, B.P. et al. 2015. Global atmospheric sulfur budget under volcanically quiescent conditions: Aerosol-chemistry-climate model predictions and validation. *J Geophys Res-Atmos* 120(1): 256–276.

Wen, H. & Carignan, J. 2007. Reviews on atmospheric selenium: emissions, speciation and fate. *Atmos Environ* 41(34): 7151–7165.

Selenium Research for Environment and Human Health: Perspectives, Technologies and Advancements – Bañuelos, Lin, Liang & Yin (eds)
© 2020 Taylor and Francis Group, London, ISBN 978-1-138-39014-0

Fractionation and distribution of soil selenium and effects of soil properties in Heilongjiang

F.Q. Chi, E.J. Kuang, J.M. Zhang, Q.R. Su & X.L. Chen
Soil Fertilizer and Environment Resources Institute, Heilongjiang Academy of Agriculture Sciences, Harbin, China
Key Laboratory of Soil Environment and Plant Nutrition of Heilongjiang Province, Harbin, China

Y.W. Zhang & Y.D. Liu
College of Resources and Environment, Northeast Agricultural University, Harbin, China

1 INTRODUCTION

Selenium (Se) is an essential trace element in humans and animals. Selenium content in the crust of the Earth is unevenly distributed, and it can be divided into available and unavailable Se. Selenium toxicity does, however, occur. In this regard, high Se content in Se-contaminated areas can lead to chronic poisoning diseases, such as alkaline disease and blindness disorder in livestock (Ohlendorf 1989). While low Se content in soil and plants can lead to white muscle disease in animals, Keshan and Kashin–Beck diseases in humans, and other Se deficiency disorders (Wang et al. 2013). Plants are an important source of Se for humans and animals, and soil is the ultimate source of Se for plants. From the perspective of environmental pathology, Tan (1982) pointed out that there is a low-selenium belt extending from North-eastern China to the southwest through the Loess Plateau and then to the southwest to the Tibetan Plateau with an average Se content of only 0.1 mg/kg. Heilongjiang in North-eastern China has abundant black soil resources and produces a major supply of commodity grains nationwide. Importantly it includes the region where the national low-Se belt starts. However, only a few studies have been conducted on the identification of different Se chemical forms in soil in this region.

2 MATERIALS AND METHODS

2.1 *Study area and sample collection*

From July to October 2013, representative soil samples were collected from the five natural geographical areas of Heilongjiang province, including Xiaoxing'anling in the North, Sanjiang Plain in the Northeast, Daxinganling in the West, mountains in the Southeast, and mountains and the Songnen Plain in the Midwest. A total of 19 typical soil types were obtained. Soil samples were collected using plum-shaped multi-point sampling method. Soil columns were collected vertically within 0–20 cm soil depth. Gravel roots and other debris in soil samples were removed. Two kg of soil were taken for each sample to the laboratory. Soil samples were air-dried, ground to 2 mm, and analysed for Se fractions and major physical and chemical properties.

2.2 *Soil sample analysis*

The classification of Se in soil was conducted according to the continuous chemical leaching technique of Qu et al. (1997) and Tan et al. (2002). Methods of total Se in soil were used according to the National Environmental Protection Standard HJ680-2103 of the People's Republic of China. Selenium was determined by using hydride generation–atomic fluorescence spectrometry. All acids used in pre-treatment were of excellent grade, and the reliability of the results was verified by the National Standards substance (soil GSS-10). Soil physical and chemical properties were measured as described by Lu (2000).

3 RESULTS AND DISCUSSION

3.1 *Selenium content and distribution in soil*

Results showed that the total soil Se content in the main soil types in Heilongjiang Province varied substantially, and the total Se content in surface soils varied from 0.069 mg/kg to 0.463 mg/kg. Ten of 19 tested samples were Se-deficient (Se content was low), whereas eight samples had medium Se content. Only one region has Se-rich soil, but its Se content was close to the lower limit value. No high nor excessive Se content was found in the soil samples.

3.2 *Contents and distribution of different Se fractions in soil*

Results showed that water-soluble Se content in the soil samples was low with great variation. However, the percentage of water-soluble Se in total Se in different soil types was 0.70–7.18%. Exchangeable

Table 1. Correlation coefficients between various Se fractions and total Se content.

Se fractions	SOL-Se	EX-Se	FMO-Se	OM-Se	RES-Se
Total Se	−0.102	0.393	0.635*	0.625*	0.720*

*Significant correlation at $p < 0.01$.

Se was similar to water-soluble Se. Ferromanganese oxide bound Se content in different soil types varied from 0.80% to 33.97%. Organic Se and residual Se were main fractions in different soil types, accounting for 8.16–50.50% and 26.32–70.90% of the total Se, respectively.

3.3 *Relationship between different Se fraction contents in different soils*

The total Se content of soils represented the potential level of Se available. The degree of Se absorption by crops in soil depended on the content of available Se. The correlation analysis showed a substantial correlation between acid-soluble Se, organic Se and residual Se in the soil and total Se content in the soil. Moreover, no substantial correlation was found between soluble Se, exchangeable Se and total Se content. This finding is similar to that of Wang et al. (2011, 2012), which was based on 15 soil types in China. The distribution of Se in soil was affected by the valence of Se and the physical and chemical properties of tested soil.

3.4 *Effects of physical and chemical properties of soil on various fractions of Se*

The relationship between soil physical and chemical properties and Se fractions in soil showed that organic carbon and cation exchange capacity in soil were negatively correlated with soluble Se, exchangeable Se and acid soluble Se in soils, whereas organic Se and residual Se were negatively correlated with organic carbon and cation exchange capacity in soil. Available Mn and Zn content in soil were positively correlated with content of residual Se in soil. A substantial positive correlation was found between available Cu in soil and acid-soluble Se. A negative correlation was found between soil sulphur content, pH, EC, and sand content, whereas a positive correlation was observed between clay content and organic matter content.

4 CONCLUSIONS

The total Se content in soil in Heilongjiang Province was generally low. Residual Se and organic Se were the main Se fractions in low Se soils, followed by acid-soluble Se. Water soluble and exchangeable Se are easily absorbed by plants. Their low contents may be the reason why Se bioavailability in soils in Heilongjiang was low. A significant correlation was found among acidic Se, organic Se, residual Se, and total Se content in soil, but the correlation between water-soluble and exchangeable Se was not substantial. Soil metal oxides, clay and sand contents were the main factors affecting the Se sequestration ability of soil and restricting the distribution of soil Se fractions. Soil pH, SOC, and other factors cannot be ignored.

REFERENCES

Lu, R.K. 2000. *Soil Chemical Analysis of Agriculture*. Beijing: China Agricultural Science and Technology Press.

Ohlendorf, H.M. 1989. Bioaccumulation and effects of selenium in wildlife. In L.W. Jacobs (ed), *Selenium in Agriculture and Environment*: 133–177. Madison, Wisconsin: American Society of Agronomy and Soil Science Society of America

Qu, J.G, Xu, B.X. & Gong, S.C. 1997 Determination of Selenium Species in Soils and Sediments by Continuous Leaching. *Chinese Journal of Environmental Chemistry* 16(3): 277–283 (in Chinese)

Tan, J.A, Zhu, W.Y, Wang, W.Y. et al. 2002. Selenium in soil and endemic diseases in China. *Sci Total Environ* 284(1): 227–235.

Tan, J.A. 1982. Keshan disease and natural environment and selenium nutrition background. *Acta Nutrimenta Sinica* 03: 175–182.

Wang, J., Li, H.R., Li, Y.H. et al. 2013. Speciation, Distribution, and Bioavailability of Soil Selenium in the Tibetan Plateau Kashin–Beck disease area - A case study in Songpan County, Sichuan Province, China. *Biol Trace Elem Res* (156): 367–375.

Wang, S.S., Liang, D.L., Wei, W. et al. 2011. The relationship between soil properties and selenium forms based on path analysis. *Chinese Journal of Soil Science* 48(4): 823–830.

Wang, S.S., Liang, D.L. & Wang, D. 2012. Selenium fractionation and speciation in agriculture soils and accumulation in corn (*Zea mays* L.) under field conditions in Shaanxi Province, China. *Sci Total Environ* 427–428: 159–164.

Selenium Research for Environment and Human Health: Perspectives, Technologies and Advancements – Bañuelos, Lin, Liang & Yin (eds)
© 2020 Taylor and Francis Group, London, ISBN 978-1-138-39014-0

Selenium distribution and speciation in seleniferous soils and controlling factors

H.-B. Qin*
State Key Laboratory of Environmental Geochemistry, Institute of Geochemistry Chinese Academy of Sciences, Guiyang, China

J.-M. Zhu
State Key Laboratory of Geological Processes and Mineral Resources, China University of Geosciences, Beijing, China

1 INTRODUCTION

Selenium (Se) is of great environmental and geochemical concern as an essential trace element for humans and a powerful proxy for paleo-ocean (Lenz & Lens 2009, Qin et al. 2012, 2013, 2017a). In the soil environment, Se naturally exists in a variety of oxidation states (VI, IV, 0, and -II) and organic forms. The study on Se speciation is critically important for understanding the mobility, bioavailability, and toxicity of Se in soils. However, the distribution, speciation, and coordination environment of Se in soils at the molecular level are still largely unknown.

Synchrotron based techniques, such as X-ray absorption fine structure (XAFS) and X-ray fluorescence (μ-XRF), are quite powerful to identify elemental speciation and local structural in solid samples (Qin et al. 2017a, b, 2019). The distribution and speciation of Se in seleniferous soils were investigated here by a combination of μ-XRF and XAFS techniques, to provide insights into the mobility and bioavailability of Se in soil environment and controlling factors at the molecular scale.

2 MATERIALS AND METHODS

Soil samples were collected from Enshi prefecture of Hubei Province, a typical seleniferous area well known for human Se poisoning in the early 1960s in China. Se K-edge XAFS spectra was obtained at beamline 1W1B at the Beijing Synchrotron Radiation Facility (BSRF), Institute of High Energy Physics (IHEP). The μ -XRF mapping and μ -XAFS experiments for the thin section samples were performed at beamline BL-4A, Photon Factory (Tsukuba, Japan).

3 RESULTS AND DISCUSSION

The XANES results showed that dominant organic Se and lesser Se(IV) were present in seleniferous agricultural soils from Enshi, and the proportion of organic Se was significantly decreased in the order of paddy soils > uncultivated soils > upland soils (Qin et al. 2017a). In contrast, Se(IV) was predominant in the soil sample from abandoned Kawazu Mine tailings in Japan, in which Fe(III) hydroxides were identified to be the host phases (Qin et al. 2017b). These findings suggested that Se speciation in agricultural soils can be significantly influenced by different cropping systems, including organic Se derived from plant litter (El Mehdawi et al. 2015, Qin et al. 2017a).

Furthermore, the distribution and speciation of Se in the soil from mine tailings was remarkably different from those for tellurium (Te), another element in Group 16 of the periodic table. A study has suggested that Te was present as a mixture of Te(VI) and Te(IV) species (Qin et al. 2017b). Although Fe(III) hydroxides were the host phases for Te(IV) and Te(VI), Te(IV) can also be retained by illite. The difference in speciation and distribution of Se and Te in soil can be clarified by their different structures of surface complexes onto Fe(III) hydroxides. Te(VI), Te(IV), and Se(IV) prefer to form stable inner-sphere complexes on the surface of Fe(III) hydroxides, whereas Se(VI) mainly forms less stable outer-sphere complexes (Harada & Takahashi 2008).

4 CONCLUSIONS

Selenium distribution and speciation are disparate in different types of soils, depending on the formed surface complexes and cropping systems. The findings are geochemical and environmental significance for better understanding the solubility, mobility, and bioavailability of Se in the surface environment.

REFERENCES

El Mehdawi, A.F., Lindblom, S.D., Cappa, J.J., Fakra, S.C. & Pilon-Smits, E.A. 2015. Do selenium hyperaccumulators affect selenium speciation in neighboring plants and soil? An X-ray microprobe analysis. *Int J Phytoremediation* 17: 753–765.

Harada, T. & Takahashi, Y. 2008, Origin of the difference in the distribution behavior of tellurium and selenium in a soil-water system. *Geochim Cosmochim Acta* 72: 1281–1294.

Lenz, M. & Lens, P.N.L. 2009. The essential toxin: The changing perception of selenium in environmental sciences. *Sci Total Environ* 407: 3620–3633.

Qin, H.B., Zhu, J.M. & Su, H. 2012. Selenium fractions in organic matter from Se-rich soils and weathered stone coal in selenosis areas of China. *Chemosphere* 86: 626–633.

Qin, H.B., Zhu, J.M., Liang, L., Wang, M.S. & Su, H. 2013. The bioavailability of selenium and risk assessment for human selenium poisoning in high-Se areas, China. *Environ Int* 52: 66–74.

Qin, H.B., Zhu, J.M., Lin, Z.Q. et al. 2017a. Selenium speciation in seleniferous agricultural soils under different cropping systems using sequential extraction and X-ray absorption spectroscopy. *Environ Pollut* 225: 361–369.

Qin, H.B., Takeichi, Y., Nitani, H., Terada, Y. & Takahashi, Y. 2017b. Tellurium distribution and speciation in contaminated soils from abandoned mine tailings: comparison with selenium. *Environ Sci Technol* 51: 6027–6035

Qin, H.B., Uesugi, S., Yang S.T. et al. 2019. Enrichment mechanisms of antimony and arsenic in marine ferromanganese oxides: insights from the structural similarity. *Geochim Cosmochim Acta* 257: 110–130.

Selenium Research for Environment and Human Health: Perspectives, Technologies and Advancements – Bañuelos, Lin, Liang & Yin (eds)
© 2020 Taylor and Francis Group, London, ISBN 978-1-138-39014-0

The role of atmospheric deposition in biogeochemical selenium cycling

L.H.E. Winkel
Swiss Federal Institute of Technology (ETH), Institute of Biogeochemistry and Pollutant Dynamics, Zurich, Switzerland
Eawag: Swiss Federal Institute of Aquatic Science and Technology, Duebendorf, Switzerland

1 INTRODUCTION

1.1 *Importance of Se to human health*

Selenium (Se) is an important micronutrient, and when it is present in selenocysteine, it is referred to as the 21st amino acid. Selenocysteine is incorporated in selenoproteins that serve a wide range of biological functions (Fairweather-Tait et al. 2011). However, Se only has a narrow range of safe dietary intake levels for humans; too low dietary intake can lead to deficiency and too high intake leads to toxicity. Generally, Se concentrations in the environment are low, and it has been estimated that up to 1 billion people around the globe have low dietary Se intakes. One of the reasons for low dietary Se levels are low content in food crops, which is due to low concentrations and bioavailability of Se in soils.

1.2 *Environmental Se cycling*

Understanding Se cycling in the environment is complex as Se is a redox sensitive element that can occur in various chemical forms in environmental compartments (e.g. soils, waters, biota, and atmosphere) (Winkel et al. 2015). Since the environmental distribution and chemical speciation of Se is closely related to environmental health issues, it is of major importance to better understand the factors that control its distribution and biogeochemical cycling. A key factor determining Se distribution in environmental compartments are its speciation, i.e. the specific forms of an element, oxidation state, and/or complex or molecular structure. Analytical activities related to identifying and/or measuring the quantities of one or more individual chemical species in a sample are referred to as speciation analysis (Templeton et al. 2000). Knowing the speciation of trace elements (TEs) provides information about the biogeochemical pathways and mobility and can be used as a predictor of bioavailability, e.g. to asses Se uptake by plants from soils.

1.3 *Climatic processes influencing Se cycling*

This talk will provide an overview of processes that control Se cycling in the natural environment. It will specifically focus on atmospheric deposition of Se, which constitutes a source of Se to the surface environment, e.g. agricultural soils and ecosystems. Processes at various spatial scales will be discussed, ranging from molecular to the global scale.

1.4 *Atmospheric Se deposition: Case study*

Climatic factors play a major role in biogeochemical Se cycling and occurrences in soils as climate controls many environmental factors. For example, rainfall is an important controlling variable of soil organic carbon contents, which is, in turn, an influential factor in the retention of Se and other trace elements in soils. Various studies indicated that atmospheric deposition, mainly in the form of wet deposition, functions as a source of Se to terrestrial environments and thus agricultural soils and food crops (Wen & Carignan 2007, Blazina et al. 2014). The atmosphere is known as a reservoir of Se and it has been estimated that 13,000–19,000 tons Se are annually cycled through the troposphere (Wen & Carignan 2007). Studies from the eighties estimated that anthropogenic sources account for around 40% of total atmospheric Se emissions and natural sources for 60% of emissions (Wen & Carignan 2007). However, there is still only scarce information available on the sources of Se to the atmosphere and to rainfall in particular. Therefore, we carried out a study investigating marine versus continental sources of Se in rainfall at two high-altitude locations in Europe, i.e. Jungfraujoch (Alps, Switzerland) and Pic du Midi (Pyrenees, France) (Suess et al. 2019). In this study, chemical speciation and precipitation sources were determined to get potential insight into sources of Se.

2 MATERIALS AND METHODS

2.1 *Chemical analyses and precipitation sources*

Rainfall was collected weekly and analyzed for total concentrations of Se, as well as sulfur (S), iodine (I), and bromine (Br) using ICP-MS/MS (Agilent 8800 ICP-QQQ). Selected precipitation samples from the time period April–September 2016 were analyzed for Se, S, I, and Br speciation (HPLC-ICP-QQQ) with a method that was initially developed to analyze

Se speciation (Dauthieu et al. 2006). In addition, dissolved organic carbon (DOC) concentrations (Shimadzu TOC-L CSH) and isotopic carbon signatures ($\delta^{13}C$, IRMS) were analyzed, and trajectory-based moisture source and loss analyses were performed to identify the precipitation directions and sources for the time periods in which samples were collected (Suess et al. 2019).

3 RESULTS AND DISCUSSION

3.1 *Link between continental rainfall and Se concentrations*

Collected data indicated mixed marine and continental source contributions for Se, as well as I, Br and S at Jungfraujoch and Pic du Midi. At both sites, continental moisture sources were found to be dominant in spring, summer and autumn, and, in these periods, concentrations of Se were highest (no samples were available for Pic du Midi in winter). The combined results from concentration analyses and moisture source diagnostics indicated that in spring-autumn, land sources are predominant (Suess et al. 2019). Furthermore, concentrations of Se correlated with DOC-$\delta^{13}C$ values, indicating that there is a link between the source of dissolved organic matter (i.e. correlations with DOC-$\delta^{13}C$ values) and these elements. This finding may indicate that biogenic land emissions are a source of Se. Also, other studies have indicated that terrestrial environments can emit biogenic Se (reviewed in Wen & Carignan 2007, Winkel et al. 2015). However, terrestrial landscapes are highly heterogeneous, and therefore more work needs to be done to understand the mechanisms of Se emissions and quantify them.

3.2 *Speciation of Se in rainfall*

Speciation analyses indicated that the main form of Se in the rainwater samples from both Jungfraujoch and Pic du Midi was selenate (SeVI), which is the most oxidized form of Se. This finding is similar to that reported in other studies of speciation in aqueous samples i.e. surface waters and soil extracts (e.g. Tolu et al. 2014, Vriens et al. 2014). The total amount of quantified species is lower than the total Se concentrations, indicating unidentified Se fractions. Currently, further research is carried out to elucidate the identity of these species.

3.3 *Further studies: Intracellular speciation*

In addition to potential biogenic land emissions, biogenic emissions from oceans are thought to be a source of atmospheric Se and ultimately to rainfall. Others have estimated that marine biogenic emissions are the main natural source of volatile Se (Wen & Carignan 2007). However, the mechanisms behind Se biomethylation by marine organisms are also largely unknown. To study marine biomethylation of Se, it is important to understand metabolic processes in phytoplankton under natural conditions. In such conditions, intracellular concentrations are low, and speciation analyses are challenging. To address this problem, new methods need to be developed, involving microwave-assisted cell disruption combined with liquid chromatography and mass spectrometry. The presentation will conclude with preliminary results for intracellular speciation and transformation of Se in marine phytoplankton.

REFERENCES

Blazina, T., Sun, Y., Voegelin, A. Lenz, M., Berg, M. & Winkel, L.H.E. 2014. Terrestrial selenium distribution in China is potentially linked to monsoonal climate. *Nat Commun* 5: 4717.

Dauthieu, M., Bueno, M., Darrouzes, J., Gilon, N. & Potin-Gautier, M. 2006. Evaluation of porous graphitic carbon stationary phase for simultaneous preconcentration and separation of organic and inorganic selenium species in "clean" water systems. *J Chromatogr A* 1114 (1): 34–39.

Fairweather-Tait, S.J., Bao, Y., Broadley, M.R. et al. 2011. Selenium in human health and disease. *Antioxid Redox Signal* 14(7): 1337–1383.

Suess, E., Aemisegger, F., Sonke, J.E., Sprenger, M., Wernli, H. & Winkel, L.H.E. 2019. Marine versus continental sources of iodine and selenium in rainfall at two European high-altitude locations. *Environ Sci Technol* 53: 1905–1917.

Templeton, D.M., Ariese, F., Cornelis, R. et al. 2000. Guidelines for terms related to chemical speciation and fractionation of elements. Definitions, structural aspects, and methodological approaches. *Pure Appl Chem* 72(8): 1453–1470.

Tolu J., Thiry, Y., Bueno, M., Jolivet, C., Potin-Gautier, M. & Le Hécho, I. 2014. Distribution and speciation of ambient selenium in contrasted soils, from mineral to organic rich. *Sci Total Environ* 479–480: 93–101.

Vriens, B., Lenz, M., Charlet, L., Berg, M. & Winkel, L.H.E. 2014. Natural wetland emissions of methylated trace elements. *Nat Commun* 5: 3035.

Wen, H. & Carignan, J. 2007. Reviews on atmospheric selenium: emissions, speciation and fate. *Atmos Environ* 41(34): 7151–7165.

Winkel, L.H.E., Vriens, B., Jones, G.D., Schneider, L.S., Pilon-Smits, E. & Banuelos, G.S. 2015. Selenium cycling across soil-plant-atmosphere interfaces: a critical review. *Nutrients* 7(6): 4199–4239.

Selenium Research for Environment and Human Health: Perspectives, Technologies and Advancements – Bañuelos, Lin, Liang & Yin (eds)
© 2020 Taylor and Francis Group, London, ISBN 978-1-138-39014-0

Volatile organic selenium in atmosphere: A mini review

Q.Q. Chen
Environmental Science, College of Life Science and Resources and Environment, Yichun University, Yichun, Jiangxi, China

X.B. Yin, Z.M. Wang, L.X. Yuan X.Q. Lu, F. Li & Z.K. Liu
School of Earth and Space Sciences, University of Science and Technology of China, Hefei, Anhui, China
Jiangsu Bio-Engineering Research Center for Selenium, Suzhou, Jiangsu, China

1 SELENIUM SPECIATIONS AND THEIR SOURCES

Selenium (Se) is an essential trace element for biological systems and is a "double-edged sword" element for biological health. Excessive and deficient levels of Se in vivo may damage biological bodies (Fordyce et al. 2000). Atmospheric Se is firstly reported as a complex system and is an important part of the biogeochemical cycle of Se. The atmospheric Se speciation is classified into particulate Se and volatile Se containing volatile inorganic and organic Se that are emitted into the atmosphere at a rate of about 13,000–19,000 tons per year (Mosher & Duce 1987). The particulate Se originated mostly from natural sources, such as sea salts, wind-blown dusts, volcanic ashes, and from anthropogenic activities that may include aerosol particles (De Santiago et al. 2014). The sources of volatile inorganic Se are volcanic and anthropogenic activities (Floor & Román-Ross 2012). The volatile inorganic Se is unstable in the atmosphere and easily transforms into the particulate Se, consisting of element Se (S^0), hydrogen selenide (H_2Se), and Se dioxide (SeO_2).

Volatile organic Se is naturally produced by organisms and microorganisms in the soil, sludge, and ocean, including dimethyl selenide (DMSe), dimethyl diselenide (DMDSe), methane selenol (MeSeH), dimethyl selenyl sulfide (DMSeS), etc. The fraction of gaseous organic Se is not clearly identified, and it has been considered as an important part of global atmospheric Se budget (Mosher & Duce 1983, 1987). This mini-review just focuses on volatile organic Se with analytical technology, generation mechanisms, and emissions of volatile organic Se in global nature systems.

2 THE ANALYTICAL TECHNOLOGY

The determination system for volatile Se is made up of the trapping subsystem and an analytical subsystem. Concentrations of volatile Se in the atmosphere are generally very low and require preconcentration. Experiments of gas trapping were generally categorized into solid sorptives (Lenz et al. 2012), mineral acids (Vriens et al. 2014a, Winkel et al. 2010), and cryotrapping (Amouroux et al. 1998). Selenium compounds involved in the gas trapping subsystem should be pretreated and transferred into specific forms for detection in laboratory. For example, gaseous Se was adsorbed on the activated carbon surface within the activated-carbon column by a small vacuum pump, and then Se is determined by continuous-flow hydride-generation atomic absorption spectrometry (Zhang & Moore 1997). Technology analyzing gaseous Se compounds includes gas trapping, pretreatment, final instrumental analysis, gas chromatography (GC), and multiple instruments (Dauchy et al. 1994). In addition, Vriens et al. (2014b) also summarized analytical techniques used for measuring volatile Se.

3 GENERATION MECHANISMS

Microbial methylation is responsible for producing volatile organic Se in terrestrial and marine systems. Methyl-transferase enzymes CH^{3+} provided by S-adenosyl methionine and methyltetrahydrfolate transfer Se atoms into amino acids SeMet and SeCys and eventually form methyl-SeMet and methyl-SeCys, respectively (Cooke & Bruland 1987). Alternatively, SeCys may be firstly converted via glutamination into γ-glutamyl-SeCys and subsequent methylation into γ-glutamyl-methyl-SeCys, and final deglutamination into methyl-SeCys (Winkel et al. 2015).

4 EMISSIONS OF GLOBAL NATURE SYSTEMS

There are few articles published in recent years that focus on wetland, grass, marine, and agricultural soil (Table 1). Atmospheric Se has also been investigated from plant, sediment-water, water-air, and soil-air interfaces. Extrapolation of the volatile Se flux (0.11–0.12 μg/m/d) in Table 1 to global wetland estimates total Se emission at $2.29–2.50 \times 10^8$ g per year. Total

Table 1. Concentrations of Se and Se species in natural systems.

Volatile Se	Areas-systems	References
62.0±3.6 mg/m^2/y	Soil-field (pickleweed)	Lin et al. 2002
16.7±1.1 mg/m^2/y	bare soil	
4.8±0.3 mg/m^2/y	saltgrass	
<12 or <25 μg/m^2/d	Soil-field (bare soil)	Bañuelos & Lin 2007
<50 μg/m^2/d	Vegetated	
5.8 μg/m^2/d	Soil-field (non-irrigated bare)	Bañuelos et al. 2005
12.33 μg/m^2/d	Irrigated bare	
21.25 μg/m^2/d	Elephant grass	
29.67 μg/m^2/d	Salado grass	
26.33 μg/m^2/d	Cordgrass	
30 μg/m^2/d	Saltgrass-turf	
19.8 μg/m^2/d	Leucaenia	
14.33 μg/m^2/d	Salado alfalfa	
31.25 μg/m^2/d	Saltgrass forage	
0.11~0.12 μg/m^2/d	Wetland-field	Vriens et al. 2014a
190~210 ng/m^2/d	Wetland-field	Vriens et al. 2014b
25~190 μg/m^{2}d	Wetland-field	Hansen et al. 1998
44~285 μg/m^2/d	Wetland-field	Gao et al. 2003
100~200 μg/m^2/y	Grassland	Haygarth et al. 1994
0.002~0.23 μg/m^2/d	Marine-field	Amouroux & Donard 1996
0.17 μg/m^2/d	Estuary-field	Amouroux & Donard 1997

volatile Se fluxes of the global grassland and the ocean were also calculated to be about 3.06×10^{11} and 3.03×10^{10} g/y using the same method. Although these results are slightly lower than the previous range of $(5–29) \times 10^{9}$ g/y published (Mosher & Duce 1987), it was a little higher than other results $(4.74–7.90) \times 10^{9}$ g/y (Weber 1999). However, the fluxes of volatile Se produced by agricultural systems cannot be estimated from scarce data currently available.

REFERENCES

Amouroux, D., Tessier, E., Pecheyran, C. & Donard, O.F.X. 1998. Sampling and probing volatile metal(loid) species in natural waters by in-situ purge and cryogenic trapping followed by gas chromatography and inductively coupled plasma mass spectrometry (P-CT-GC-ICP/MS). *Anal Chim Acta* 377: 241–254.

Amouroux, D. & Donard, O.F.X. 1996. Maritime emission of selenium to the atmosphere in eastern Mediterranean seas. *Geophys Res Lett* 23: 1777–1780.

Amouroux, D. & Donard, O.F.X. 1997. Evasion of selenium to the atmosphere via biomethylation processes in the Gironde estuary, France. *Mar Chem* 58: 173–188.

Bañuelos, G.S. & Lin, Z.Q. 2007. Acceleration of selenium volatilization in seleniferous agricultural drainage sediments amended with methionine and casein. *Environ Pollut* 150:306–312.

Bañuelos, G.S. Lin, Z.Q. Arroyo I. & Terry, N. 2005. Selenium volatilization in vegetated agricultural drainage sediment from the San Luis Drain, Central California. *Chemosphere* 60: 1203–1213.

De Santiago, A. Longo, A.F., Ingall, E.D. et al. 2014. Characterization of Selenium in Ambient Aerosols and Primary Emission Sources. *Environ Sci Technol* 48: 8988–8994

Cooke, T.D. & Bruland, K.W. 1987. Aquatic chemistry of selenium-Evidence of biomethylation. *Environ Sci Technol* 21: 1214–1219.

Dauchy, X., Potin-Gautier, M.A. & Astruc, M. 1994. Analytical methods for the speciation of selenium compounds: a review. *Fresenius J Anal Chem* 348: 792–805.

Floor, G.H. & Román-Ross, G. 2012. Selenium in volcanic environments: A review. *Appl Geochem* 27: 517–531.

Fordyce, F.M. Zhang, G.D., Green, K. & Liu, X.P. 2000. Soil, grain and water chemistry in relation to human selenium-responsive diseases in Enshi District, China. *Appl Geochem* 15: 117–132.

Gao, S., Tanjia, K.K., Lin, Z.Q., Terry, N. & Peters, D.W. 2003. Selenium removal and mass balance in a constructed flow-through wetland system. *J Environ Qual* 32: 1557–1570.

Hansen, D., Duda, P.J., Zayed, A. & Terry, N. 1998. Selenium removal by constructed wetlands: Role of biological volatilization. *Environ Sci Technol* 32: 591–597.

Haygarth, P.M., Fowler, D., Stürup, S., Davison, B.M. & Tone, K.C. 1994. Determination of gaseous and particulate selenium over a rural grassland in the UK. *Atmos Environ* 28(22): 3655–3663

Lin, Z.Q., Cervinka, V., Pickering, I.J., Zayed, A. & Terry, N. 2002. Managing selenium-contaminated agricultural drainage water by the integrated on-farm drainage management system: Role of selenium volatilization. *Water Res* 36: 3150–3160.

Lenz, M., Floor, G.H., Winkel, L.H.E., Román-Ross, G. & Corvini, P.F.X. 2012. Online preconcentration-IC-ICP-MS for selenium quantification and speciation at ultratraces. *Environ Sci Technol* 46: 11988–11994.

Mosher, B.W. & Duce, R.A. 1987. A global atmospheric selenium budget. *J Geophys Res* 92(D11): 13289–13298.

Mosher, B.W. & Duce, R.A. 1983. Vapour phase and particulate selenium in marine atmosphere. *J Geophys Res* 88: 6761–6768.

Vriens, B., Lenz, M., Charlet, L., Berg, M. & Winkel, L.H.E. 2014a. Natural wetland emissions of methylated trace elements. *Nat Commun* 5: 3035.

Vriens, B., Ammann, A.A., Hagendorfer, H., Lenz, M. & Winkel, L.H.E. 2014b. Quantification of methylated selenium, sulfur, and arsenic in the environment. *PLOS One* 9: e102906

Weber, J.H. 1999. Volatile hydride and methyl compounds of selected elements formed in the marine environment. *Marine Chem* 65: 67–75.

Winkel, L.H.E., Feldmann, J. & Meharg, A.A. 2010. Quantitative and qualitative trapping of volatile methylated selenium species entrained through nitric acid. *Environ Sci Technol* 44: 382–387.

Winkel, L.H.E., Vriens, B., Jones, G.D., Schneider, L.S., Pilon- Smits, E. & Bañuelos, G.S. 2015. Selenium cycling across soil-plant-atmosphere interfaces: A critical review. *Nutrients* 7: 4199–4239.

Zhang, Y. & Moore, J.N. 1997. Environmental conditions controlling selenium volatilization from a wetland system. *Environ Sci Technol* 31: 511–517.

Selenium Research for Environment and Human Health: Perspectives, Technologies and Advancements – Bañuelos, Lin, Liang & Yin (eds)
© 2020 Taylor and Francis Group, London, ISBN 978-1-138-39014-0

Evaluation standards of selenium-rich soil in Guanzhong in Shaanxi Province

R. Ren, X. Chao & J.P. Chen
Shaanxi Hydrological Engineering and Environmental Geology Survey Center, Xi'an, China

F. Yuan*
Xing Zhi College of Xi'an University of Finance and Economics, Xi'an, China

1 INTRODUCTION

Selenium (Se) plays an irreplaceable key role in nutrition, including lipids, proteins, and other microelements (Xiang et al. 2016). The lack of selenium-rich soils (Rayman 2012) and the benefits of Se for human health may increase consumer demand for developing natural Se-rich agricultural products. Meanwhile, there is increased value in selenium-rich soils for their ability to develop natural selenium-rich agricultural products (Hu et al. 2010), which can safely and effectively transfer Se to human beings through the food chain. However, there is no unified standard in China to define selenium-rich soil. This study is based on the multi-target geochemistry survey in Shaanxi Province, which collected surface soil Se data over an area of 39500 km^2, which systematically sorted out data on wheat grain Se content, rhizosphere soil Se content, and pH. There were two projects: (1) Comprehensive study on the development and utilization of alkaline Se-rich soil in Guanzhong region of Shaanxi Province and (2) Geochemical investigation and evaluation of Se-rich area in Guanzhong-Tianshui economic zone (Guanzhong basin).

2 MATERIALS AND METHODS

2.1 *Sample collection*

The surface soil in Guanzhong area is basically alkaline (pH > 7.5), therefore, the evaluation standard of Se-rich soil in this research is only for alkaline soil. A grid method was used in the multi-objective regional geochemical survey in Shaanxi Province. The surface (0–20 cm) soil sampling density was 1 point/km^2, and every 4 km^2 samples were combined and analyzed, a total of 10,114 surface soil data were obtained. The Mei flower point method was used for multi-point sampling of wheat samples, and 544 groups of wheat and corresponding rhizosphere soil samples were collected according to different soil Se content intervals. The samples were measured for Se by atomic fluorescence spectrophotometry (AFS).

2.2 *Modelling procedure*

The procedure was as follows: One-way analysis of variance-correlation analysis-regression analysis. Firstly, crop sample data were grouped according to soil Se content, and then SPSS 22.0 software was used for one-way ANOVA to test whether soil Se content has significant effect on wheat Se content. If the soil Se content had a significant impact on the Se content of crops, then the correlation between indicators could be tested by bivariate correlation analysis. When soil Se and crop Se showed significant linear correlation, a regression model was established.

3 RESULTS AND DISCUSSION

3.1 *Soil Se geochemistry statistical classification*

According to the principle of geochemistry statistics, the grading boundary of Se content in surface soil was determined by two methods: Cumulative frequency method and iterative elimination method (Jia 2013). The statistical values of low value area, low background area, and background limit obtained by the two methods are basically the same, and the lower limit of abnormal area is 0.22 mg/kg. According to the results of regional geochemical statistics, 0.22 mg/kg is determined to be the lower limit of Se anomaly in surface soil in Guanzhong area.

3.2 *Wheat and soil Se membership function model*

3.2.1 *One-way ANOVA*

After the heterogeneity test and normal distribution test of 544 wheat and root soil samples collected in the survey area, 535 sets of data were retained for one-way ANOVA. Data were divided into 6 groups according to the Se content in rhizosphere soil. One-way ANOVA was used to test whether the soil Se content had a significant impact on the Se content in wheat, and multiple comparisons were applied to analyze the differences among different data groups. The results showed that the mean values of Groups 1 and 2 and other groups were significantly different, and the

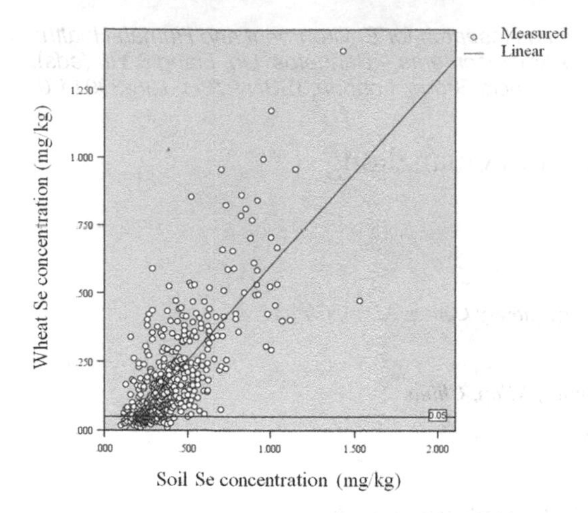

Figure 1. Correlation map of soil-wheat Se content.

mean values of Se content in wheat were significantly different from the latter groups. We concluded that the Se content of soil in Guanzhong area had a significant effect on Se content of wheat.

3.2.2 *Correlation analysis of Se content in soil and wheat*

Bivariate correlation analysis was used to test the correlation between indicators. The results showed that the correlation coefficient between the Se content in soil and wheat was 0.750, and the P value of the correlation coefficient test was approximately 0, (less than the significance level of 0.01). Hence, the correlation coefficient between the Se content in soil and the Se content in wheat was positive strong. Therefore, we concluded that there is a significant linear correlation between Se content in wheat and soil in Guanzhong region.

3.2.3 *Determination of critical value of soil selenium*

Scatter plot was drawn in the rectangular coordinate system (Fig. 1). The independent variable (X) was soil Se content, and the dependent variable (Y) was wheat Se content. We see from the correlation diagram that the soil Se content and wheat Se content showed a linear correlation, and the linear regression equation was established. The expression of the model is ($R = 0.563$, $p < 0.01$). According to this regression equation, when Y (Se content of wheat) is 0.05 mg/kg, the corresponding X (Se content of soil) can be determined to be 0.21 mg/kg.

3.3 *Experimental verification*

According to the geochemical anomalies of Se in surface soil and the results of the membership function of Se in wheat and soil, it is reasonable to define the evaluation standard of alkaline selenium-rich soil in Guanzhong region of Shaanxi Province as 0.22 mg/kg.

The data of 535 wheat samples collected in Guanzhong were statistically calculated. Among the 476 wheat samples, rhizosphere soil Se content was 0.22 mg/kg higher than the critical value, the Se content of 397 wheat samples was 0.05 mg/kg, which was higher than the standard of Se-rich value, and the Se-rich rate of the samples was 83.40%, which was calculated statistically according to the administrative division and soil type. The Se-rich rate of wheat could basically reach more than 80%, indicating that the standard was universal and representative.

4 CONCLUSIONS

According to the standard content of Se-rich wheat in Shaanxi Province (DB 61/T 556-2018), the membership function model of Se in wheat and soil has been established. The evaluation standard of alkaline Se-rich soil in Guanzhong area of Shaanxi Province is the Se content of surface soil ≥ 0.22 mg/kg. 83.40% natural wheat produced by Se-rich soil defined by this content can reach the standard of Se-rich wheat in Shaanxi Province.

REFERENCES

Hu, Y.H. Wang, J.E. Cai, Z.H. Song, M.Y. Kang, Z.J. & Yang, T.Z. 2010. Content, distribution and influencing factors of selenium in soil of Jiashan area, Northern Zhejiang Province. *Gelogical Science and Technology Information* 29(6): 84–88. (in Chinese)

Jia, S.A. 2013. Evaluation standards and genesis of selenium-rich soil in Anhui Province. *Resources Survey and Environment* 34(2): 133–137. (in Chinese)

Rayman, M.P. 2012. Selenium and human health. *Lancet* 379: 1256–1268.

Susanne, E.G. Trine, A.S. Anne, F.O. & Ivar, A. 2007. Plant availability of inorganic and organic selenium fertilizer as influenced by soil organic matter content and pH. *Nutr Cycl Agroecosys* 79(3): 221–231.

Xiang, J.Q. Zhu, Y.F. & Yin, H.Q. 2016. Current Status and thoughts on the development and utilization of selenium resources in Enshi autonomous prefecture. *Hubei Agricultural Science* 55(17): 4366–4550. (in Chinese)

Selenium Research for Environment and Human Health: Perspectives, Technologies and Advancements – Bañuelos, Lin, Liang & Yin (eds)
© 2020 Taylor and Francis Group, London, ISBN 978-1-138-39014-0

Smart technology drives for Tibetan and Yungui plateau selenium-enriched agriculture in China

S.S. Zuo, D. Wu & W.L. Wu*
College of Resources and Environmental Sciences, China Agricultural University, Beijing, China

1 INTRODUCTION

1.1 *Selenium in Tibet*

Selenium (Se) plays a crucial role in our diet, and it can help prevent the Kaschin-Beck disease (KBD) and affect immune responses. Although KBD is presently controlled in most regions of China, it is still active and severe in the Tibetan Plateau (Chen et al. 2015), especially in the north area of the Yalu Zangbu River. The average total Se content of the Tibetan Plateau is much lower than the inland of China. Therefore, there has been increased attention given to local people's health in Tibet, and to creating a smart-technology-driven Se-enrichment strategy for the Tibetan plateau.

1.2 *Selenium-enriched organic Puer-tea in Yunnan*

There is over one thousand years of history on growing Puer-tea in China. As a kind of fermented tea, Puer-tea is popular because of its potential health beneficial properties of scavenging free radicals, maintaining beauty, binding of heavy metals, and fighting bacteria and hyperlipidemia (Mo et al. 2008). Because it is grown in a unique geographical environment, high quality and abundant tea plants can be produced, there is tremendous Yunnan support for developing organic Puer-tea. However, there is a lack of information known about the specific characteristics of organic Se-enriched Puer-tea.

2 MATERIALS AND METHODS

2.1 *Selenium-enriched methods in Tibet*

Firstly, it is necessary to establish scientific national certification standards for Se-enriched agricultural products. Secondly, we must use the following Se-enriched technologies to improve the Se content of agricultural products: (1) introduce, improve, and domesticate high-quality germplasm resources, such as highland barley and potato; (2) use biological Se-enriched carbon fertilizer or green manures as fertilizer; (3) spray Se-enriched fertilizer on the leaves; (4) apply microorganisms that can assimilate Se into the agricultural products, such as employment of yeast that can enrich Qingke barley beer with Se or be used to cultivate Seenriched mushrooms; and (5) research and apply bio-nano Se materials. Based on the above technologies, we can build a modern demonstration base of Se-enriched agricultural products that will rapidly drive the production of Se enriched agricultural products of the whole Tibetan plateau.

2.2 *Strategies for the development of organic Se-enriched Puer-tea in Yunnan*

The first step is to choose an ecological organic tea garden as a model field. Selenium-enriched methods for Puer-tea were described above. During these processes, it is necessary to build some smart infrastructures that use cloud computing and visualization technology to construct an "agricultural information cloud" (Zhao et al. 2012), and to provide high quality information to meet the standard production of organic Se-enriched Puer-tea.

3 RESULTS AND DISCUSSION

3.1 *Build a Se-enriched ecological industry chain*

As Figure 1 shows, developing a Se-enriched ecological industry chain mainly includes the production of Se-enriched agricultural products, development of both Se enrichment processes, and the commercial activity after harvest of Se-enriched products. The key point of the total chain is based upon Se-enriched technology. Yin et al. (2012) reported that mixing biological carbon Se fertilizers with another slow release fertilizer can efficiently enrich and improve production of potato with Se. Garcia-Franco et al. (2015) revealed that green manure is good for soil organic carbon sequestration. Therefore, the combination of a biological carbon Se fertilizer and green manure is probably an ideal option for organically producing Se-enriched agricultural products. In addition, traditional measurements of Se content are commonly performed under laboratory conditions, which are time-consuming and inefficient. Hence, hence on-line and real-time application of a portable instrument is required, preferably if it can connect with our smart phone.

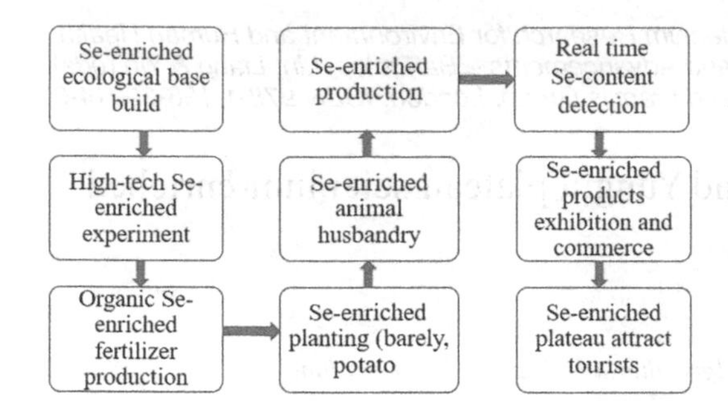

Figure 1. Selenium-enriched ecological industry chain in Tibet.

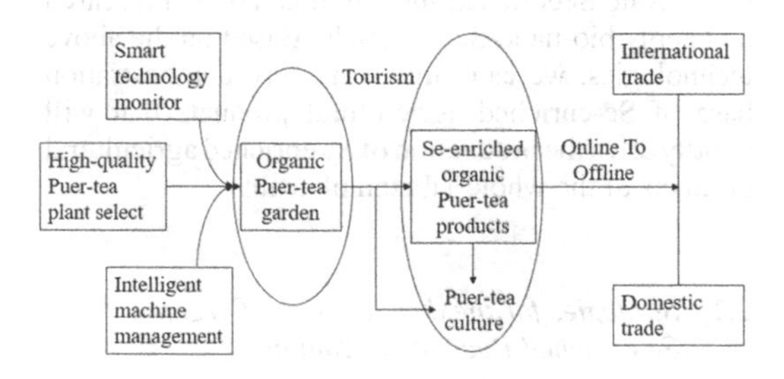

Figure 2. Selenium enriched organic Puer-tea industry chains in Yunnan.

3.2 *Build a Se-enriched organic Puer-tea industry chain*

The Se-enriched organic Puer-tea industry chain is illustrated in Figure 2. Organic Puer-tea planting gardens should strengthen every link between various production technologies, especially by smart technology innovation, resulting in the development of the Se-enriched organic Puer-tea products geared for specific market-oriented policies. Moreover, the adoption of internet to promote the sales of Se-enriched organic Puer-tea products not only increases the income of farmers, company, and government, but also enables consumers to appreciate the Se-enriched organic Puer-tea growing process and tea products.

3.3 *Formation of a smart plateau Se-enriched agriculture development system*

Plateau-intelligent agriculture of Se-enriched industrial chain is shown in Figure 3, which includes the Se-enriched technology in farming and animal husbandry, processing industry, tourism, logistics, commerce, and trade. A smart garden will be constructed by the standard processing of Se-enriched agricultural

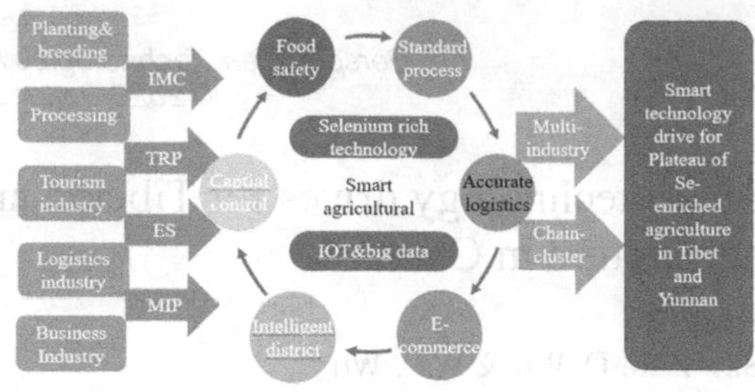

Figure 3. A smart plateau Se-enriched agriculture system. Brand-integrated marketing communication (IMC); technological research promotion (TRP); E-commerce services (ES); model innovation promotion (MIP); Internet of things (IOT).

products. Food safety and accurate logistics will be more accurate using "internet" and the large amount of available agricultural data. The health and happiness of local people can be greatly improved after the industrial chain is constructed.

4 CONCLUSIONS

We constructed two plateaus of Se-enriched industrial chain in Tibet and Yunnan. By developing a Se rich smart agricultural technology, the transformation of traditional extensive plateau agriculture into organic and functional agriculture will be realized in the nearby future.

REFERENCES

Chen, Z., Li, H., Yang, L. et al. 2015. Hair selenium levels of school children in Kashin–Beck disease endemic areas in Tibet, China. *Biol Trace Elem Res* 168(1): 25–32.

Garcia-Franco, N., Albaladejo, J., Almagro, M. et al. 2015. Beneficial effects of reduced tillage and green manure on soil aggregation and stabilization of organic carbon in a Mediterranean agroecosystem. *Soil Till Res* 153: 66–75.

Mo, H., Zhu, Y. & Chen, Z. 2008. Microbial fermented tea– a potential source of natural food preservatives. *Trends Food Sci. Technol* 19(3): 124–130.

Yin, J.Y., Geng, Z.C., Meng, L.J. et al. 2012. Effects of different selenium fertilizers on the yield, selenium contents and qualities of potatoes. *Journal of Northwest A & F University (Natural Science Edition)* 40(9): 122–127. (in Chinese)

Zhao, X., Liao, G.P., Shi, X.H. et al. 2012. Construction of agricultural service mode in IOT and cloud computing environment. *Journal of Agricultural Mechanization Research* 4:142–147. (in Chinese)

Selenium Research for Environment and Human Health: Perspectives, Technologies and Advancements – Bañuelos, Lin, Liang & Yin (eds)
© 2020 Taylor and Francis Group, London, ISBN 978-1-138-39014-0

The variation pattern of selenium in geological history in relation to regional variation in topical environment

K.L. Luo & S.X. Zhang
Institute of Geographic Sciences and Natural Resources Research, Chinese Academy of Sciences, Beijing, China

1 INTRODUCTION

The distribution of selenium (Se) in the topical environment is extremely variable with large regional differences (Tan 1991). During Earth's geological periods of 4.6 billion years, various types of rocks and sediments were formed on surface of Earth, which were deposited during various geological periods. In the surface environment of China's mainland, the true Quaternary Holocene sediments account for about 20% of the sediments (Ma & Liu 2002, Deng 2007), while the present sediments are mainly distributed along the coastal rivers, modern riverbeds, Aeolian deposits at the bottom of lakes, and in the modern desert. Nearly 80% of the land surface are covered by ancient rocks and sediments deposited during the geological periods, especially in the mountainous, hilly, and highland areas, which have been weathered in a denuded surface environment. Hence, the surface environment has inherited variability.

However, due to the differences in climate, geographical environment, and provenance in different geological periods, rocks formed in different geological periods with different levels of trace elements, even in the same lithological rock beds. In the study of the environmental background of the local landmarks and the ecological environment, geographical and environmental scientists have long recognized the relationship between the content of trace elements in bedrock in the endemic area and the content of trace elements in soil and water (Luo et al. 2001, 2002, 2004). However, there is still a lack of research in the distribution and variation pattern of Se in geological history and its relationship with the regional differentiation of Se in the surface environment.

2 MATERIALS AND METHODS

To investigate the distribution and variation patterns of Se in geological periods and their relationships with the regional differentiation of Se in the surface environment, concentrations of Se were analyzed in about 3000 samples including bedrock, soil, rice, corn, and from stratum sections with different geological periods and topical environments in the stratum distribution area in different tectonic units of China.

3 RESULTS AND DISCUSSION

Our results show that the distribution of Se in a geological period is non-uniform and discontinuous, and most of the stratigraphic intervals are close to the crustal Clark value, which is about 0.02–0.10 mg/kg. The enrichment and lack of Se is nearly the same in the same lithology and marine sediment of the same geological period and in the same tectonic units, but Se content varies greatly in the same lithology and terrestrial deposits. The patterns of enrichment and lack of Se varies greatly in same geological period, in different tectonic units, and even in the same lithology and marine sedimentary.

The difference of original Se content in the strata is the main factor controlling the Se content and its spatial distribution and differentiation in soil, water, and crops in the topical environment. The distribution of Se in the environment is similar to the Se enrichment or losses occurring throughout geological history in a non-uniform and discontinuous distribution.

We found that the increasing phases and strong enriched strata of Se (> 10 mg/kg) are consistent with a series of major biological evolution sequences occurring in all geological periods. The enrichment of Se during the geohistory is closely related to geological events and biological events. The main enrichment horizon occurred mainly during the geological transition period.

The most concentrated (average Se content > 10 mg/kg) stratigraphic horizon is the Ediacaran and early Cambrian; both are critical periods for the evolution of multicellular life from ancient to recent times. Two rock beds about 14–15 m thick interbedded in Middle – Upper Permian, followed by upper of Lower Cretaceous in China, with a Se content varying from 0.02 to 257 mg/kg in those three intervals.

The average Se content in the stratigraphic sequences of these three geological periods intervals are 22 mg/kg, 148 mg/kg, and 12 mg/kg from old to young formations, respectively. The highest Se content of Se-rich layer among them is located in a 15 m thick layer, which interbedded in the Changhsingian Stage of the Lopingian Series and upper Capitanian Series of Permian located in the Yangtze platform. But this strong enriched Se-rich layer is only 15 m thick in the Yangtze platform and 35 m in southeastern China. The thickest Se-rich stratum is the Lujiaping Formation of Ediacaran-Cambrian period in the Qinling Mountain in Central China, with a thickness of about 1,000 m and about 300 m thick in the Yangtze platform. High concentrations of Se in soil in Enshi (>5 mg/kg) and Ziyang (>10 mg/kg) have resulted in Se toxic effects to humans and animals (Luo et al. 2004). These two geological sources of Se-rich stratum are considered Se-rich areas in southern China.

The Albian carbonaceous siliceous rocks of upper of Lower Cretaceous are about 50 meters thick in Tibet. Similar to the Utah region in western USA, the Albian carbonaceous siliceous rocks are the main sources of Se in those Se-rich areas in Tibet and Utah regions.

The lowest Se stratigraphic horizon (<0.05 mg/kg) and low Se distribution areas (soil < 0.20 mg/kg) are mainly associated with Precambrian Archean and Proterozoic granitic gneiss strata and are related to the granite rock mass in different geological periods and distribution areas. The Mesozoic and Cenozoic terrestrial deposited strata and their sources are around granitic gneiss strata or granite rock. Furthermore, the junction area of tectonic activity zones since the Cenozoic, or the terrestrial sedimentary strata, were deposited at the margins of the platform. For example, China's well-known "Selenium-deficiency Zone" located N 30–40° from west Yunnan in southwestern China to Heilongjiang in northeast China, is situated in the distribution area of Mesozoic and Cenozoic continental strata at the junction of the second-order terrace and the third-order terrace in China. It contains thick Tertiary and Quaternary mantle deposits or terrestrial sediments and their distribution areas are located in front of orogenic belts or the margins of the platform.

4 CONCLUSIONS

Globally, regardless of the geological history and modern times, the endogenic geological process is the dominant factor for the regional variation of Se distribution in the surface environment, while the exogenic geological process is a secondary dominant factor. The regional differentiation of Se on the surface of the earth is a comprehensive product resulted from 4.6 billion years of geological interaction with diversity and inheritance.

ACKNOWLEDGEMENT

This study is supported by the National Natural Science Foundation of China (No. 41172310, 41472322, & 41877299).

REFERENCES

Deng, Q.D. 2007. *Chinese Active Tectonics Map* (1:4000000): 1–2. Beijing: Earthquake Publishing House.

Luo, K.L. 2003. Determination of the stratigraphic era of selenium poisoning area in southern Shaanxi. *Geol Rev* 49(4): 383–388.

Luo, K.L., Pan, Y.T., Wang, W.Y. & Tan, J.A. 2001. Selenium content and distribution pattern in the Palaeozoic strata in the southern Qinling Mountains. *Geol Rev* 47(2): 211–217.

Luo, K.L., Tan, J.A., Wang, W.Y., Xiang, L.H. & Li, D.Z. 2002. Preliminary study on the chemical activity of selenium in early Paleozoic strata and stone coal in Daba mountain area. *Acta Sci Circumst* 22(1): 86–91.

Luo, K.L., Xu, L.R., Tan, J.A., Wang, D.H. & Xiang, L.H. 2004. Selenium source in the selenosis area of the Daba region, South Qinling Mountain, China. *Environ Geol* 45(3): 426–432.

Ma, L. & Liu, N. 2002. *Geological Atlas of China*: 9-12. Beijing: Geological Publishing House.

Tan, J.A. 1991. *Atlas of Endemic Diseases and the Environment of the People's Republic of China*. Beijing: Science Press.

Uptake and accumulation of Se

Selenium Research for Environment and Human Health: Perspectives, Technologies and Advancements – Bañuelos, Lin, Liang & Yin (eds)
© 2020 Taylor and Francis Group, London, ISBN 978-1-138-39014-0

Unravelling the complex trait of Se hyperaccumulation: Advances in research on potential candidate genes involved

M. Schiavon
Department of Agricultural Biotechnology, University of Padova, Legnaro (PD), Italy

Y. Jiang
College of Agronomy and Biotechnology, China Agricultural University, Beijing, China

M. Pilon, L.W. Lima & E.A.H. Pilon-Smits
Department of Biology, Colorado State University, Fort Collins, Colorado, USA

1 INTRODUCTION

Plants that hyperaccumulate selenium (Se) contain more than 1 mg Se/g DW in their aboveground tissues (Schiavon & Pilon-Smits 2017). The capacity to hyperaccumulate and hypertolerate Se by these species possibly developed as a result of convergent evolution of selective transporters and biochemical pathways in distinct angiosperm clades during geological periods when Se in soil was widely abundant and more widespread than today (White 2015). Some of the mechanisms evolved by Se hyperaccumulators are those involved in the regulation of specific Se/sulfur (S) transporters, methylation of Se-amino acids, rupture of Se-cysteine (Se-Cys) into alanine and elemental Se, and volatilization of Se-organic compounds (Schiavon & Pilon-Smits 2017, Lima et al. 2018). Additional mechanisms might include oxidation and transamination of selenocysteine (Secys), and conversion of SeCys to selenomethionine (SeMet) (Zhou et al. 2018).

The advent of RNA-Seq technology, which permits both *de novo* transcriptome assembly and profiling of global gene expression without the need for a reference genome, has provided a novel tool for the analysis of Se hyperaccumulators and improved our understanding of the molecular basis of Se hyperaccumulation in plants. Here, we describe an RNA-seq study conducted in the Se hyperaccumulator *Stanleya pinnata* and non-accumulator *Stanleya elata* grown with or without selenate with the aim to identify and characterize candidate genes responsible for the traits of Se hyperaccumulation and hypertolarance. Attention was then paid to two genes involved in Se/S uptake and assimilation.

2 MATERIALS AND METHODS

Stanleya pinnata and *Stanleya elata* plants were grown with or without 20 μM selenate for three weeks. A transcriptome-wide study via RNA-seq (Illumina HiSeq-2000) was further performed to highlight differences in root and shoot gene expression levels between the two species. Results of this study were compared with those obtained from another recent work to find out commonalities and differences between Se hyperaccumulators and suggest potential future avenues of research in this field.

3 RESULTS AND DISCUSSION

The study has identified a number of genes either up-regulated by Se or overexpressed in the hyperaccumulator compared to the non-accumulator, principally involved in antioxidant systems (e.g. glutathione-related genes and peroxidases), plant defense, jasmonic acid/salicylic acid/ethylene signaling, S acquisition and assimilation.

Genes involved in sulfate/selenate transport and assimilation were among the most differentially expressed between the two species (Figs 1, 2). Such differences in gene expression for S transporters, especially of SULTR1.2 (Fig. 2), were then supported by kinetic studies of Se uptake, while the gene coding for ATP-sulfurylase isoform ATPS2 was in deep characterized and its expression was found for an "abnormal" transcript, whose product is purely cytosolic, with a different C terminus. Higher ATPS2 gene expression correlated with greater ATPS activity in plants. Mature SpATPS2 was also expressed in *Escherichia coli* and purified. In a reverse assay, SpATPS2 was inhibited by both sulfate and selenate, but the effect of selenate was less pronounced, thus suggesting a greater affinity of this protein for Se over S.

Our results were in line with a study conducted on the hyperaccumulator *Cardamine hupingshanensis*, where genes functioning in S assimilation were overexpressed at high Se concentrations, as well as genes implied in degradation pathways, metal ion binding, and storage (Zhou et al. 2018).

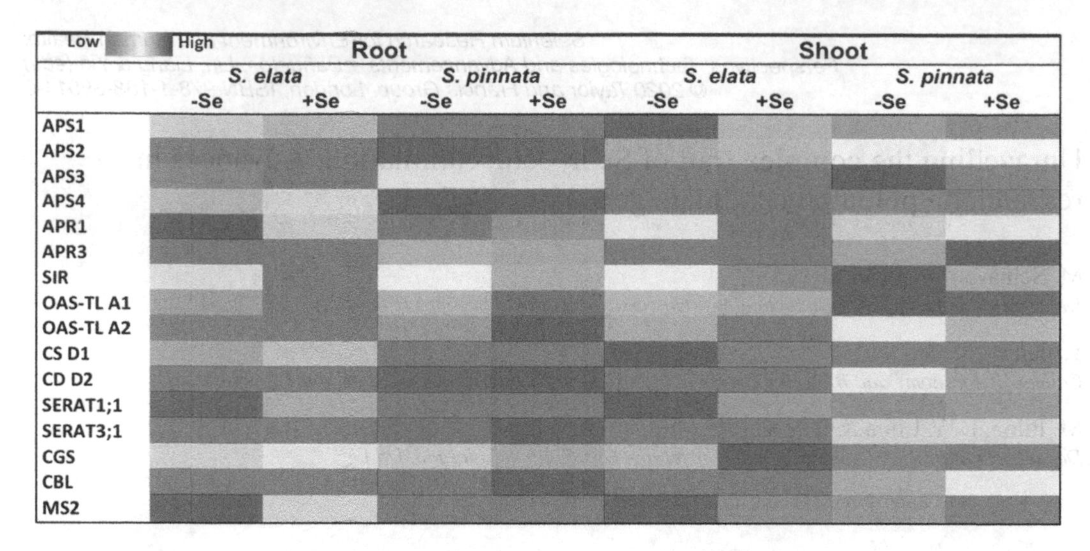

Figure 1. Heat Map of genes involved in S/Se assimilation expression in roots and shoots of *S. pinnata* and *S. elata* plants. Values used for Heat Map generation are the mean TMM-normalized RPKM values of 3 replicates per treatment. For each gene, different colors compared expression between treatments, organ and species.

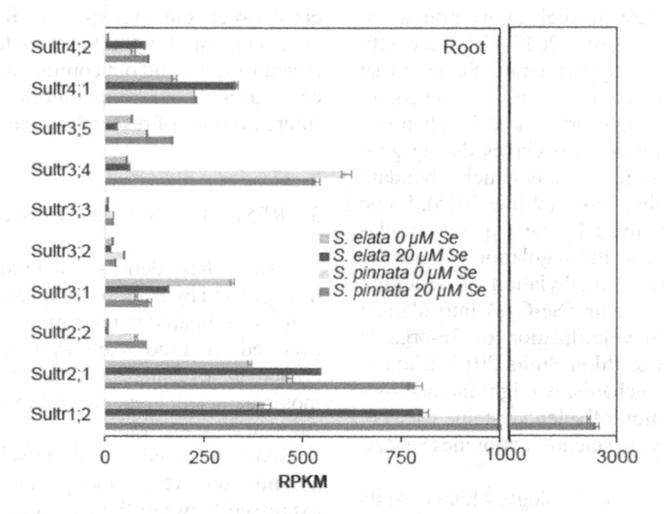

Figure 2. Expression levels (TMM-normalized RPKM values) of sulfate transporter (Sultr) genes in roots of *S. pinnata* and *S. elata* grown on 0 or 20 μM sodium selenate. S. Root expression levels of Sultr genes (mean RPKM ± SD, n = 3).

4 CONCLUSIONS

Overall, we can hypothesize that Se hyperaccumulation and hypertolerance in *S. pinnata* may be mediated by constitutively high expression of genes involved in sulfate/selenate uptake and assimilation, associated with elevate expression of defense- and stress-related genes. ATPS2 gene *from S. pinnata*, in particular, looks an intriguing candidate for explaining elevate Se assimilation rates and tolerance in this hyperaccumulator, while SULTR1.2 could be responsible for extremely high rates of Se uptake.

REFERENCES

Lima, L.W., Pilon-Smits, E.A.H. & Schiavon, M. 2018. Mechanisms of selenium hyperaccumulation in plants: A survey of molecular, biochemical and ecological cues. *Biochim Biophys Acta Gen Subj* 1862: 2343–2353.

Schiavon, M. & Pilon-Smits, E.A.H. 2017. The fascinating facets of plant selenium accumulation – biochemistry, physiology, evolution and ecology. *New Phytol* 213(4): 1582–1596.

Zhou, Y., Tang, Q., Wu, M. et al. 2018. Comparative transcriptomics provides novel insights into the mechanisms of selenium tolerance in the hyperaccumulator plant *Cardamine hupingshanensis*. *Sci Rep* 8(1): 2789.

Selenium Research for Environment and Human Health: Perspectives, Technologies and Advancements – Bañuelos, Lin, Liang & Yin (eds)
© 2020 Taylor and Francis Group, London, ISBN 978-1-138-39014-0

The genetics of selenium accumulation by plants

P.J. White & K. Neugebauer
The James Hutton Institute, Invergowrie, Dundee, UK

1 INTRODUCTION

Although selenium (Se) is an essential element for many prokaryotes, archaebacteria and eukaryotes, it is not required by either fungi or higher plants (White 2016). Nevertheless, Se is taken up, translocated and metabolised by higher plants because of its chemical similarity to sulphur (S). Tissue Se concentrations in an individual plant are largely determined by the Se phytoavailability in the substrate in which the plant is growing and the plant's capacity for S uptake, translocation, and assimilation (White 2016, 2018a, b). Since excessive Se accumulation in plant tissues can be toxic, the maximum tissue Se concentration in a living plant is determined by its ability to tolerate Se physiologically (White 2016). Most plants that grow on non-seleniferous soils cannot tolerate tissue Se concentrations greater than 10 to 100 mg Se/kg dry matter (DM). However, some plant species have evolved greater Se tolerance and can accumulate more than 100 mg Se/kg DM in their tissues. Plant species that can colonise both non-seleniferous and seleniferous soils are termed "Se indicator" plants, whereas those that only grow on seleniferous soils are termed "Se accumulator" plants. Some Se accumulator plants can have tissue Se concentrations of 1000 to 15,000 mg Se/kg DM when growing their native environment and are termed "Se hyperaccumulator" plants. However, fewer than 60 species have been reported that hyperaccumulate Se (White 2016). There is considerable genetic variation within all plant species in both Se accumulation and tissue Se tolerance. It is thought that this genetic variation (1) led to the evolution of plant species that colonised seleniferous soils (White 2016, 2018a) and (2) might be used in agriculture to develop crop genotypes with greater Se concentrations in their edible portions to improve the nutrition of humans and their livestock (White & Broadley 2009, White 2016). This article first reviews the genetic variation in Se accumulation between plant species and then the genetic variation in Se accumulation that occurs within plant species.

2 GENETIC VARIATION IN SELENIUM ACCUMULATION AMONG SPECIES

There is considerable variation in shoot Se concentrations among angiosperm species grown in the same

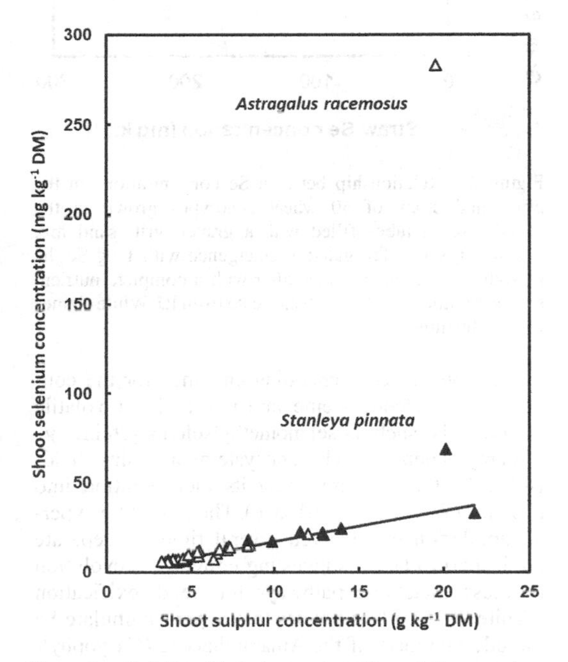

Figure 1. Relationship between shoot Se and S concentrations in 39 species of flowering plants grown hydroponically with 0.91 mM sulphate and 0.63 μM selenate. The line indicates a shoot Se/S quotient of 1.725 mg Se per g S. Closed symbols indicate species from the Brassicales. *Stanleya pinnata* (Brassicaceae) and *Astragalus racemosus* (Fabaceae) are Se hyperaccumulator plants. Data from White et al. (2007).

environment (Fig. 1, White et al. 2007, 2016, 2018a, b). Among most plant species, except Se hyperaccumulators, there is a strong stoichiometric relationship between shoot Se and S concentrations, and plants that accumulate more S, such as alliums and brassicas, also accumulate more Se (White 2016, 2018a, b). This is because Se and S share common pathways for uptake by roots, translocation between organs, and metabolism within the plant. However, the shoot Se/S quotient of Se hyperaccumulators is far greater than that of other plant species grown in the same environment (Fig. 1, White et al. 2007). The ability of Se hyperaccumulator species to accumulate, and tolerate, large tissue Se concentrations is related to the constitutive expression of genes encoding Se transport proteins, constitutive expression of genes involved in the assimilation of selenate, and the

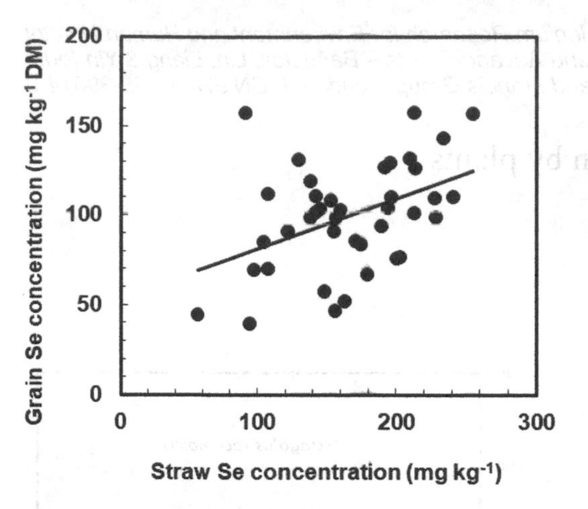

Figure 2. Relationship between Se concentrations in the grain and straw of 40 wheat genotypes grown in the glasshouse in tubes filled with a gravel: grit: sand mix (40:40:40, v:v:v) fertigated at emergence with 10 g Se /ha as sodium selenate and thereafter with a complete nutrient solution without Se. Unpublished data from P.J. White (James Hutton Institute).

expression of genes encoding enzymes for the conversion of selenocysteine into non-toxic or volatile compounds, such as selenomethylselenocysteine, γ-glutamyl selenomethylselenocysteine and dimethyldiselenide, thereby preventing its incorporation into proteins (White 2016, 2018a, b). The trait of Se hyperaccumulation has evolved several times in separate angiosperm clades, suggesting convergent evolution of these metabolic pathways for Se detoxification (White 2016). Plant species that hyperaccumulate Se include members of the Amaranthaceae (Caryophyllales), Asteraceae (Asterales), Brassicaceae (Brassicales), Fabaceae (Fabales), Orobanchaceae (Lamiales) and Rubiaceae (Gentianales).

3 GENETIC VARIATION IN SELENIUM ACCUMULATION WITHIN SPECIES

In addition to variation in the ability of different species to accumulate Se, there is considerable variation in tissue Se concentration among genotypes of a species growing in the same environment (White 2016). For example, shoot Se concentrations differ among genotypes of both Se hyperaccumulator species, such as *Stanleya pinnata* and *Symphyotrichum ericoides*, and non-accumulator species, such as arabidopsis, tall fescue and various crops.

This genetic variation has been used to identify genes affecting Se uptake, translocation, assimilation and volatilisation in several plant species and might be used to develop crop cultivars with greater tolerance and accumulation of Se in their tissues (White & Broadley 2009, White 2016, 2018b). In addition to the genetic variation in the Se concentrations of leafy vegetables, such as onion, kale, broccoli, cauliflower, Chinese cabbage, Indian mustard, chicory leaves and lettuce, genetic variation has also been reported in grain Se concentration of cereals, including bread wheat (Fig. 2), durum wheat, rice, millet, barley, and oats, in seed Se concentration of legumes, including common bean, field pea, chickpea, lentil, mung bean, and soybean, and in Se concentration of other horticultural crops, including tomato, pepper and potatoes.

4 CONCLUSIONS

There is considerable genetic variation in the ability to tolerate and accumulate Se in tissues both among and within plant species. Differences in Se metabolism underlie the contrasting abilities of plant species to tolerate and accumulate Se in their tissues and led to the evolution of the seleniferous flora. Genetic variation in Se accumulation within plant species has led to the identification of genes impacting Se accumulation that might enable the development of edible crops with greater Se concentrations to improve the nutrition of humans and their livestock.

REFERENCES

White P.J. & Broadley, M.R. 2009. Biofortification of crops with seven mineral elements often lacking in human diets – iron, zinc, copper, calcium, magnesium, selenium and iodine. *New Phytol* 182: 49–84.

White, P.J., Bowen, H.C., Marshall, B. & Broadley, M.R. 2007. Extraordinarily high leaf selenium to sulphur ratios define 'Se-accumulator' plants. *Ann Bot* 100: 111–118.

White, P.J. 2016. Selenium accumulation by plants. *Ann Bot* 117: 217–235.

White, P.J. 2018a. Selenium metabolism in plants. *Biochim Biophys Acta - Gen Subj* 1862: 2333-2342.

White, P.J. 2018b. Selenium in soil and crops. In B. Michalke (ed.), *Selenium*: 29–50. Cham: Springer.

Selenium Research for Environment and Human Health: Perspectives, Technologies and Advancements – Bañuelos, Lin, Liang & Yin (eds)
© 2020 Taylor and Francis Group, London, ISBN 978-1-138-39014-0

Effects of different kinds of hormones on selenium accumulation in rice

Z.H. Dai, Y. Yuan, H.L. Huang, M. Rizwan & S.X. Tu
College of Resources and Environment, Huazhong Agricultural University, Wuhan, China
Microelement Research Center, Huazhong Agricultural University, Wuhan, China

1 INTRODUCTION

Selenium (Se) is an essential element for human beings and animals. Different selenium-enriched agricultural products have different Se-enriched standards, and the regulation of Se content has always been a research hotspot.

We know that the hormones are widely used in the agricultural production. Auxins are arguably the most important signaling molecules in plants and have a profound impact on plant growth and development (Weijers & Wagner 2016). Gibberellic acid (GA3) is an important plant growth hormone which can accelerate stalk and leaf growth, improve seed shooting and increase fructification yield (Tang *et al.* 2000). The role of ethylene in defense responses to pathogens is widely recognized (Dubois et al. 2018). The 6-benzyl aminopurine (6BA) is one kind of cytokinin which can inhibiting the decomposition of chlorophyll, nucleic acid and protein in plant leaves, keeping green and preventing aging (Wojtania & Skrzypek 2014). Brassinosteroids (BRs) are steroidal plant hormones that are widely distributed in lower to higher plants, and the brassinolide (BL) exhibits the highest biological activity among naturally occurring BRs (Kim et al. 2000). Salicylic acid (SA) is a phenolic compound which can regulate plant physiological functions, such as seed germination, photosynthesis, respiration, growth and flowering (Rivas & Plasencia 2011). Plant growth regulator methyl jasmonate (MeJA) is a member of jasmonate group, which can regulates many aspects of plant growth and development. MeJA is known as a signaling molecule which plays a role in many biotic and abiotic stress responses.

At present, the development of Se-enriched agricultural products is the most popular and effective way to meet human nutritional needs for Se. While the hormones are inevitably used in the production of Se-enriched agricultural products, the effects of applying different hormones on Se accumulation in rice is not clear. Hence, the objective of this study was to investigate the relationship between Se accumulation and different hormones in rice.

2 MATERIALS AND METHODS

A hydroponic experiment was conducted in the greenhouse of the College of Resources and Environment in Huazhong Agricultural University in Wuhan, China. Rice (*Oryza sativa* L., cv. yangliangyou 6) seeds were surface sterilized in 10% H_2O_2 for 10 min, then rinsed with deionized water (ddH_2O) thoroughly, and soaked in ddH_2O for one night. Thereafter, the rice seeds were placed in plastic trays for germination with 30°C in the growth cabinet. Until the three-leaf stage, the rice seedlings were transplanted into the greenhouse at 27°C and subject to lighting for 14 h per day. The rice seedlings were cultured with 1/2 international (Yoshida) rice nutrient solution, and the nutrient solution was changed every 3 days. After 7 days, uniform-sized rice seedlings were selected and transplanted into plastic pots. The Na_2SeO_3, indole-3-acetic acid (IAA), GA3, Ethylene, 6BA, BL, SA, and MeJA were added to full strength international rice nutrient solution in different treatments. In this experiment, 8 groups of treatments were designed, and each group had 5 different levels. Each treatment was replicated three times. The concentration of Se used in this experiment was 5 μM. During the growth of rice seedlings for 14 days, the nutrient solutions were replaced every 3 days, and pH value was adjusted to 6.0 with NaOH or HCl. All chemicals were purchased from Sigma-Aldrich unless stated otherwise.

The total Se concentrations in rice tissues were measured by Hydride Generation-Atomic Fluorescence Spectrometry (HG-AFS) (Titian AFS-8220, Beijing).

3 RESULTS AND DISCUSSION

Concentrations of Se in rice roots and shoots were significantly influenced by different kinds and levels of hormones. The highest Se concentration was found in the single Se treatment.

In the IAA treatment group, the mean concentration of Se in roots and shoots was 55.53 mg/kg and 10.53 mg/kg, respectively. With an increasing level of IAA, the Se content in rice roots and shoots decreased gradually.

The average concentration of Se in roots and shoots was 60.08 mg/kg and 9.42 mg/kg in the GA3 treatment group, respectively. There was a significant 46.39% and 15.74% decrease of Se concentration in rice roots and shoots in the highest GA3 treatment compared with single Se treatment.

All ethylene treatments decreased the Se concentration in rice seedlings, and the mean content of Se in roots and shoots was 60.08 mg/kg and 9.88 mg/kg, respectively.

Applying 6BA effectively decreased the Se concentration in rice seedlings, and the average Se concentration in roots and shoots was 56.98 mg/kg and 10.40 mg/kg.

In the BL treatment group, all treatments decreased the Se content. With an increasing level of BL, the Se content in rice roots and shoots gradually decreased. In the highest BL treatment, the Se content in rice roots and shoots decreased by 6.36% and 25.41% compared with single Se treatment.

All SA and MeJA treatments decreased the Se concentrations in rice seedlings. These two kinds of hormones were the most effective way to decrease Se content in rice.

4 CONCLUSIONS

The results of this study clearly demonstrate that all the hormones used in this study can decrease the Se accumulation in rice seedlings. The SA and MeJA may be useful for reducing Se levels in rice, if Se content in rice exceeds rice consumption standard.

REFERENCES

Dubois, M., Lisa, V.D.B. & Inzé, D. 2018. The pivotal role of ethylene in plant growth. *Trends Plant Sci* 23(4): 311–323.

Kim, T.W., Chang, S.C., Choo, J. et al. 2000. Brassinolide and [26,28-$^{2}H_{6}$] brassinolide are differently demethylated by loss of C-26 and C-28, respectively, in *Marchantia polymorpha*. *Plant Cell Physiol* 41(10):1171–1174.

Rivas-San Vicente, M. & Plasencia, J. 2011. Salicylic acid beyond defence: its role in plant growth and development. *J Exp Bot* 62(10): 3321–3338.

Tang, Z., Zhou, R. & Duan, Z. 2000. Separation of gibberellic acid (GA3) by macroporous adsorption resin. *J Chem Technol Biot* 75(8): 695–700.

Weijers, D. & Wagner, D. 2016. Transcriptional responses to the auxin hormone. *Annu Rev Plant Biol* 67(1): 539–574.

Wojtania, A. & Skrzypek, E. 2014. Effects of cytokinins on antioxidant enzymes in in vitro grown shoots of *Pelargonium hortorum* L.H. Bayley. *Acta Agrobot* 67(4): 33–42.

Selenium Research for Environment and Human Health: Perspectives, Technologies and Advancements – Bañuelos, Lin, Liang & Yin (eds)
© 2020 Taylor and Francis Group, London, ISBN 978-1-138-39014-0

The relationship between sulfur and selenium in the plant-soil system

X.W. Liu, Z.Q. Zhao & X.F. Deng
Microelement Research Center, Huazhong Agricultural University, Wuhan, China

1 INTRODUCTION

Selenite and selenate are the major selenium (Se) species in soil that can be efficiently utilized by plants (Ellis & Salt 2003). Currently, the plant uptake mechanisms of selenite are not well understood. Some studies postulate that it enters the plant via a phosphate transporter (Li et al. 2008), and phosphate inhibits crop absorption of selenite. However, pot experiments have shown that phosphate application reduces Se content in crops through the dilution effect produced by crop growth, which did not seem to reduce Se accumulation (Mora et al. 2008). Since selenate and sulfate have the same mechanism of transport, there are significant inter-element antagonism effects between them. In contrast, field experiments showed that sulfate facilitated selenate uptake in wheat and buckwheat (Golob et al. 2016, Stroud et al. 2010). Therefore, there is a necessity to systematically clarify the regulatory effect of sulfur (S) on crop uptake of selenite and selenate in the production practice. It's noteworthy that the bioavailability of Se in the human body is closely associated with Se speciation in the crops; therefore, the effect of S on Se speciation in the edible parts of crop should be of concern.

2 MATERIALS AND METHODS

2.1 *Experimental design*

Experiment 1: Rape was chosen as the experimental plant material. Sulfur was applied at the rates of 0 and 60 kg/ha as elemental S or as magnesium sulfate. The application of Se at the rate of 60 g/ha was applied as sodium selenite or sodium selenate. A no-S and no-Se treatment was used as the control.

Experiment 2: Rape seedlings grew in hydroponic solution. The treatments began four weeks after the seedlings were planted. Different levels of S were used in the experiment at rates of 0.5, 1, 2, and 4 mmol/L; Se was applied at a constant rate of 10 μmol/L. Rape was exposed to the Se and S treatments for 24 h, using sodium selenite, sodium selenate, and magnesium sulfate compounds.

Experiment 3: Eight kg of air-dried and sieved aqueous soil was placed into pots. Elemental S was used as the S source, and the application rates were 0, 150, and 300 mg/kg. The Se source was sodium selenite, and it was applied at 0 and 15 mg/kg. Rape was harvested on the 70th day after sowing. Soil samples were randomly taken from 5 points with a 3-cm diameter punch and then mixed. Extraction of Se fractions in soil is described in Liu et al. (2015).

Experiment 4: Twenty kg of air-dried and sieved yellow-brown and calcareous alluvial soil was placed into pots. Selenite and selenate were used as the Se sources, and the application rates were 2 mg/kg. The S source was elemental S, and it was applied at 0 and 100 mg/kg. Soybean was harvested on the 100th day after sowing.

2.2 *Statistical analysis*

One-way ANOVA was carried out with multiple comparisons using Duncan's test to compare the means of different treatments at $p \leq 0.05$. All statistical analyzes were performed using the SPSS 17.0 statistical package (SPSS Inc., Chicago, IL, USA).

3 RESULTS AND DISCUSSION

3.1 *Effects of S on Se concentration in field trials*

Irrespective of the species of Se fertilizers, S application significantly inhibited the Se uptake, and there was no significant difference in the inhibitory effect between elemental S and sulfate. For selenate-treated plants, the two S fertilizers caused a consistent reduction (approximately 40%) in the Se concentration of seeds. The reduction in seed Se concentration (by approximately 25%) was also consistent in selenite-treated plants. These results show that S application caused a significantly reduction in Se concentration of seeds in selenate-treated plants compared with selenite-treated plants.

3.2 *Effects of S on Se absorption in solution culture*

The results for the selenite treatment showed that there was no significant impact on the Se concentrations in rape shoots with any rate of S application (Fig. 1). In contrast, with the selenate treatment, sulfate significantly reduced the Se concentrations in shoots. These reductions did not decrease further with an increase of sulfate concentration to 2 mmol/L. Overall, sulfate

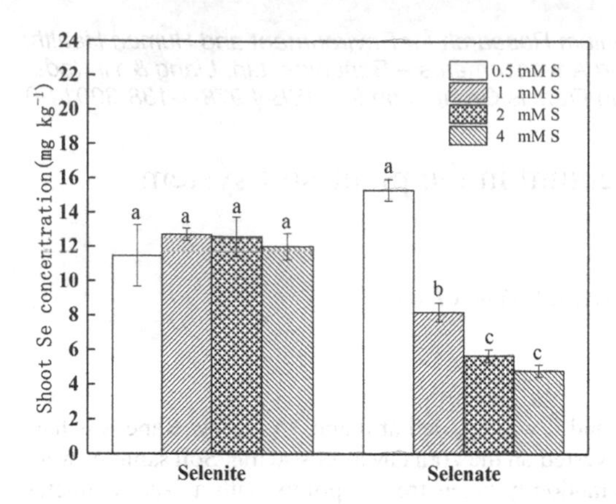

Figure 1. Effects of sulfate on the Se concentration in shoots of rape seedlings grown in solution culture treated with selenite or selenate.

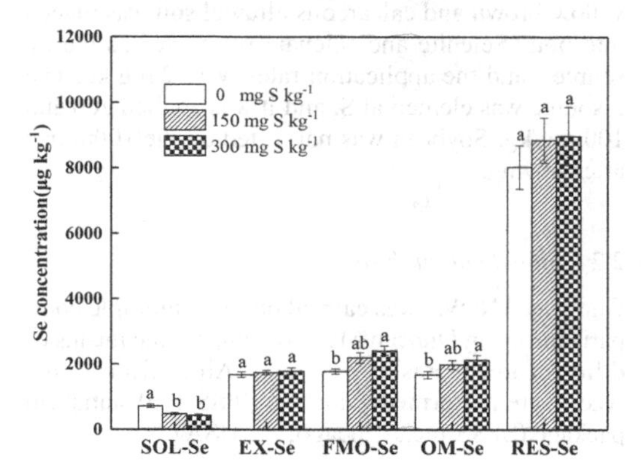

Figure 2. Selenium concentrations in different fractions of the soil with increasing S application rates.

barely affected selenite-Se uptake by the plants, but reduced selenate-Se uptake.

3.3 *Effects of S on soil Se fractionation*

Selenium fractions and species in soil are more relevant to Se uptake and accumulation in crops than the total Se in soil. The results showed that the most of Se in the soil treated with selenite can be adsorbed or fixed at the seedling stage of rape (Fig. 2). Soluble and exchangeable Se (available Se) in soil making up approximately 10-15% of the total Se. Fe-Mn oxide-bound and organic bound Se (ineffective Se) accounted for 25-30% of the total Se. The application of S promoting soluble Se transformation to Fe-Mn oxide-bound Se, organic matter-bound Se and residual Se, hence reducing the uptake of Se by rape in soil culture.

3.4 *Effects of S on Se speciation in Se-enriched soybean*

SeMet was the main Se species in soybean seeds (>90%) grown under selenite or selenate treatment. Selenium in soybean seeds exists in the organic form of

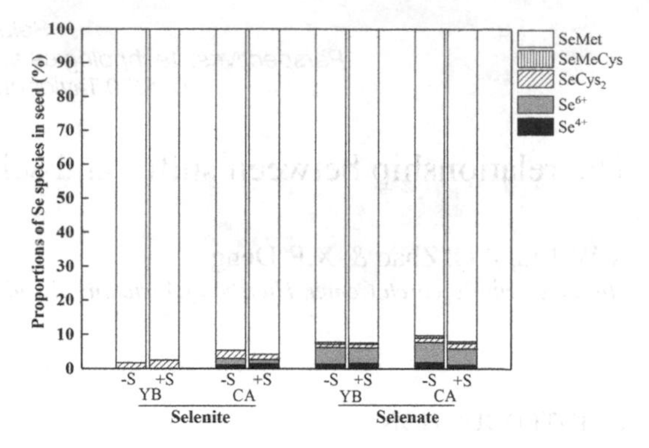

Figure 3. Effects of S on Se species in soybean grain.

SeMet and $SeCys_2$ in yellow-brown soil, while about 3% the inorganic Se^{4+} and Se^{6+} in calcareous alluvial soil (Fig. 3). Sulfur application also had no significant effect on the transformation of Se form in soybean seeds produced from the yellow-brown soil and calcareous alluvial soil. Overall, S had no significant effect on the transformation of Se forms in soybean grains treated with selenite and selenate.

4 CONCLUSIONS

Application of S and Se fertilizers significantly caused antagonisms between Se and S and inhibited the uptake of Se by roots, and also decreased the availability of Se in soil. Concentrations of Se were reduced in crops treated with selenite or selenate; however, there was little effect on the speciation of Se in soybean grain.

REFERENCES

Ellis, D.R. & Salt, D.E. 2003. Plants, selenium and human health. *Curr Opin Plant Biol* 6: 273–279.

Golob, A., Gadžo, D., Stibilj, V. et al. 2016. Sulphur interferes with selenium accumulation in Tartary buckwheat plants. *Plant Physiol Bioch* 108: 32–36.

Liu, X.W., Zhao, Z.Q., Duan, B.H. et al. 2015. Effect of applied sulphur on the uptake by wheat of selenium applied as selenite. *Plant Soil* 386: 35–45.

Li, H.F., McGrath, S.P. & Zhao, F.J. 2008. Selenium uptake, translocation and speciation in wheat supplied with selenate or selenite. *New Phytol* 178: 92–102.

Mora, M.L., Pinilla, L., Rosas, A. & Cartes, P. 2008. Selenium uptake and its influence on the antioxidative system of white clover as affected by lime and phosphorus fertilization. *Plant Soil* 303: 139–149.

Stroud, J.L., Li, H.F., Lopez-Bellido, F.J. et al. 2010. Impact of sulphur fertilisation on crop response to selenium fertilisation. *Plant Soil* 332: 31–40.

Selenium Research for Environment and Human Health: Perspectives, Technologies and Advancements – Bañuelos, Lin, Liang & Yin (eds)
© 2020 Taylor and Francis Group, London, ISBN 978-1-138-39014-0

Selenium in agriculture soils and its accumulation in crops in southern Jiangxi, China

M.L. Tang, H. Tian & Z.Y. Bao
Faculty of Materials Science and Chemistry, China University of Geosciences, Wuhan, Hubei, China

L.Y. Yao & M. Ma
Zhejiang Institute, China University of Geosciences, Hangzhou, Zhejiang, China

1 INTRODUCTION

Selenium (Se) is an essential micronutrient for humans and animals, with a narrow range between its nutrient requirement and its toxicity (Rayman 2000). Selenium consumption for humans is supplied mainly by the food chain. In soil, Se exists mainly in four valence states: 2−, 0, 4+, and 6+). Selenate (SeO_4^{2-}) and selenite (SeO_3^{2-}) are highly soluble and Se^{4+} is mainly adsorbed to charged surfaces of clay minerals, whereas elemental Se (Se^0) and selenide (Se^{2-}) are poorly mobile, and are represented as stable and insoluble forms (Hartikainen 2005).

Several field and pot experiments have been conducted to investigate soil parameters affecting Se accumulation in crops without Se fertilization. Researchers found that soil total Se content, KH_2PO_4 extractable Se, KH_2PO_4 extractable S, $CaCl_2$ extractable Se, $CaCl_2$ extractable dissolved organic carbon, organic carbon content, and pH influence crop Se accumulation (Zhao et al. 2005, Stroud et al. 2012, De Temmerman et al. 2014, Supriatin et al. 2016). Regression models have been established based on the above-mentioned soil parameters to determine Se content in crops.

The objectives of this study were (1) to investigate the Se contents in local agriculture products and corresponding rhizosphere soils in Southern Jiangxi Province, China; (2) to identify soil parameters affecting Se uptake in crops; and (3) to establish the best fitting regression models to predict Se concentrations in edible part of crops.

2 MATERIALS AND METHODS

2.1 *Field sampling and sample preparation*

In July 2017, sampling was performed in Ganzhou City, Jiangxi Province, China. Rice seeds (*Oryza sativa* L.), peanut seeds (*Arachis hypogaea* Linn.), and tea leaves (*Camellia sinensis* (L.) O. Kuntze) and corresponding rhizosphere soils were sampled. Each soil or plant sample was mixed with 4 subsamples. Soil samples were air-dried, removed from roots and ground to pass through a 0.74 μm nylon sieve. Crop samples were hulled and washed with running tap water and then with ultrapure water. The washed crop samples were then oven-dried at 60°C to a constant weight. All the crop samples were powdered, sealed and stored in refrigerator at −20°C.

2.2 *Chemical analysis*

Soil pH was measured using a pH meter. Soil total organic carbon (TOC) was determined by total organic carbon analyzer. Soil organic matter (SOM) = TOC × 1.724 (Rowell 2014). Plant Se and soil-bioavailable Se was extracted by 0.5 mol/L $NaHCO_3$ and determined by AFS.

2.3 *Statistical analysis*

Statistical analyses were performed with Minitab 16. Soil and crop Se concentrations were analyzed by ANOVA. Best subset regression models were derived to correlate crop Se and soil parameters.

3 RESULTS AND DISCUSSION

3.1 *Selenium in soils and crops*

Table 1 summarizes the total Se in the edible part of crops, and total Se and bioavailable Se of different cultivated soils in Southern Jiangxi. 14.49% of paddy soils, 26.32% of peanut soils, and 88.26% of tea soils met the level of 0.4–3.0 mg/kg for selenium-rich soil proposed by Tan (1989). The average of total Se and bioavailable Se followed the order of tea soil > peanut soil > paddy soil. Similar average percentages of bioavailable Se to total Se had been found within paddy soil (16.47%), peanut soil (17.77%) and tea soil (17.71%).

The Se content of rice and peanut seeds in this area was generally low, with an average of 0.05 mg/kg. In this study 22.46% of rice, and 21.05% of peanuts meet the standard of Se-rich crops (0.07–0.30 mg/kg)

Table 1. Total Se in crops, total Se and bioavailable Se in corresponding soils of Southern Jiangxi.*

Crops	N	Se in Crops (mg/kg)	Soils	
			Se (mg/kg)	Bio-Se (μg/kg)
Rice	138	0.05^{B} (0.01-0.21)	0.29^{Bb} (0.11-1.06)	46.03^{C} (8.53-142.4)
Peanut	95	0.05^{B} (0.01-0.32)	0.36^{Ba} (0.08-2.71)	61.86^{B} (4.96-333.9)
Tea	17	0.10^{A} (0.04-0.24)	0.55^{A} (0.32-0.79)	94.26^{A} (73.0-120.3)

* The data are presented as mean concentrations. Values in brackets are ranges of the parameters. Means below the quantification limit (LOQ) were replaced with LOQ/2. Mean values with the different letter indicate significant differences at $p < 0.05$ following Turkey's HSD multiple range test.

Table 2. Best subset regression equations of the type (C-Se, mg/kg) = a + b × (Se, mg/kg) + c × (SOM, %) + d × pH**.

	(a)	Se (b)	SOM (c)	pH (d)	R^2_{adj}	P
R-Se	0.025	0.092	−0.012	0.007	0.306	<0.001
P-Se	0.045	0.105	−0.010	–	0.531	<0.001
T-Se	−0.509	0.316	−0.052	0.118	0.366	0.030

** C-Se, R-Se, P-Se, and T-Se refer to crop Se, rice Se, peanut Se and tea Se respectively. Samples for regression analyses included 128 of rice, 68 of peanut and 17 of tea. The crop Se below the quantification limit were not included.

according to local standards established in Jiangxi Province (DBD36/T 566-2009), while none of the tea samples met the standard of 0.5–3.0 mg/kg for tea.

3.2 *Regression models for Se in crops*

To identify which soil properties determine crop Se content and to establish a model for prediction, best subset regression models were performed between crop Se and soil parameters including total Se, SOM, and pH (Table 2). To simplify the model and make it practical for the Geochemical Survey of Land Quality in China, bioavailable Se was not included, and $\log_{10}$ transformation was not conducted in models.

Due to the limited variability of parameters measured in the soils sampled, the equations can explain 53.1% of the variation at maximum but all were significant at $p < 0.05$. As the main source of plant uptake, total soil Se and bioavailable represented as Se^{6+} and Se^{4+} determined crop Se to a certain extent (Favorito et al. 2017). Rice and tea Se accumulation mainly increased at elevated soil pH. Wang et al. (2017) concluded that increasing soil pH inhibited the aging of selenate and thus improved Se bioavailability. In contrast, decreasing soil pH elevated soil H^+ concentration and reduced the negative charge on the soil surface at the same time, making the acid radical anion selenite more easily adsorbed and fixed onto the soil surface, resulting in a lower Se bioavailability (Liu et al. 2016). Soil organic matter played a dual role in regulating soil Se bioavailability. The immobilization of Se by SOM of both biotic and abiotic mechanisms reduced Se bioavailability, whereas the decomposition of OM-bound Se provided more bioavailable Se in soil solution (Li et al. 2017). In this study, the higher SOM content mainly reduced soil Se bioavailability, resulting in lower Se accumulation in crops in Southern Jiangxi.

4 CONCLUSIONS

The majority of cultivated soils in Southern Jiangxi are rather low in Se, resulting in low crop Se contents. Based on regression analyses, we concluded that soil Se, bioavailable Se, organic matter and pH are important factors determining crop Se in Southern Jiangxi.

REFERENCES

De Temmerman, L., Waegeneers, N., Céline, T. et al. 2014. Selenium content of Belgian cultivated soils and its uptake by field crops and vegetables. *Sci Total Environ* 468–469: 77–82.

Favorito, J.E., Eick, M.J., Grossl, P.R. et al. 2017. Selenium geochemistry in reclaimed phosphate mine soils and its relationship with plant bioavailability. *Plant Soil* 418(1–2): 1–15.

Hartikainen, H. 2005. Biogeochemistry of selenium and its impact on food chain quality and human health. *J Trace Elem Med Biol* 18: 309–318.

Li, Z., Liang, D., Peng, Q. et al. 2017. Interaction between selenium and soil organic matter and its impact on soil selenium bioavailability: A review. *Geoderma 295*: 69–79.

Liu, J., Peng, Q., Liang, D.L. et al. 2016. Effects of aging on the fraction distribution and bioavailability of selenium in three different soils. *Chemosphere* 144: 2351–2359.

Rayman, M.P. 2000. The importance of selenium to human health. *Lancet* 356(9225): 233–241.

Rowell, D.L. 2014. *Soil Science: Methods and Applications*. London: Routledge.

Stroud, J.L., Broadley, M.R., Foot, I. et al. 2010. Soil factors affecting selenium concentration in wheat grain and the fate and speciation of Se fertilisers applied to soil. *Plant Soil* 332(1–2): 19–30.

Supriatin, S., Weng, L. & Comans, R.N.J. 2016. Selenium-rich dissolved organic matter determines selenium uptake in wheat grown on low-selenium arable land soils. *Plant Soil* 408(1–2): 73–94.

Tan, J.A. 1989. *The Atlas of Endemic Diseases and Their Environments in the People's Republic of China*. Beijing: Science Press.

Wang, D., Zhou, F., Yang, W. et al. 2017. Selenate redistribution during aging in different Chinese soils and the dominant influential factors. *Chemosphere* 182: 284–292.

Zhao, C., Ren, J., Xue, C. et al. 2005. Study on the relationship between soil selenium and plant selenium uptake. *Plant Soil* 277(1–2): 197–206.

Selenium Research for Environment and Human Health: Perspectives, Technologies and Advancements – Bañuelos, Lin, Liang & Yin (eds)
© 2020 Taylor and Francis Group, London, ISBN 978-1-138-39014-0

Studies on adsorption kinetics of selenium by different materials

J.Y. Li, H. Tian & C.H. Wei
Engineering Research Center of Nano-Geo Materials of Ministry of Education, Faculty of Materials Science and Chemistry, China University of Geosciences, Wuhan, Hubei, China

Z.Y. Bao
Engineering Research Center of Nano-Geo Materials of Ministry of Education, Faculty of Materials Science and Chemistry, China University of Geosciences, Wuhan, Hubei, China
Zhejiang Institute, China University of Geosciences (Wuhan), Hangzhou, China

1 INTRODUCTION

Selenium (Se) is an essential trace element for human beings and animals (Xia 2013). Diet is the main way for selenium supplementation. Thus, more attention has been given to Se-biofortified agriculture. Selenate (SeVI) and selenite (SeIV) are the main effective forms of Se-enriched nutrient fortifier for soil, which can result in rapid release, easy loss, and incomplete utilization for plants in soil (Liu et al. 2010). Thus, it is of great practical significance to develop a slow-release Se-enriched fertilizer.

Adsorption material has a strong adsorption capacity for some elemental forms under certain physical and chemical conditions. The adsorbed elements can be fixed on the surface of the materials and then slowly released under natural conditions (Ke et al. 2005). Similarly, the addition of adsorption material to Se fertilizer can bring two benefits for plants: (1) effectively control the bio-available Se content in soil at unit time; (2) ensure a long-term supply of Se element during the growth process of crop (Ni & Fan 2018). In other words, these are key mechanisms for the sustained controlled release fertilizer. Therefore, the selection of adsorption materials, adsorption capacity, adsorption, and release are of great significance for the successful development of slow-release Se fertilizer.

Langmuir adsorption isotherm equation (Zhao 2005):

$$q = q_0 \frac{KC_e}{1+KC_e} \tag{1}$$

where

q: adsorbing capacity
q_0: saturated adsorption capacity
K: the adsorption constant
C_e: adsorbent concentration

Freundlich adsorption isotherm equation:

$$q = KC_e^{1/n} \tag{2}$$

In this paper, in order to obtain the best slow-release adsorption materials, two materials (fly ash, DS18) are selected as the research objects. Adsorption simulation experiments with Se(IV) were then conducted to investigate their maximum adsorption capacity and adsorption/release rules.

2 MATERIALS AND METHODS

2.1 *The instrument condition*

The solution Se concentration was measured by Hydride Generation-Atomic Fluorescence Spectrometer (HG-AFS). In order to obtain a high analytical sensitivity, precision and accuracy, the instrument conditions were optimized as followed: The negative high voltage 270 V, the lamp current 80 ma, the atomization device height 8 mm, the carrier gas flow rate 300 ml/min, the shielding gas flow rate 800 ml/min, reading time 10 s, and delay time 2 s.

2.2 *Experiments*

Selenite (50 mg/L and 1000 mg/L) solutions were prepared by dissolving a specific quantity of $NaSeO_3$ in ultrapure water. Firstly, 2 g fly ash were weighed and transferred into a 50 ml polyethylene tube, and 50 ml selenite solutions (50 mg/L) were added to the tubes. Secondly, 1 g DS18 new material was weighed and transferred into a 1000 ml polyethylene tube and 800 ml selenite solutions (1000 mg/L) were added to the tubes.

We designed different condition experiments to determine their optimal adsorption time for each material and the best suitable material. Then, the desorption trend of the best material was discussed.

3 RESULTS AND DISCUSSION

3.1 *The adsorption isotherm of Se(IV) for fly ash*

As shown in Table 1, with increased adsorption time, the adsorption concentration of Se(IV) on fly ash was also improved from 14.1 μg/L to 496.3 μg/L.

Table 1. The adsorption capacity of Se on fly ash.

Time (h)	Adsorbed Se Concentration (μg/L)	The Adsorption Capacity (μg/g)
0.5	14.1	35
1	16.4	41
2	35.8	90
3	99.6	249
6	139	348
9	192	481
24	403	1009
48	496	1241

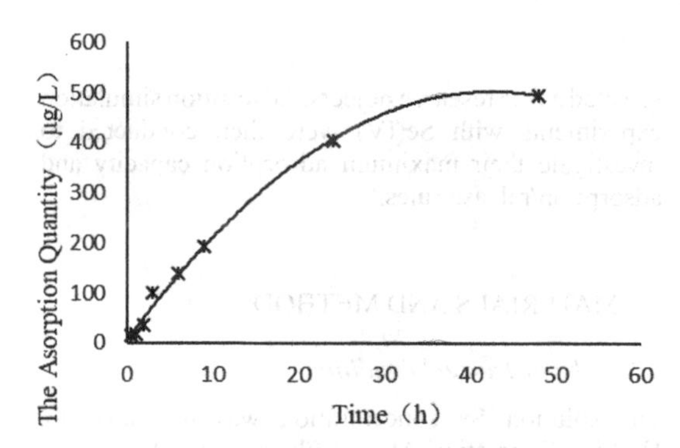

Figure 1. Time-adsorption quantity diagram for fly ash adsorbing Se(IV).

Origin was used to fit the experimental data, and the Langmuir adsorption isotherm equation was obtained as follows (Fig. 1).

The adsorption capacity of fly ash for Se(IV) presents a rapid increase trend and tends to be stable after 24 hours later at temperature of 35°C (Fig. 1). The Langmuir adsorption isotherm equation was also fitted with the software of Origin 8, and the fitted equation is $q = 496 \times (0.06C_e/1 + 0.06C_e)$, $R = 0.998$, which shows very good agreement to the Langmuir adsorption isotherm curve. Besides, the maximal adsorption capacity is calculated as 1241 μg/g.

3.2 *Experiment on the maximum adsorption capacity of DS18 material*

At room temperature, new material DS18 can complete the adsorption process in 5 minutes and almost no Se was left in the residual solution. As calculated from the experimental data (not shown), DS18 could adsorb as much as 6.4 mg/g Se(IV) onto their surface, which showed a much higher adsorption capacity for Se(IV) than that of fly ash.

4 CONCLUSIONS

Our results show that new material DS18 can absorb much more Se(IV) than fly ash, the ratio value between these two materials can reach 1.6×10^3. We suggest that the new material DS18 can be a potential Se carrier for controlled-release Se fertilizer despite the absence of an experimental study on Se desorption. This study provided data support for development of slow-release Se fertilizer and its application guidance.

REFERENCES

Ke, G.J., Yang, X. F., Peng, H. et al. 2005. Research progress on activity mechanism of chemically activated fly ash. *Acta Coal Sinica* 30(3): 366–370.

Liu, M., Chen, W.M. & Zhang, Z.H. 2010. Occurrence status of selenium and research status of the sulfur selenium ratio. *Mineral Deposits* 29(s1): 1105–1106. (in Chinese)

Xia, Y.M. 2013. Selenium. *J Nutr* 35(3): 223–226.

Zhao, Z.G. 2005. *Application Principle of Adsorption*. Beijing: Chemical Industry Press.

Selenium Research for Environment and Human Health: Perspectives, Technologies and Advancements – Bañuelos, Lin, Liang & Yin (eds)
© 2020 Taylor and Francis Group, London, ISBN 978-1-138-39014-0

Effects of selenium treatment on sulfur nutrition and metabolism

F.E.M. Santiago, M. Tian, P.F. Boldrin & L. Li
Robert W. Holley Center for Agriculture and Health, USDA-ARS, Cornell University, Ithaca, New York, USA

1 INTRODUCTION

Selenium (Se) is an essential micronutrient and has multiple health benefits to humans. Selenium biofortification in crops not only helps combat Se deficiency problems, but also can provide bioactive Se compounds in some cases to reduce the risk of cancer.

Sulfur (S) is an essential nutrient for plants and has diverse functions related to plant growth and development. Selenium as a S analogue shares the S uptake, translocation, and assimilation pathways in plants (White 2018). Thus, it is expected that Se fertilization for biofortification affects S nutrition and metabolites in plants. While antagonistic effects on S levels are often noticed when Se is supplied at high dosages, interestingly, Se application at low concentrations enhances S nutrition and metabolism in crops. We evaluated the effects of Se treatments on S nutrition and metabolites in a number of plant species and examined the basis underlying Se-induced S accumulation in plants (Boldrin et al. 2016, Tian et al. 2018, Santiago et al., in prep.). Moreover, we provided evidence for the protective role of S in reducing Se toxicity in plants, as Se can be toxic to plants even at low dosages (Tian et al. 2017).

2 MATERIALS AND METHODS

2.1 *Plant materials and Se treatments*

Seeds of Se accumulators broccoli and arugula, as well as non-selenium accumulators wheat and lettuce, were germinated and grown hydroponically in a greenhouse with a 14-h light and 10-h dark photoperiod. Na_2SeO_4 at various concentrations (0, 5, 10, 20 μM) were applied in nutrient solutions for one or two weeks before plant tissues were harvested and analyzed.

2.2 *Mineral and metabolite analysis*

Total Se and S contents in dry leaf or root samples of these plants were determined using an inductively-coupled plasma (ICP) emission spectrometer. Sulfur-containing compounds, i.e. glucosinolates, in Se accumulators were extracted and analyzed using a Waters UPLC.

2.3 *RNA isolation and quantitative real-time PCR*

Total RNA was isolated from plant tissues using Trizol reagent. The total RNA was reversely transcribed into cDNA and used for quantitative real-time PCR analysis with gene-specific primers.

2.4 *Analysis of antioxidant enzymes and related compounds*

Antioxidant enzymes were extracted from plant leaf tissues and their activities were measured spectrophotometrically. *In Situ* ROS detection, membrane permeability, and lipid peroxidation assays were also carried out.

3 RESULTS AND DISCUSSION

3.1 *Low dosages of Se treatments promote plant growth*

Plant biomasses in Se accumulators broccoli and arugula, as well as in non-selenium accumulators wheat and lettuce, increased following Se treatments at low dosages. Selenate was more effective than selenite in promoting plant growth at the same dosages. In general, increased plant growth was observed when the supplied Na_2SeO_4 concentrations were up to 10 μM for non-selenium accumulators and to 20 μM for Se accumulators under hydroponic growth conditions. The growth-stimulated effect was noticed at lower concentrations for selenite than selenate. Interestingly, while selenite is more toxic to plants under normal S nutrition conditions, much high growth inhibitory effects were found in selenate than selenite treated plants under S deficiency conditions.

3.2 *Low dosages of Se supplements enhance S levels*

Low-dosage Se treatments were found to effectively enhance total S levels in tissues of both Se accumulators and non-accumulators. While selenate fertilization enhances S accumulation in both aboveground tissues and roots, selenite fertilization shows little effect on S level in roots. Much higher promoting effects on growth were normally found with selenate than selenite treatments.

3.3 *Selenium treatments mimic sulfur deficiency*

Selenate uptake uses sulfate transporters, which are upregulated under S deficiency. Selenate fertilization greatly increased the expression of *Sultr1:1* transporter at both transcriptional and translational levels, suggesting that the increased S accumulation by selenate treatment is due to its action in mimicking S starvation to stimulate S uptake. Unexpectedly, we recently observed that Arabidopsis *sultr1:1* knockout line has enhanced plant growth in comparison with wild type and *sultr1:2* knockout line.

3.4 *Sulfur nutrition status affects Se toxicity to plants*

Under low S conditions, plant growth is dramatically suppressed when treated with either selenate or selenite in comparison with untreated, indicating S protects plants from Se toxicity. The protection was discovered to be most likely through reducing non-specific integration of Se into proteins, decreasing cell membrane damage, and mediating redox enzyme activities. Thus, adequate S nutrition is critically important to prevent Se toxicity during Se biofortification in crops.

3.5 *Selenium fertilization affects glucosinolate synthesis*

Glucosinolates are sulfur-containing chemopreventive compounds, and their levels in crops are greatly affected by Se status. Se at a dosage with no apparent detrimental effect to plant growth can reduce glucosinolate levels. The Se-conferred glucosinolate reduction was found not to necessarily be associated with plant S nutrition status, but rather with suppressed expression of genes involved in glucosinolate biosynthesis as well as other cellular processes.

3.6 *Se accumulators have high redox capacity*

Our recent comparative study of Se accumulators and non-selenium accumulators reveals Se accumulators have relatively higher redox capacity than non-selenium accumulators, which may explain why the accumulator crops can tolerate high levels of Se.

4 CONCLUSIONS

Different dosages and forms of Se fertilization exert divergent effects on plant growth and S nutrition status in crops. Our studies support that plant growth and nutrition benefit from low dosages of Se fertilization and provide information for the basis underlying Se-induced S accumulation in crops. Further studies should offer more mechanistic insights into our understanding of the effects of Se fertilization on S nutrition and metabolism.

REFERENCES

Boldrin, P.F., de Figueiredo, M.A., Yang, Y. et al. 2016. Selenium promotes sulfur accumulation and plant growth in wheat (*Triticum aestivum*). *Physiol Plant* 158: 80–91.

Tian, M., Hui, M., Thannhauser, T.W., Pan, S. & Li, L. 2017. Selenium-induced toxicity is counteracted by sulfur in broccoli (*Brassica oleracea* L. var. italic). *Front Plant Sci* 8: 1425.

Tian, M., Yang, Y., Avila, F.W. et al. 2018. Effects of selenium supplementation on glucosinolate biosynthesis in broccoli. *J Agr Food Chem* 66: 8036–8044.

White, P.J. 2018. Selenium metabolism in plants. *BBA-Gen Subj* 1862: 2333–2342.

Selenium Research for Environment and Human Health: Perspectives, Technologies and Advancements – Bañuelos, Lin, Liang & Yin (eds)

Effect of soil properties and contact time on selenium transfer to wheat

C. Ramkissoon
University of Adelaide, South Australia, Australia
University of Nottingham, Nottingham, UK

F. Degryse & M.J. McLaughlin
University of Adelaide, South Australia, Australia

S.D. Young & E.H. Bailey
University of Nottingham, Nottingham, UK

1 INTRODUCTION

To assess the fate of residually added selenium (Se) fertilizer in soil over time, many studies have used chemical extraction methods to quantify "available" Se (Dhillon et al. 2005, Keskinen et al. 2009). Few studies have used plant evaluation trials to assess the bioavailability of the added Se in different soils (Supriatin et al. 2015), and even fewer have assessed changes in bioavailability of Se in soil over time (Li et al. 2016). Our experiment aimed to assess how Se transfers from soil to plant over a total period of 12 months. By using both chemical extraction methods and a pot experiment, we investigated (1) the chemical extraction method which gave the most consistent prediction for plant Se uptake and (2) the effect of soil properties on time-dependent Se fixation. Results from this study can be used in conjunction with others to provide a more comprehensive understanding of residual fate and bioavailability of Se.

2 MATERIALS AND METHODS

Eight soils varying in physiochemical properties were spiked with 0.5 mg/kg Se in the form of sodium selenate (Na_2SeO_4) at different time periods (0, 1, 2, 3, and 10 months) and incubated at 25°C for the duration of the experiment. Soil Se was fractionated by sequential extraction procedures into soluble Se (Se_{sol}) by $CaCl_2$ extraction, adsorbed Se (Se_{ads}) by KH_2PO_4 extraction and organically bound Se (Se_{OM}) by TMAH extraction, and analysed by ICPOES after hydride generation (HG-ICPOES). As a result of the low recovery of Se_{OM} by this method, a single TMAH extraction was repeated on all soil samples, and the extracts were analysed for Se by ICPMS. The organically bound Se fraction was calculated as the difference between total Se extracted by TMAH and Sesol + Seads. In parallel, 250 g of the incubated soil were used to set up a pot trial, in which wheat was grown for 8 weeks under controlled conditions. After harvest, plants were dried, ground, and acid digested prior to total Se analysis (Se_{plant}) by HG-ICPOES.

3 RESULTS AND DISCUSSION

The soluble Se fraction decreased with incubation time at different rates depending on the soil type. For soils with high Se sorption capacity, such as Kingaroy (52% clay and highest Al/Fe oxides contents), up to 73% of the added Se was fixed into non-labile forms (Se not extractable by $CaCl_2$) within 24 hr (t = 0 months). A rapid decrease in Se_{sol} (>80% loss) was also observed in the alkaline calcareous soil within the first month, suggesting rapid precipitation of Se into carbonates. The Se_{ads} fraction averaged at 5-15% of the added Se over the period of 0-10 months, with small, statistically insignificant changes over time for most soils, except Kingaroy (Se_{ads} consistently accounted for > 50% of the added Se over time). This observation suggests that the adsorbed Se fraction did not explain the loss of Se solubility over time, and that mechanisms other than surface adsorption of Se onto mineral oxides or clay particles might be responsible for Se fixation.

The measured soluble fraction (Se_{sol}) gave the best prediction for Se uptake (r = 0.77) (Fig. 1) while $Se_{sol} + Se_{ads}$ did not improve that correlation (r = 0.75).

To assess long-term bioavailability of Se in different soils, a transfer factor (TF) was calculated as the Se_{plant} (mg/kg) divided by total "available Se", defined as the $Se_{sol} + Se_{ads}$ (mg/kg). Transfer factors varied significantly among soils, from 28.9 to 238 within the first month, with the highest TF observed in alkaline soils, probably due to the predominance of selenate ions. Apart from the sandy soils where no significant change in TF was observed over time, alkaline soils were observed to fix Se to a greater extent with incubation time compared to acidic ones, regardless of pH (Fig. 2). Mechanisms explaining this observation will be discussed.

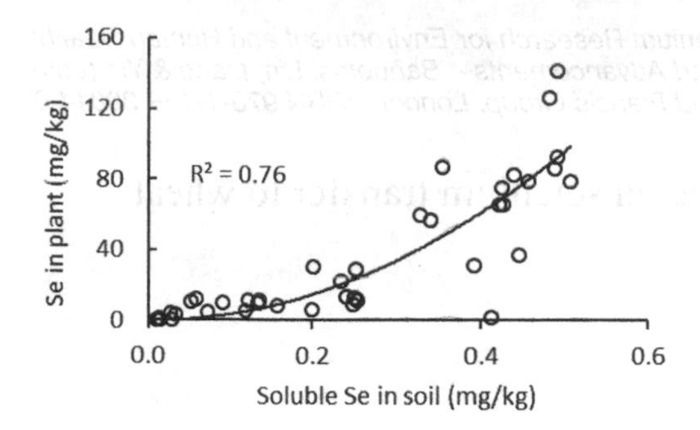

Figure 1. Relationship between soil soluble Se and plant Se concentration.

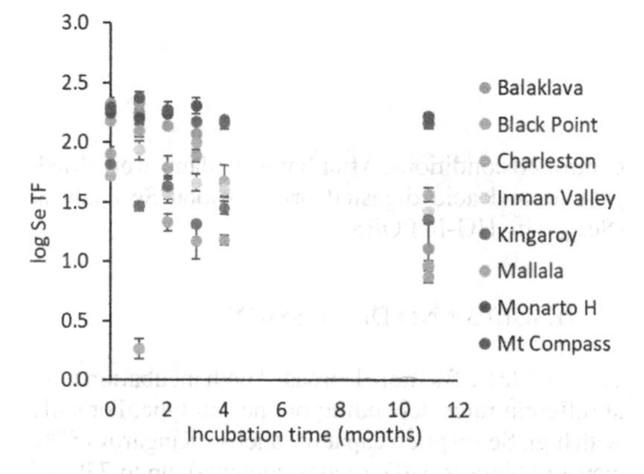

Figure 2. Selenium transfer factor ($log_{10}TF$) for wheat, using "available Se" concentration as the denominator, for 8 contrasting soils over a total soil-Se contact time of 11 months.

4 CONCLUSIONS

The bioavailability of Se was influenced by soil properties such as pH and soil texture, as well as incubation time. By investigating the effect of soil parameters and contact time on the availability of Se added to soil, this study sheds light on the fate of Se fertilizer used in crop biofortification and has possible implications for environmental safety.

REFERENCES

Dhillon, K.S., Rani, N. & Dhillon, S.K. 2005. Evaluation of different extractants for the estimation of bioavailable selenium in seleniferous soils of Northwest India. *Aust J Soil Res* 43 (5): 639–645.

Keskinen, R., Ekholm, P., Yli-Halla, M. & Hartikainen, H. 2009. Efficiency of different methods in extracting selenium from agricultural soils of Finland. *Geoderma* 153 (1): 87–93.

Supriatin, S., Weng, L. & Comans, R.N.J. 2015. Selenium speciation and extractability in Dutch agricultural soils. *Sci Total Environ* 532: 368–382.

Li, J., Peng, Q., Liang, D. et al. 2016. Effects of aging on the fraction distribution and bioavailability of selenium in three different soils. *Chemosphere* 144: 2351–2359.

Selenium Research for Environment and Human Health: Perspectives, Technologies and Advancements – Bañuelos, Lin, Liang & Yin (eds)
© 2020 Taylor and Francis Group, London, ISBN 978-1-138-39014-0

Screening and identification of soil selenium-enriched bacteria in Guangxi, China

Q. Liao, Y.X. Liu, Y. Xing, P.X. Liang, L.P. Pan, J.P. Chen & Z.P. Jiang*

Agricultural Resource and Environmental Research Institute, Guangxi Academy of Agricultural Sciences, Nanning, China
Selenium Enriched Agriculture Research Center of Guangxi, Nanning, China

1 INTRODUCTION

Selenium (Se) is an essential micronutrient for humans and its function includes preventing diseases, improving health, and delaying aging. The area of Se-rich soil in Guangxi is 2.12 million hm^2, and the maximum content of Se in soil is 2.29 mg/kg. However, the Se-rich soils in Guangxi are mostly acidic, and the Se in the soils is mostly insoluble complex iron selenite, leading to the low availability of Se and directly affecting its utilization. The bioavailability of Se in soil is significantly affected by environmental microorganisms which can change the form and valence of Se through their activity. The tolerance of soil microorganisms to Se was bacteria > fungi > actinomycetes (Liao et al. 2017). Soil bacteria can transform inorganic Se and grow well in high Se environment. Therefore, in this study bacteria were isolated from Se-enriched soil. The Se-tolerant bacteria were then screened by Se-containing medium, and the Se-enriched strains were screened by Se-enriched test. Lastly, the high-efficiency Se-transforming bacteria were identified. This research provides information about using specific bacteria for the improving the utilization of Se resources in soil and promoting the production of Se-enriched agricultural commodities.

2 MATERIALS AND METHODS

2.1 *Medium and reagent*

Medium: nutrient agar medium and nutrient broth.

Reagent: 20 mg/mL Na_2SeO_3 solution, phosphate buffered saline (PBS), and 0.1 mol/L KH_2PO_4-K_2HPO_4 solution.

2.2 *Isolation and purification of strains*

Soil samples were collected from several Se-rich areas in Guangxi, and 10 g of soil samples were added to 90 mL of sterile PBS with 10 glass beads. The suspension was collected and centrifuged for 15 min at 5000 r/min. The deposit was suspended at 10 mL of PBS and stored at 4°C. Two mL of stored liquid were added to 100 mL nutrient broth medium containing 100 μg Se/mL. Selenium-tolerant bacteria were obtained on nutrient agar medium by spread plate and streak plate methods.

2.3 *Measurement of Se-enrichment capacity of strains*

The strains were respectively transferred to a triangular flask containing 100 mL nutrient broth medium with 5% inoculation rate and cultured in shaker. The suspensions were centrifuged for 30 min at 4000 r/min. The supernatants were digested, and the values of fluorescence intensity were measured by atomic fluorescence spectrometer. The concentration of Se in the solution was obtained from the standard curve. The Se enrichment capacity of the strain was calculated from the formula of Se conversion rate. Se conversion rate (%) = [(total Se content – residual Se content)/total Se content] × 100.

2.4 *Strain identification*

The genomic DNA of Se-enriched bacterial was extracted and used as template. Fragments of 16S rRNA gene were amplified by universal bacterial primers (27F/1492R). The amplified products were purified by gel recovery kit and sequenced by ABI 3730 sequencer. The sequence was compared with the sequence in GenBank database. The phylogenetic tree was drawn with MEGA5.0 to identify the species.

2.5 *Bioactivation of Se in soil*

Selenium activation test was carried out according to Long's method (Long et al. 2017). Selenium content was measured by atomic fluorescence spectrometry.

3 RESULTS AND DISCUSSION

3.1 *Screening of Se-tolerant bacteria*

A total of 32 strains of Se-tolerant bacteria with different colonial characteristics were obtained. Selenium-tolerant bacteria existed in red and lateritic red soils of sandy shale parent material in different Se-rich areas of Guangxi.

Through analyzing of the growth curve of these strains (Fig. 1), four strains with short culture time

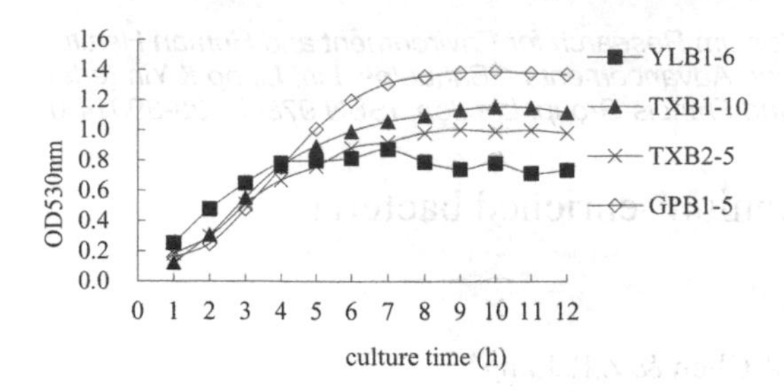

Figure 1. Growth curves of Se-tolerant bacteria.

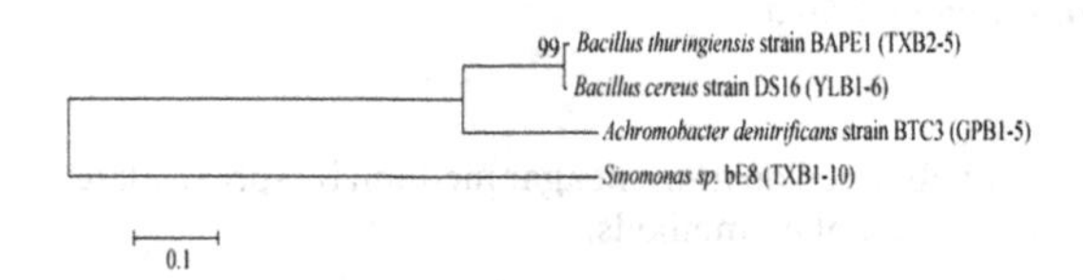

Figure 2. Phylogenetic tree based on 16S rDNA sequences.

(<12 h) and similar logarithmic growth phases were screened for further study. These four strains were named as YLB1-6, TXB1-10, TXB2-5, and GPB1-5.

3.2 *Selenium-enrichment capacity of strains*

Selenium-enriched bacteria normally have a Se conversion rate of 50%–80%. In our study, the Se conversion rate of YLB1-6, TXB1-10, TXB2-5, and GPB1-5 were 74.22%, 66.05%, 55.31%, and 63.30%, respectively, showing strong Se-enriched ability.

3.3 *Identification of Se-enriched bacteria*

After the 16S rDNA sequencing and BLAST analysis, YLB1-6 was identified as *Bacillus cereus*, TXB1-10 as *Sinomonas sp.*, TXB2-5 as *Bacillus thuringiensis*, and GPB1-5 as *Achromobacter denitrificans* (Fig. 2). Se-enriched bacterial genera reported till now include *Bacillus, Stenotrophomonas, Enterobacter,* and *Pseudomonas*. In this study, *Sinomonas* and*Achromobacter* were first reported to be Se-enriched bacteria, which increased the number of species to the Se-enriched strains database.

3.4 *Bio-activation of Se in soil by strains*

Water-soluble Se and exchangeable Se are the main forms of available Se in soil. Table 1 showed that exchangeable Se was the main effective form of Se, while water soluble Se occupied a small proportion of available Se. There was no significant difference between water soluble Se and CK, but exchangeable Se was increased by treating with different bacteria

Table 1. Concentrations of water-soluble Se and exchangeable Se in treated soils.

Treatments	Water-soluble Se (mg/kg)	Exchangeable Se (mg/kg)
CK	0.005 a†	0.053 bB†
YLB1-6	0.006 a	0.112 aA
TXB1-10	0.004 a	0.101 aA
TXB2-5	0.006 a	0.085 aAB
GPB1-5	0.008 a	0.088 aAB

†Different upper-case and lower-case letters in the same column indicate that the difference is statistically significant at $p < 0.01$ and $p < 0.05$, respectively.

dosage. In this regard, selenite adsorbed on the surface of oxides and clay minerals and can easily bind to the peptide and amino acids secreted by the bacterial strains. In addition, the strains secrete extracellular phosphatase that can dissolve insoluble phosphorus in the soil, and the Se inside the phosphorus is activated and released. The activating mechanism needs to be further studied. Microbial metabolism is an important way to activate soil Se, and it shows great potential to improve the bioavailability of Se in soil. Therefore, Se-enriched bacteria, as a potential tool for Se bio-activation, have a practical application value in Se-enriched crop production.

4 CONCLUSIONS

Four bacterial strains with strong Se-enriched capability were screened from Se-enriched areas in Guangxi. They were named as YLB1-6, TXB1-10, TXB2-5, and GPB1-5. 16S rDNA sequence analysis and phylogenetic tree analysis showed that these 4 strains were identified as *Bacillus cereus*, *Sinomonas sp.*, *Bacillus thuringiensis*, and *Achromobacter denitrificans*, respectively. The Se conversion rates of these Se-enriched bacteria were 55.3% - 74.2%. The exchangeable Se content in soil was significantly increased by treating with various strains, which played a strong role in activating Se in soil.

REFERENCES

Liao, Q., Liu, Y.X., Xing, Y. et al. 2018. Screening and identification of Se-enriched bacteria from Se-rich soils in Guangxi. *Soils* 50(6): 1203–1207.

Long, Y.C., Chen, X. & Zhou, S.Q. 2017. Isolation, identification and assessment on selenium biofortification of siderophore-producing rhizobacteria. *Curr Biotechnol* 7(5): 402–408.

Selenium Research for Environment and Human Health: Perspectives, Technologies and Advancements – Bañuelos, Lin, Liang & Yin (eds)
© 2020 Taylor and Francis Group, London, ISBN 978-1-138-39014-0

The effect of soil type on selenium uptake and recovery by a maize crop

A.D.C. Chilimba
AGRISO Consultants and Ngolojere Investments, Zomba, Malawi

S.D. Young & M.R. Broadley
University of Nottingham, School of Biosciences, Sutton Bonington Campus, Loughborough, UK

1 INTRODUCTION

Agronomic biofortification can increase selenium (Se) concentration in crops and hence dietary intake of Se (Broadley et al. 2010, Eurola, 2005, Lyons et al. 2005). However, more information on the fate and transformation of Se added to the soil-plant system is required (Keskinen et al. 2009, Keskinen et al. 2010) to maximize the efficiency of Se use in biofortification programs (Keskinen et al. 2010). Current understanding suggests that Se recovery by crops is inefficient, and applied Se is likely to be rapidly leached as soluble selenate, adsorbed as selenite, or immobilized into organic forms. Chilimba et al. (2012a,b) reported Se recovery of 6.5% at Chitedze research station and 10.8% at Mbawa research station, which seemed to indicate that Se recovery could vary with soil types. This paper is aimed at establishing whether Se uptake and recovery is influenced by soil type.

2 MATERIALS AND METHODS

Experiments were conducted during the 2008-2010 growing seasons at six different sites in Malawi: Bvumbwe, Chitedze, Makoka, Chitala, Mbawa, and Ngabu Research Stations. The soil types were: Lixisols at Mbawa, Luvisols at Chitedze, Makoka, and Bvumbwe, and Vertisols at Chitala and Ngabu. The experimental treatments consisted of eight levels of Se application (0, 5, 10, 15, 25, 50, 75, and 100 g/ha) in maize crop. The experimental plots were laid out in a randomized complete block design with three replicates. To ensure even application to the crop, $Na_2SeO_{4(aq)}$ was applied as a high-volume drench (667 L/ha of water) using a knapsack sprayer, with the operator wearing personal protective equipment of overalls, boots, face-shield, and nitrile gloves (Chilimba et al. 2012a, b). Whole grains and stover samples were collected, dried, and milled. Selenium (^{78}Se) analysis was conducted by ICP-MS (X-SeriesII, Thermo Fisher Scientific Inc., Waltham, MA, USA). Total Se accumulated by grain and stover was computed by adding grain and stover Se.

3 RESULTS AND DISCUSSION

Selenium uptake and recovery was the highest with vertisols (Ngabu and Chitala) followed by Lixisols (Mbawa) and Luvisols (Bvumbwe, Makoka, and Chitedze). Selenium uptake increased with Se application at all sites, but Se recovery did not vary with the rate of application (Fig. 1). Selenium recovery was 35.7% and 27.9% in Vertisols, 15.7% in Lixisols, and 12.3, 14.9, and 11.1% in Luvisols (Fig. 2) for Chitedze, Makoka and Bvumbwe, respectively. Luvisols have high base saturation, dominated by high activity clays with high cation exchange capacity (CEC). Consequently, they

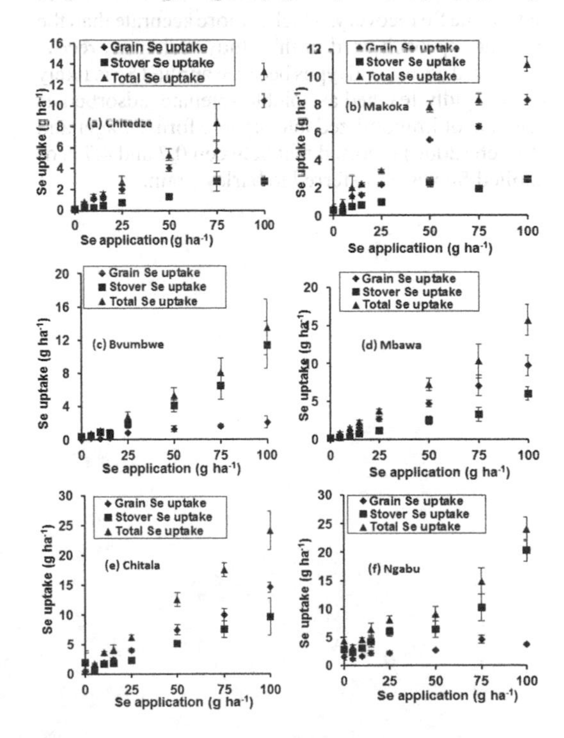

Figure 1. Effect of Se application on Se uptake at (a) Chitedze, (b) Makoka, (c) Bvumbwe-Luvisols; (d) Mbawa-Lixisols; (e) Chitala, (f) Ngabu-Vertisols. Error bars = standard errror of the mean.

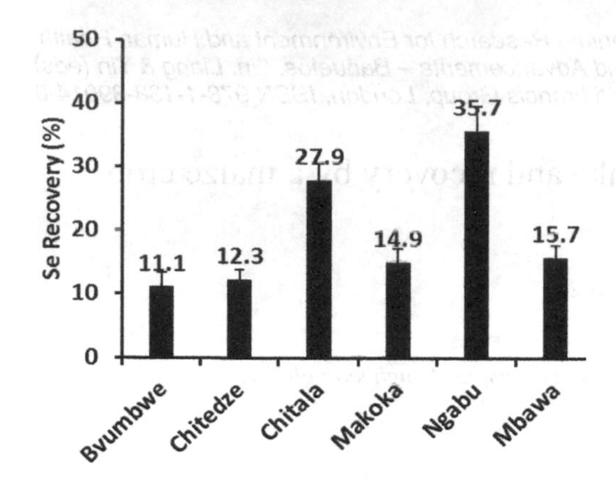

Figure 2. Effect of soil types on Se recovery ((a) Chitedze, (b) Makoka, (c) Bvumbwe-Luvisols; (d) Mbawa-Lixisols; (e) Chitala, (f) Ngabu-Vertisols. Error bars = standard errror of the mean.

likely fix more Se than Lixisols, which is dominated by low activity clay with low CEC. This CEC difference explains why Se recovery is higher in Lixisols than in Luvisols. Chilimba et al. (2012a, b) observed Se recovery rates of between 6.5 and 10.8% in maize grain after applying 10 g Se/ha. The results reported in this paper are in agreement with Chilimba et al. (2012a, b). Selenium uptake increased with Se application at all sites. Chilimba et al. (2012a, b) reported Se recovery of 6.5 to 10.8% when ^{77}Se was used to determine Se recovery, which is more accurate than the current approach used in this study. Selenium recovery varied across soil types because applied Se is likely to be rapidly leached as soluble selenate, adsorbed as selenite, or immobilized into organic forms. Sager and Hoesch (2006) reported that between 0.7 and 4.7% of applied Se was transferred to barley grain.

4 CONCLUSIONS

Soil types influenced Se uptake and recovery, and therefore soil properties have a significant effect on Se uptake and Se recovery. There is a need to investigate further and identify the critical soil properties that have the greatest effect on Se uptake. Vertisols and Lixisols seemed to have greater Se uptake and recovery compared to Luvisols.

REFERENCES

Broadley, M.R., Alcock, J., Alford, J. et al. 2010. Selenium biofortification of high-yielding winter wheat (*Triticum aestivum* L.) by liquid or granular Se fertilisation. *Plant Soil* 332: 5–18.

Chilimba, A.D.C., Young, S.D., Black, C.R., Meacham, M.C., Lammel, J. & Broadley, M.R. 2012. Assessing residual availability of selenium applied to maize crops in Malawi. *Field Crops Res* 134: 11–18.

Eurola, M. 2005. *Proceedings: Twenty Year's of Selenium Fertilization.* Helsinki: MTT Agrifood Research, Finland.

Keskenin, R., Turakainen, M. & Hartikainen, H. 2010. Plant availability of soil selenite additions and selenium distribution within wheat and ryegrass. *Plant Soil* 333: 301–313.

Keskinen, R., Ekholm, P., Yli-Halla, M. & Hartikainen, H. 2009. Ef?ciency of different Methods in extracting selenium from agricultural soils of Finland. *Geoderma* 153: 87–93.

Lyons, G., Ortiz-Monasterio, I., Stangoulis, J. & Graham, R. 2005. Selenium concentration in wheat grain: Is there sufficient genotypic variation to use in breeding? *Plant Soil* 269: 269–380.

Sager, M. & Hoesch, J. 2006. Selenium uptake in cereals grown in lower Austria. *J Cent Eur Agric* 7: 71–77.

Selenium Research for Environment and Human Health: Perspectives, Technologies and Advancements – Bañuelos, Lin, Liang & Yin (eds)
© 2020 Taylor and Francis Group, London, ISBN 978-1-138-39014-0

Inoculation of AM fungi enhanced rhizosphere soil Se bioavailability of winter wheat

J. Li, W.J. Xing, W.Q. Luo & F.Y. Wu*
College of Natural Resources and Environment, Northwest A&F University, Yangling, Shaanxi, China

1 INTRODUCTION

Selenium (Se) is an essential micronutrient for humans and animals. Others have assessed that 39–61% of the Chinese population has an inadequate intake of Se (26–34 μg/d). The average daily Se intake is inadequate for human health due to the low Se concentrations in food products like wheat (Liu et al. 2016). Low Se content in food products is due to low Se availability in soils. Therefore, research on soil Se bioavailability is of great significance.

Arbuscular mycorrhizal fungi (AMF) are an important group of soil fungi, forming symbiotic associations with roots of the majority of plant species (Smith & Read 2008). Larsen et al. (2006) found that mycorrhizal inoculation increased Se content by ten times in garlic, whereas Munier-Lamy et al. (2007) demonstrated that mycorrhizal inoculation decreased soil Se uptake by ryegrass up to 30%. To the best of our knowledge, there is still very limited information about the interactions between AMF and Se fractions in rhizosphere soils of host plants. Therefore, the objectives of the present study were: (1) to investigate the effects of AMF inoculation on Se fractions in rhizosphere soils of wheat and (2) to reveal the correlation between Se fractions in the rhizosphere soil and selected soil properties.

2 MATERIALS AND METHODS

2.1 *Pot experiment*

Uncontaminated farmland soil was collected from a farm in Yangling, Shaanxi Province, China. Soil was air-dried and sieved through a 2-mm mill. Two species of AMF, *Funneliformis mosseae* (BGC BJ05A) (*F.m*) and *G. versiforme* (BGC GD01B) (*G.v*), were purchased from Beijing Academy of Agriculture and Forestry Sciences, China and used for pot trials. The soil was autoclaved (121°C for 120 min). Se(IV) and Se(VI) were added separately as a solution of Na_2SeO_3 or Na_2SeO_4 and mixed with the soil at a concentration of 2.5 mg/kg, respectively. Soil without Se addition was used as the control. Triplicate pots were prepared for each treatment. Each pot received a mixture of 3 kg soil, and 200 g AM fungal inoculum for mycorrhizal treatment or sterilized AM fungal inoculum (200 g) were added into the pots for non-mycorrhizal treatment. After 12 weeks, rhizosphere soils were collected and analyzed for various chemical properties and Se fractions.

2.2 *Determination of DOC, SOM, EE-GRSP, and T-GRSP*

The dissolved organic carbon (DOC) content was determined as methods as described by Jones et al. (2006), and soil organic matter was determined with heated $K_2Cr_2O_7$ oxidation and $FeSO_4$ titration. Glomalin-related soil protein (GRSP) was extracted from treated soils as total GRSP (T-GRSP) and easily extractable GRSP (EE-GRSP), as described by Bedini et al. (2007).

2.3 *Determination of Se fractions in soils*

Soil samples were analyzed for Se fractions using the five-step sequential extraction method described by Ali et al. (2017). Total Se in soil was digested by an acid mixture (HNO_3+$HClO_4$, 3:2, v/v) at 165°C on an electrothermal digestion furnace (LabTech, Beijing, China), until the soil appeared white. Concentrations of Se was analyzed using an atomic fluorescence spectrophotometer (AFS-9780, Beijing JiTian Instruments) with hydride generation.

3 RESULTS AND DISCUSSION

3.1 *Distribution characteristics of different Se fractions in soil*

In the soil without Se addition, the inoculation of *F. mosseae* and *G. versiforme* significantly ($p < 0.05$) increased available Se (SOL-Se + EXC-Se) content by 22.34% and 24.76%, respectively (Fig. 1). In Se(IV)- and Se(VI)-spiked soils, the inoculation of *F.m* significantly ($p < 0.05$) enhanced available Se (SOL-Se + EXC-Se) content by 30.21% and 69.75%, respectively, compared with the soil without mycorrhizal inoculation.

Selenium mainly existed in stably oxidizable fractions in the control, in which FMO-Se and OM-Se

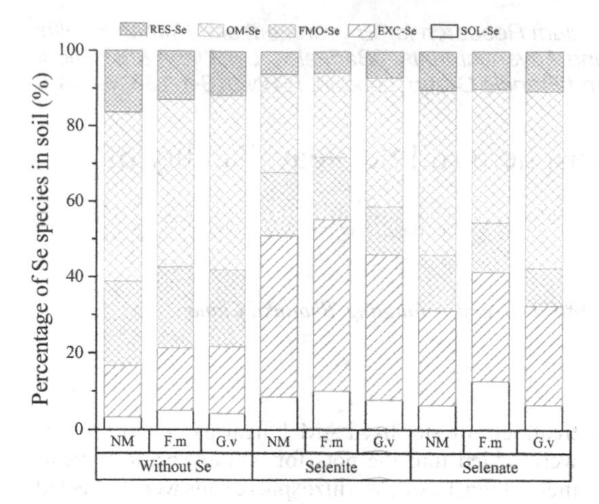

Figure 1. The proportion of different Se fractions in Se-spiked soils when uninoculated/inoculated with *Funneliformis mosseae* (*F.m*) or *Glomus versiforme* (*G.v*) (mean ± SE, n = 3). SOL-Se: soluble Se; EXC-Se: exchangeable and carbonate-bound Se; FMO-Se: Fe/Mn oxides-bound; OM-Se: organic matter-bound; RES-Se: residual Se.

accounted for 22.2% and 44.84% of the total soil Se, respectively, while SOL-Se and EXC-Se took up only 3.32% and 13.36%, respectively (Fig. 1). Compared with control, the dominant Se fraction was EXC-Se in the Se(IV) treatment, and the proportion of EXC-Se increased by 24.86–31.73%. The proportion of EXC-Se increased by 11.45–15.29% in selenate spiked soils compared with control.

3.2 *Effects of AMF on DOC, EE-GRSP, and T-GRSP in rhizosphere soils of wheat*

Root colonization in wheat were 80.0–95.56% with Se(IV) and Se(VI) treatment. There was no significant difference in the colonization rate between the two species of AMF (data not shown). Inoculation with two species of AMF significantly ($p < 0.05$) increased soil EE-GRSP and DOC in all treatments. T-GRSP and SOM significantly ($p < 0.05$) improved under Se(IV) or Se(VI) treatments (Table 1).

3.3 *Correlation between soil Se fraction and selected soil properties*

Redundancy analysis (RDA) was carried on to investigate the correlation between the soil Se fractions and soil properties (data not show). Interestingly, we observed that DOC and EE-GRSP contents were significantly positively correlated with available Se in mycorrhizal rhizosphere soils, while they were negatively correlated with the pH value, indicating the significant role of DOC and EE-GRSP in the activated of soil available Se.

Table 1. Change of soil properties in non-mycorrhizal and mycorrhizal wheat.

Se (mg/kg)	Inoculation	EE-GRSP (mg/g)	T-GRSP (mg/g)	DOC (mg/kg)
CK	NM	0.828±0.06 b†	3.589±0.20 b	55.86±2.17 b
	G.m	0.907±0.01 a	4.362±0.12 ab	62.18±2.22 a
	G.v	0.968±0.02 a	4.629±0.75 a	64.36±2.75 a
Selenite (2.5)	NM	0.921±0.05 b	3.748±0.15 b	56.73±1.67 c
	G.m	1.015±0.03 a	4.462±0.21 a	66.78±4.80 b
	G.v	1.085±0.05 a	4.392±0.41 a	78.06±4.03 a
Selenate (2.5)	NM	0.905±0.05 b	3.921±0.25 b	56.83±2.88 b
	G.m	1.072±0.08 a	4.704±0.38 a	70.97±3.78 a
	G.v	1.074±0.10 a	4.489±0.06 a	69.13±3.80 a

†Different letters mean significant difference at $p < 0.05$.

4 CONCLUSIONS

Application of Se (IV) or Se (VI) can significantly affect the proportion and distribution of Se fractions in rhizosphere soils of wheat. In all treatments, the inoculation of *F.m* significantly ($p < 0.05$) enhanced available Se (SOL-Se + EXC-Se) content compared to soils without mycorrhizal inoculation. In addition, DOC and EE-GRSP contents were significantly and positively correlated with available Se in rhizosphere soils. In-depth studies must be conducted to elucidate the molecular mechanism of AMF affecting Se speciation in soil and to ultimately produce Se-rich wheat.

REFERENCES

Al, F., Peng, Q., Wang, D. et al. 2017. Effects of selenite and selenate application on distribution and transformation of selenium fractions in soil and its bioavailability for wheat (*Triticum aestivum* L.). *Environ Sci Pollut Res Int* 9: 1–11.

Bedini, S., Avio, L., Argese, E. et al. 2007. Effects of long-term land use on arbuscular mycorrhizal fungi and glomalin-related soil protein. *Agric. Ecosyst Environ* 2: 463–466.

Larsen, E.H., Lobinski, R., Burger-Meÿer, K. et al. 2006. Uptake and speciation of selenium in garlic cultivated in soil amended with symbiotic fungi (mycorrhiza) and selenate. *Anal Bioanal Chem* 385(6): 1098–1108.

Liu, H., Yue, E.Y., Zhao, H.W. et al. 2016. Selenium content of wheat grain and its regulation in different wheat production regions of China. *Sci Agric Sin* 49: 1715–1728.

Jones, D.L. & Willett, V.B. 2006. Experimental evaluation of methods to quantify dissolved organic nitrogen (DON) and dissolved organic carbon (DOC) in soil. *Soil Biol Biochem* 38(5): 991–999.

Munier-Lamy, C., Deneux-Mustin, S., Mustin, C. et al. 2007. Selenium bioavailability and uptake as affected by four different plants in a loamy clay soil with particular attention to mycorrhizae inoculated ryegrass. *J Environ Radioact* 97(2–3): 148–158.

Smith, S.E. & Read, D.J. 2008. Mycorrhizal symbiosis. *Q Rev Biol* 3(3): 273–281.

Selenium Research for Environment and Human Health: Perspectives, Technologies and Advancements – Bañuelos, Lin, Liang & Yin (eds)
© 2020 Taylor and Francis Group, London, ISBN 978-1-138-39014-0

Uptake competition between selenium and sulphur fertilizers in sequential harvests of ryegrass: A stable isotope study

L. Jiang, S.D. Young, E.H. Bailey, M.R. Broadley & N.S. Graham
School of Biosciences, Sutton Bonington Campus, University of Nottingham, Leicestershire, UK

S.P. McGrath
Sustainable Agriculture Sciences, Rothamsted Research, Harpenden, UK

1 INTRODUCTION

Recently, the demand for sulphur (S) fertilizers has increased due to declining S emissions from coal burning and subsequent deposition on agricultural land (Webb et al. 2016). However, application of sulphate promotes competition between selenate and sulphate for crop uptake, as both share the same transporters in plant roots (Terry et al. 2000). In our experiment, enriched selenium (Se) isotopes $^{74}Se^{VI}$ and $^{77}Se^{IV}$ were used to trace the fate of Se in (1) sequential ryegrass harvests, (2) the soil pore water, and (3) soil sorbed Se fractions. Treatments tested included a series of slow- and fast-release S fertilizers (i.e. polyhalite).

2 MATERIALS AND METHODS

2.1 *Soil and pots*

Topsoil, arable Wick series, a sandy loam (pH = 6.62, soluble S = 17.2 ± 0.2 mg/kg; soluble P = 4.0 ± 0.1 mg/kg), was collected from Sutton Bonington farm near Loughborough, UK (52°449′48.6″N, 1° 14′24.2″W). Dry soil was sieved to <4 mm, fully mixed, and then re-packed in 17.4-cm high and 14.8-cm diameter pots.

2.2 *Isotopic Se solution preparation*

Solutions of Se $K_2^{74}SeO_4$ ($^{74}Se^{VI}$) and $K_2^{77}SeO_3$ ($^{77}Se^{IV}$) were made from elemental ^{74}Se and ^{77}Se powder (99.2% isotopic enrichment) purchased from Isoflex, USA.

2.3 *Sulphur fertilizers*

Five sulphate fertilizers were selected in this experiment based on a previous soil column leaching experiment that investigated the solubility and S release rates in a wide range of marketable sulphate fertilizers. Our current experiment sub-selects fertilizers depending on representatives of high, medium, and low sulphur release rates, including $MgSO_4.H_2O$ (Mg) and Patentkali® (SMP) (i.e. *high*); PotashpluS 37 (Potash 37) (i.e. *medium*); Polysulphate® (polyhalite, Poly) and $CaSO_4.2H_2O$ (Gypsum) (i.e. *low*).

2.4 *Ryegrass growth and fertilizers application*

An Italian ryegrass (*Lolium multiflorum,* Bb2235A), was sown (~5 g per pot) in a glasshouse (20°C, 16 h day light, 14°C night 8 h dark) for seven months with five replicated per treatment (described below) and five harvests (Cut 0 – Cut 4). The Se and S fertilizers were applied at the same time, two weeks after Cut 0 was harvested. Ryegrass was harvested by cutting shoot and leaf ~1 cm above soil surface. Each isotopic Se species was applied at 20 g/ha as a joint $^{74}Se^{VI}$&$^{77}Se^{IV}$ treatment in two ways: (1) "fertigation" on the soil surface in 120 ml (FG) and (2) liquid placement as three drops of 0.5 ml each (LP). Sulphur fertilizers were applied at 60 kg/ha S, except for gypsum, which was applied at 1320 kg/ha S. The growth time between each cut was ~40 days. Nutrient differences were made up by relevant chloride solutions due to different cation mixes in the S fertilizers.

2.5 *Rhizosphere and residual soil samples*

After Cut 0, a rhizon pore water sampler was placed into each pot (5 cm porous, 12 cm PVC tubing, female Luer lock, produced by Rhizosphere Research Products, The Netherlands) for sampling soil rhizosphere solution. Rhizosamplers were collected at the same time as each grass harvest, and then frozen at −20°C prior to elemental analysis using ICP-MS. Major anions were analyzed using IC, and Se speciation was determined using LC-ICP-MS with digested grass leaf samples. After the final grass cut, soil was extracted by sequential extraction methods for determining soil residual Se.

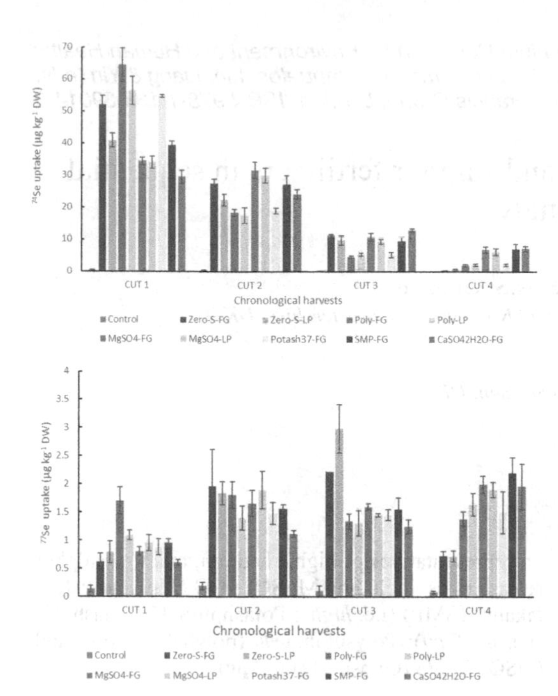

Figure 1. Forage grass Se uptake under different S fertilizers with jointly applied $^{74}Se^{VI}$ (top) and $^{77}Se^{IV}$ (bottom). Selenium application methods were fertigation (FG) and liquid placement (LP). Sulphur fertilizers were applied as 60 kg/ha S except gypsum was applied as 1320 kg/ ha S.

3 RESULTS AND DISCUSSION

3.1 *The effect of S fertilizers on Se isotope concentrations in sequential harvests*

The Se isotopic composition was analyzed in the harvested grass samples (Fig. 1). Uptake of selenite (^{77}Se) (b) was significantly less than that of selenate (^{74}Se) (a) ($p < 0.05$) from Cut 1 to Cut 3, although the difference between the two species narrowed over sequential harvests. Slow-release S increased $^{74}Se^{VI}$ uptake in Cut 1 and then suppressed uptake at Cut 2. In contrast, the fast-release fertilizers inhibited $^{74}Se^{VI}$ only in Cut 1. Sulphur did not affect uptake of $^{77}Se^{IV}$, which was less variable across sequential cuts. However this small variability may be explained by morphology of roots and phosphate uptake (Li et al. 2008). Greater level of applied gypsum significantly ($p < 0.05$) initially inhibited $^{74}Se^{VI}$ uptake; however, in all S applications, $^{74}Se^{VI}$ uptake was enhanced in comparison with Zero-S in Cut 4. This result may be explained by gene up-regulation in response to a change in S status over the first three harvests (White et al. 2004).

Selenium isotope transfer to the grass (% recovery) from soil to plant showed that $^{74}Se^{VI}$ was lower than 5 %, and $^{77}Se^{IV}$ was lower than 0.25% in each harvest. Nevertheless, $^{74}Se^{VI}$ TF was reduced to ~0.5% at the end Cut 4 from ~4.5% at Cut 1, but $^{77}Se^{IV}$ was relatively stable across all cuts. The similar leaf Se concentrations implied that $^{74}Se^{VI}$ was initially available but was gradually *fixed* in the soil (Mikkelsen & Wan 1990). In contrast, $^{77}Se^{IV}$ was likely immediately adsorbed after application in solution to the soil.

4 CONCLUSIONS

Slow-release S may only be beneficial for grass Se status initially and less beneficial for the later harvests. Applied selenate is more phytoavailable than selenite, however, its residual availability is reduced to a very low level within a short time. Besides, applied S seems to enhance residual selenate phytoavailability, especially when soil environmental S status is decreased in later harvests. In contrast, applied selenite is rapidly adsorbed in soil and can be a stable long-term Se pool available for plant uptake albeit at a comparatively low level of availability.

REFERENCES

Li, H.F., McGrath, S.P. & Zhao, F.J. 2008. Selenium uptake, translocation and speciation in wheat supplied with selenate or selenite. *New Phytol* 178: 92–102.

Mikkelsen, R.L. & Wan, H.F. 1990. The effect of selenium on sulfur uptake by barley and rice. *Plant Soil* 121: 151–153.

Terry, N., Zayed, A., de Souza, M. & Tarun, A. 2000. Selenium in higher plants. *Annu Rev Plant Biol* 51: 401–432.

Webb, J., Jephcote, C., Fraser, A. et al. 2016. Do UK crops and grassland require greater inputs of sulphur fertilizer in response to recent and forecast reductions in sulphur emissions and deposition? *Soil Use Manag* 32: 3–16.

White, P.J., Bowen, H.C., Parmaguru, P. et al. 2004. Interactions between selenium and sulphur nutrition in Arabidopsis thaliana. *J Exp Bot* 55: 1927–1937.

Selenium Research for Environment and Human Health: Perspectives, Technologies and Advancements – Bañuelos, Lin, Liang & Yin (eds)
© 2020 Taylor and Francis Group, London, ISBN 978-1-138-39014-0

Effects of soil selenate and selenite on selenium uptake and speciation in wheat (*Triticum aestivum* L.)

M. Wang, F. Ali & D.L. Liang*
College of Natural Resources and Environment, Northwest A&F University, Yangling, Shaanxi, China

1 INTRODUCTION

Selenium (Se) is arguably one of the most interesting elements because it is both essential and toxic for most organisms, with a very narrow window between deficiency and toxicity compared with other trace elements (Schiavon & Pilon-Smits 2017). Selenium deficiency is considered the fourth most serious deficiency of minerals after iron (Fe), zinc (Zn), and iodine (I) (Boldrin et al. 2013). At present, Se biofortification in edible parts of crops may be an effective approach to reducing Se deficiency in the food chain (Zhou et al. 2018). Various countries have successfully implemented Se biofortification strategies, such as Finland (Broadley et al. 2007), the UK (Lyons 2010), and New Zealand (Curtin et al. 2006).

Selenate and selenite, two predominant forms of soil Se, have been successfully applied for crop Se biofortification in various crops (Zhou et al. 2018). The uptake and accumulation of exogenous Se in crops is, however, different (Boldrin et al. 2013). Moreover, the bioavailability of Se to humans and animals largely depends on the speciation of Se in the edible parts rather than the total concentration of Se (Schiavon & Pilon-Smits 2017). Wheat (*Triticum aestivum* L.) cereal is the main staple food of more than one-third of the world's population. Based on this, a field biofortification experiment was conducted to evaluate the soil application of different Se species (selenite and selenate) at different rates (20 g/ha and 100 g/ha) on Se concentration and speciation in different parts of wheat.

2 MATERIALS AND METHODS

Field experiments were conducted on wheat (*Triticum aestivum* L.) on October 6, 2016 in the experimental field located at Northwest A&F University in Yang Ling, Shaanxi Province, China (34° 20′ N, 108° 24′ E at 521 m altitude). The total Se of the Lou soil from 0-20 cm was 0.052 mg/kg. The experiment was performed using a randomized complete block design with plot area of 3.8 m × 1.4 m and 30 cm row spacing of wheat. Two kinds of Se fertilizers (selenite and selenate) and two rates of Se (20 and 100 g/ha) were used in the experiment. Soil without added Se was designated as control (CK). A total of five treatments with three replications was set up per treatment. Nitrogen 100 kg/ha (urea) and phosphorus 80 kg/ha (triple superphosphate) were applied as basal fertilizers to all plots before sowing, and another nitrogen 100 kg/ha (urea) application was applied at jointing stage of wheat. Other measures of agronomic management followed local practices. The date of wheat harvesting was June 5, 2017.

After harvest, ten plants were randomly selected from each treatment to measure biomass, yield, and grain yield, then the plant was divided into five parts: root, stem, leaf, glume, and grain. Samples of each treatment were divided into two parts, one part was measured for Se concentration after sample was dried and ground by hydride generation atomic fluorescence spectrophotometry (HG-AFS), while the other part was stored at −80°C and then measured for grain Se speciation by high-performance liquid chromatography inductively coupled plasma mass spectrometry (HPLC-ICP-MS). Statistical analysis was conducted with SPSS 20.0.

3 RESULTS AND DISCUSSION

3.1 *Effect of selenate and selenite on Se uptake in different parts of wheat*

No significant differences were observed in biomass and grain yield of wheat for all treatments (data not presented). Significant differences were observed on Se concentrations of different parts of wheat (Fig. 1).

Overall, Se distribution in selenite-treated wheat was in the descending order of leaf > root > glume > stem > grain, and it ranged as leaf > glume > grain > stem > root with selenate treatment.

Selenium concentration in root with selenite treatment was significantly ($p < 0.05$) increased by 0.52 times and 1.11 times compared with selenate applied at 20 and 100 g/ha treatment, respectively. These results are consistent with the results of Boldrin et al. (2013). In contrast, the Se content of leaves treated with selenate (20 and 100 g/ha) was 1.74 and 2.35 times higher than that treated with selenite, respectively. No significant difference was observed in the stem of any treatment. Concentrations of Se in glumes

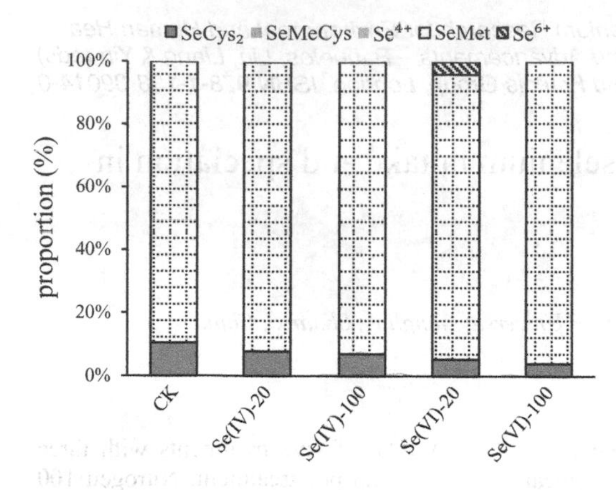

Figure 1. Effects of selenate and selenite on Se concentration in wheat parts.

in the selenate treatment of 100 g Se/ha were 1.98 times higher than those in the selenite treatment. Compared with selenite, the Se concentration in wheat grain treated with selenate (20 and 100 g/ha) was significantly ($p < 0.05$) increased by 33% and 71%, respectively. This result indicated that selenate was easily transported to the grain (the edible part), while selenite tended to accumulate in the root of wheat (Li et al. 2017).

3.2 *Effects of selenate and selenite on grain Se speciation*

All the extraction efficiency rates were above 82% (data not shown) for different Se species in wheat grain, which indicated the enzymatic hydrolysis of protease XIV can efficiently extract the Se compounds in wheat grain. The main Se speciation of wheat grain was selenomethionine (SeMet) for all treatments, accounting for about 92.3% of total Se (Fig. 2). This species of Se was followed by selenocysteine ($SeCys_2$), which accounted for about 6% of total Se in wheat. Organic Se is more beneficial to humans than inorganic Se (Schiavon & Pilon-Smits 2017). Our results indicated that wheat has a strong ability to convert inorganic Se from wheat into plant organic Se. Notably, compared with selenite treatment, the percentage of $SeCys_2$ decreased 37% when selenite was applied to the soil.

4 CONCLUSIONS

Soil application of Se can reliably and effectively increase the Se content of wheat. At the same time, Se from soil application of selenite tended to accumulate in wheat root, while soil application of selenate was more efficient at increasing Se content in wheat

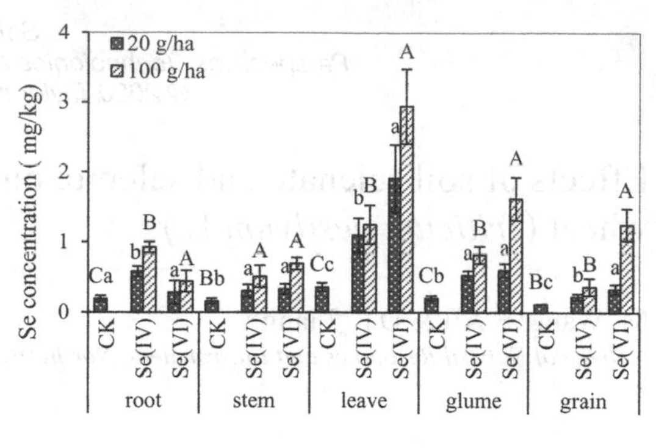

Figure 2. Effects of selenate and selenite on grain Se speciation of wheat.

grain. SeMet was the main organic Se species in wheat grain, accounting for about 92.3% of total Se. We suggest that soil application of selenate at 20 g/ha was the most efficient for Se biofortification in wheat in Lou soil.

REFERENCES

Boldrin, P.F., Faquin, V., Ramos, S.J., Bpldrin, K.V.F., Avila, F.W. & Guilherme, L.R.G. 2013. Soil and foliar application of selenium in rice biofortification. *J Food Compos & Anal* 31(2): 238–244.

Broadley, M.R., White, P.J., Bryson, R.J. et al. 2007. Biofortification of UK food crops with selenium. *Proc Nutr Soc* 65(2): 169–181.

Curtin, D., Hanson, R., Lindley, T.N. & Butler, R.C. 2006. Selenium concentration in wheat (*Triticum aestivum*) grain as influenced by method, rate, and timing of sodium selenate application. *New Zeal J Crop Hort* 34(4): 329–339.

Li, X., Wu, Y., Li, B., Yang, Y. & Yang, Y. 2017. Selenium accumulation characteristics and biofortification potentiality in turnip (*Brassica rapa* var. *rapa*) supplied with selenite or selenate. *Frontiers Plant Sci* 8: 2207.

Lyons, G. 2010. Selenium in cereals: improving the efficiency of agronomic biofortification in the UK. *Plant Soil* 332(1-2): 1–4.

Schiavon, M. & Pilon-Smits, E.A.H. 2017. The fascinating facets of plant selenium accumulation-biochemistry, physiology, evolution and ecology. *New Phytol* 213(4): 1582–1596.

Zhou, F., Yang, W.X., Wang, M.K. et al. 2018. Effects of selenium application on Se content and speciation in *Lentinula edodes*. *Food Chem* 265: 182–188.

Selenium Research for Environment and Human Health: Perspectives, Technologies and Advancements – Bañuelos, Lin, Liang & Yin (eds)
 ISBN 978-1-138-39014-0

Effect of cow manure amendment and root-induced changes on Se fractionation and plant uptake by Indian mustard

Q.T. Dinh & T.A.T. Tran
College of Natural Resources and Environment, Northwest A&F University, Yangling, Shaanxi, China
Faculty of Natural Science, Thu Dau Mot University, Thu Dau Mot City, Binh Duong, Vietnam

F. Zhou, H. Zhai & D.L. Liang
College of Natural Resources and Environment, Northwest A&F University, Yangling, Shaanxi, China

1 INTRODUCTION

Selenium (Se) is an essential micronutrient for human and animal health. The Se status in a human population depends on the daily dietary Se intake, which is in part governed by the amount of Se available that enters the food chain from the soil. Applying cow manure is a traditional method to maintain or increase soil fertility for agricultural production. The application of cow manure affects the transformation of Se fraction in soil, which in turn affects its bioavailability. Until now, only a few studies have been conducted on the effects of cow manure on the bioavailability of Se in soil, and they have been limited to bulk soil. Moreover, the rhizosphere itself was also different from the bulk soil, which may have a significant effect on the fractionation, speciation and, eventually, the bioavailability of Se. Therefore, the objective of this study was to gain a better insight into how the application of manure amendments to selenium-contaminated soil affects the rhizosphere response of Indian mustard (*Brassica juncea* (L.) *Czern. et Coss*) and the speciation and the bioavailability of Se.

2 MATERIALS AND METHODS

2.1 *Experiment design*

Soil used in the experiment was collected from the surface layer (0–20-cm depth) of a farm located at the Northwest A&F University in Shaanxi Province, China. The soil type is Eum-Orthic Anthrosol that represents a typical agricultural soil with an old cultivation history in Northern China. The relevant physicochemical properties are as follows: Soil pH 7.99, organic matter (OM) 14.34 g/g, and total Se 0.19 mg/kg. The cow manure was collected from farmland at Hanzhong district, Shaanxi, China.

The soil was artificially contaminated with 1 mg/kg DW of Se using a Na_2SeO_3 solution. Four cow manure application rates were selected as 0, 2%, 4%, and 6%. Each treatment was replicated three times and soils were maintained at 70% water holding capacity. After 30 days of homogenization and equilibration, ten seeds of Indian mustard were sown into each pot. After germinating for a week, the seedlings were thinned to 3 per pot. The plants were harvested 120 d after planting. At harvest, rhizosphere and bulk soil was recovered by gently shaking and scraping the roots and then air-dried. Plant shoots, tubers, and roots were rinsed with distilled water and air-dried.

2.2 *Chemical analysis*

Selenium fraction analysis in soil was adopted from the modified method described by Qin et al. (2012); the modified method took into account not only Se fractionation but also Se species in soil. The Se concentration in plant tissue was determined by an atomic fluorescence spectrophotometer (AFS-9780, Beijing Titan Instruments Co., Ltd.) after digestion with 4:1 (v/v) HNO_3-$HClO_4$ at 170°C.

3 RESULTS AND DISCUSSION

3.1 *Distribution of Se fraction in the rhizosphere and bulk soil*

Cow manure application depressed the percentage of SOL-Se and EXC-Se, while it increased the pools of OM-Se and RES-Se. Similar trends were observed for the Se concentrations in the different fractions. For example, at the 2% cow manure amendment treatment, the SOL-Se and EXC-Se concentrations were 96.76 and 237.68 μg/kg in the rhizosphere soil and 73.81 and 118.55 μg/kg in bulk soil, respectively. In contrast, OM-Se concentrations in bulk soil were significantly higher than those in the rhizosphere soil ($p < 0.05$). Compared to the rhizosphere soil, OM-Se concentration was 1.80–2.47-fold higher than that in bulk soil. Among all treatments, the percentages of available Se were higher in the rhizosphere than bulk soil ($p < 0.05$), which were 1.42-fold in non-manure treatment, and 1.74–2.10-fold in cow manure treatment.

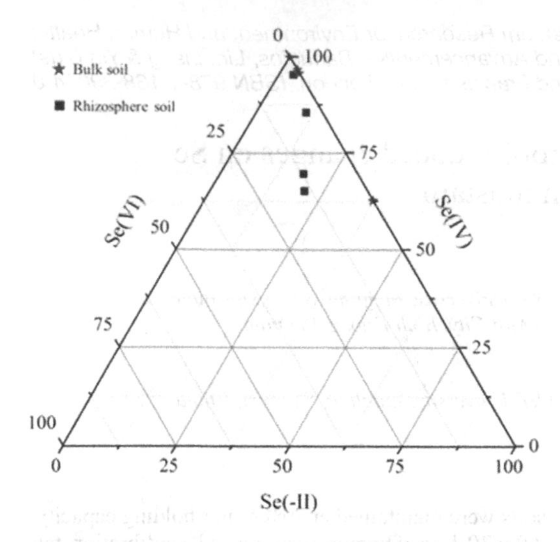

Figure 1. Ternary diagram of the average Se speciation from rhizosphere and bulk soil oxidation state maps.

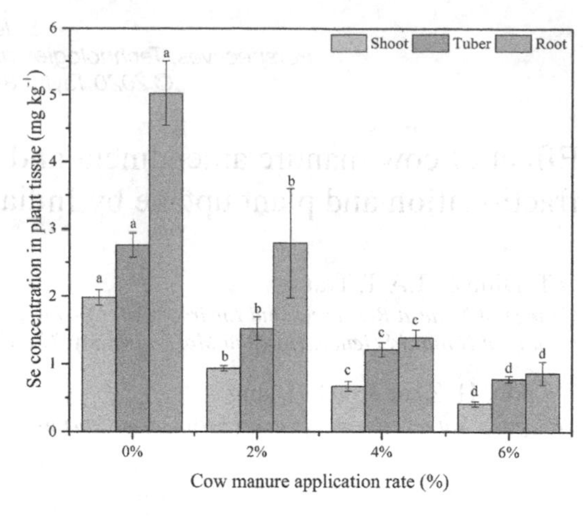

Figure 2. Selenium concentration in plant tissues. Different letters indicate significant differences ($p < 0.05$).

This difference indicates that the rhizosphere process leads to the mobilization of Se and thus increases the bioavailability of Se in soil. In other words, the rhizosphere process has somewhat limited cow manure's effect of reducing the Se availability.

3.2 *Interactive effects of manure and plant root on Se species*

Besides the fractionation, Se bioavailability is also governed by the speciation of Se in soil (Dinh et al. 2018). Se(IV) was the predominant species in the rhizosphere and bulk soil after harvest, ranging from 65.14-95.31% and 62.82-100%, respectively. Soil oxidation state maps typically had more Se(IV) than Se(VI) and Se(-II). The proportion of Se(-II) increased with increasing cow manure application rate, indicating cow manure application can promote the reduction of Se to a lower valence state. With the initial application of selenite, Se(VI) was not observed in bulk soil, while it was detected in the rhizosphere soil (Fig. 1). Thus, it is assumed that a rhizosphere process is causing unavailable forms of Se (e.g. Se[-II] and Se[IV]) to be oxidized and mobilized.

3.3 *Selenium uptake by plant*

The Se concentration in Indian mustard tissues dramatically decreased with increasing manure amendments ($p < 0.05$) (Fig. 2). Compared to the single Se(IV) treatment without the addition of manure, the Se concentration in shoots, tubers, and roots were reduced by 52.9–79.2%, 44.8–71.7%, and 44.5–82.8% in cow manure treatments, respectively. This observation suggested that the Se immobilization process promoted by cow manure amendment plays a dominant role over mineralization by plant roots in determining the bioavailability of Se in soil-plant system.

4 CONCLUSIONS

The application of cow manure to the soil significantly reduced the bioavailability of Se due to liming effect, which subsequently promoted the Se shift into less available pools, whereas the rhizosphere process leads to the mobilization of Se, and the bioavailability of Se was greatly increased. However, applying cow manure generally decreased Se bioavailability, resulting in a decrease in the amount of Se accumulated by Indian mustard.

REFERENCES

Dinh, Q.T., Wang, M., Tran, T.A.T. et al. 2018. Bioavailability of selenium in soil-plant system and a regulatory approach. *Crit Rev Env Sci Tech* 49: 443–517.

Qin, H.B., Zhu, J.M. & Su, H. 2012. Selenium fractions in organic matter from Se-rich soils and weathered stone coal in selenosis areas of China. *Chemosphere* 86: 626–633.

Selenium Research for Environment and Human Health: Perspectives, Technologies and Advancements – Bañuelos, Lin, Liang & Yin (eds)
 ISBN 978-1-138-39014-0

Selenium bioaccessibility in selenium-enriched *Lentinula edodes*

F. Zhou, N.N. Liu, Y. Liu & D.L. Liang
College of Natural Resources and Environment, Northwest A&F University, Yangling, Shaanxi, China
Key Laboratory of Plant Nutrition and the Agri-environment in Northwest China, Ministry of Agriculture, Yangling, Shaanxi, China

1 INTRODUCTION

Selenium (Se) is an essential micronutrient for mammals and certain algae, although the difference between deficient, essential, and toxic doses of Se is small. To ensure the adequate daily Se intake while avoiding the toxic effects of excessive intake, it is important to obtain a safe and efficient dietary Se supplement. Assunção et al. (2014) have shown that a variety of edible fungi have strong ability to accumulate Se, and eventually convert exogenous Se into organic forms (mainly as SeMet). Therefore, edible fungi can be an important food crop for Se biofortification.

In view of the double effects of Se on health, both total Se content and the food safety of the Se biofortified products need to be evaluated. Studies by Maseko et al. (2013) and Egressy-Molnár et al. (2016) demonstrated that application of exogenous Se can significantly improve (27–399 times) the Se content in mushroom compared with control. Based on this result, it is thus clear that the excessive application of exogenous Se may result in health risk for Se-enriched products. Bioaccessibility represents the concentration of nutrients or contaminants in foods, which can be digested by human gastrointestinal tract and further absorbed through the intestinal mucosa (Bhatia et al. 2013). The *in vitro* simulated gastrointestinal digestion test, which has the advantages of low cost, simple operation, and easy control of conditions, is a good approach to determine the element bioaccessibility under laboratory conditions and evaluate the health risk of Se-enriched products.

Lentinula edodes was selected as the material of Se biofortification. Exogenous Se supplements were injected into cultivation substrate to produce Se-enriched *L. edodes* fruit bodies. Total Se concentration and Se bioaccessibility were determined, and the health risk was preliminarily evaluated based on these indexes.

2 MATERIALS AND METHODS

2.1 *Cultivation of selenium-enriched* L. edodes

One week before fruiting, sodium selenite (Na_2SeO_3), sodium selenate (Na_2SeO_4), or Se-enriched yeast water solution was uniformly injected into the substrate to maintain a Se concentration of 5 mg Se/kg in the substrate. Two flushes of Se-enriched *L. edodes* fruiting bodies were harvested on the 49th and 70th days after the Se application. The fruiting bodies were cut into pieces, mixed evenly, freeze-dried, ground, and stored at −20°C.

2.2 *Determination of total Se concentration and Se bioaccessibility*

The fruiting bodies were digested in HNO_3 and $HClO_4$ mixture solution (4:1, v/v). Selenate Se was reduced with concentrated HCl, and then Se was determined using HG-AFS. PBET method was used for the determination of Se bioaccessibility in *L. edodes*. Bioaccessibility of Se is expressed by the percentage of total Se in digestive juice of gastric or gastrointestinal phase in total Se of samples.

3 RESULTS AND DISCUSSION

3.1 *Total Se in fruit bodies of* L. edodes

The Se content of the control fruit body of *L. edodes* harvested on Day 49 and Day 70 was 0.46 mg/kg (data not shown). The total Se of fruit bodies treated with Na_2SeO_3, Na_2SeO_4, and Se yeast was 53.1–89.7 mg/kg, 32.4–33.5 mg/kg, and 25.2–35.3 mg/kg, respectively (Fig. 1).

The *L. edodes* fruit body had a strong ability to accumulate exogenous Se, which was consistent with the findings of Rzymski et al. (2016).

Total Se in *L. edodes* treated with selenite and Se yeast was significantly reduced ($p < 0.05$) from Day 49 to Day 70 but was non-significantly increased in the selenate treatment (Fig. 1). This result might be partly due to strong adsorption of Se(IV) to substrate and mycelium. The Se-yeast treatment resulted in similar levels of Se accumulation in the fruit body. In contrast, the highest Se accumulation in the fruit body was observed from the selenite treatment at two different growth stages.

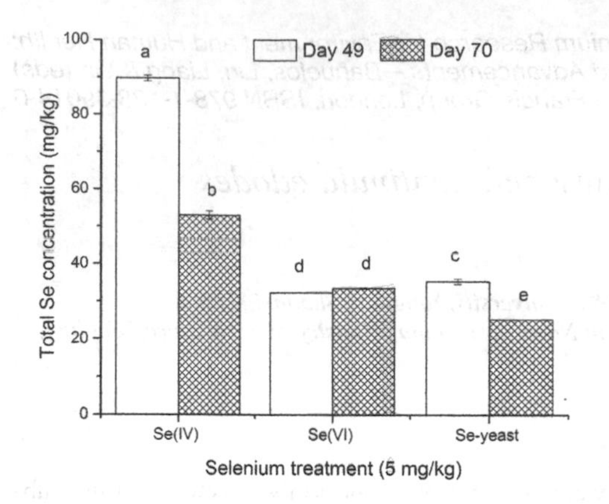

Figure 1. Selenium concentration of *L. edodes* treated with selenite, selenate, and Se yeast. Means with different letters are not statistically significant ($p > 0.05$).

Table 1. Selenium bioaccessibility in gastric and gastrointestinal fractions of *L. edodes*.

Harvest time	Se Treatment	Se bioaccessibility in gastric fraction (%)	Se bioaccessibility in gastrointestinal fraction (%)
Day 49	Se(IV)	36.0b	78.1b
	Se(VI)	43.9a	73.5c
	Se-yeast	42.3a	80.1ab
Day 70	Se(IV)	37.0b	82.6a
	Se(VI)	44.8a	84.1a
	Se-yeast	43.8a	79.9ab

3.2 *Selenium bioaccessibility in fruit bodies of* L. edodes

In gastric digestion stage, Se bioaccessibility of *L. edodes* fruit bodies treated with selenite was 36%–37%, which was lower than that those in selenate and Se yeast treatments (44%–45% and 42%–44%, respectively) (Table 1). Moreover, no significant difference was observed between Se bioaccessibility in fruit bodies harvested at Day 49 and Day 70. The Se bioaccessibility in gastric digests were lower than that in gastrointestinal digests in each Se treatment. With the increase of the application time of Se, Se bioaccessibility in gastrointestinal digests were significantly increased in selenite (from 78% to 83%) and selenate (from 74% to 84%) treatments ($p < 0.05$). However, there was almost no change in the fruit bodies treated with Se yeast (about 80%). This result is consistent with the range of Se bioaccessibility in Se-enriched mushrooms (60%–80%) produced *in vitro* by Bhatia et al. (2013), and higher than that reported in various cereals and green leafy vegetables (Khanam & Platel 2016). Therefore, Se-enriched *L. edodes*, produced by applying Se supplements into substrate, is a good Se dietary source.

Based on the bioaccessibility of Se, the maximum daily Se intake through the consumption of the Se-enriched *L. edodes* (daily intake is 30 g fresh weight with water content is 90%) obtained from selenite, selenate, and Se yeast treatment was 210.1 μg, 84.6 μg, and 84.8 μg, respectively. Consequently, no health risk is associated with consuming the Se-enriched *L. edodes* obtained in this study.

4 CONCLUSIONS

The application of the three Se supplements (selenite, selenate, and Se yeast) can all significantly improve the Se concentrations in *L. edodes* fruit bodies. The highest utilization of Se was observed with the selenite treatment. The Se bioaccessibility (80–84%) was high in selenium-enriched fruit bodies harvested from each Se treatment. Importantly, no health risk would be associated with the consumption of the selenium-enriched *L. edodes*. Therefore, it can be considered a good Se dietary source.

REFERENCES

Assunção, L.S., Silva, M.C.S., Fernandez, M.G. et al. 2014. Speciation of selenium in *Pleurotus ostreatus* and *Lentinula edodes* mushrooms. *J Biotechnol Lett* 5(1): 79–86.

Bhatia, P., Aureli, F., D'Amato, M. et al. 2013. Selenium bioaccessibility and speciation in biofortified Pleurotus mushrooms grown on selenium-rich agricultural residues. *Food Chem* 140(1-2): 225–230.

Egressy-Molnár, O., Ouerdane, L., Győrfi, J. & Dernovics, M. 2016. Analogy in selenium enrichment and selenium speciation between selenized yeast *Saccharomyces cerevisiae* and *Hericium erinaceus* (lion's mane mushroom). *LWT-Food Sci Technol* 68: 306–312.

Khanam, A. & Platel, K. 2016. Bioaccessibility of selenium, selenomethionine and selenocysteine from foods and influence of heat processing on the same. *Food Chem* 194: 1293.

Maseko, T., Callahan, D.L., Dunshea, F.R., Doronila, A., Kolev, S. D. & Ng, K. 2013. Chemical characterisation and speciation of organic selenium in cultivated selenium-enriched *Agaricus bisporus*. *Food Chem* 141(4): 3681–3687.

Rzymski, P., Mleczek, M., Niedzielski, P., Siwulski, M. & Gąsecka, M. 2016. Cultivation of *Agaricus bisporus* enriched with selenium, zinc and copper. *J Sci Food Agr* 97(3): 923–928.

Selenium Research for Environment and Human Health: Perspectives, Technologies and Advancements – Bañuelos, Lin, Liang & Yin (eds)
© 2020 Taylor and Francis Group, London, ISBN 978-1-138-39014-0

Selenium content in soils and rice from Guangxi, China

H.W. Zang
School of Plant Protection, Anhui Agricultural University, Hefei, Anhui, China
Jiangsu Bio-Engineering Research Center for Selenium, Suzhou, Jiangsu, China

Z.D. Long, L.X. Yuan & Z.M. Wang
Jiangsu Bio-Engineering Research Center for Selenium, Suzhou, Jiangsu, China

X.B. Yin
School of Earth and Space Sciences, University of Science and Technology of China, Hefei, Anhui, China
Jiangsu Bio-Engineering Research Center for Selenium, Suzhou, Jiangsu, China

Q.Q. Chen
Environmental Science, College of Life Science and Resources and Environment, Yichun University, Yichun, Jiangxi, China

1 INTRODUCTION

Selenium (Se) is an essential nutrient for animals and humans and is related to many human health problems. Selenium deficiency in humans is related to endemic Kashin-Beck and Keshan diseases in China (Rayman 2000), while excessive Se can be toxic and leads to human DNA damage, cancer, and mutagenesis (Foster et al. 1995).

A close relationship has been found between available forms of trace elements in soil, such as Se, boron, nickel, and molybdenum, and the number of humans living to over 90 years old in Rugao County, Jiangsu Province, China (Huang et al. 2009, Lv et al. 2008). As an important natural source and pool of Se, soil Se content affects Se levels in the complete ecosystem and impacts human health indirectly through the food chain (Li et al. 2008). Guangxi City, China is a natural lab to study the relationship between Se intake and human health since there are over 33 thousand km^2 area with selenium-rich soils and about 10 longevity villages. To better understand the relationship between the concentrations of Se in environment and human health, soil and rice seedling samples were collected in Pubei County and Guiping County, Guangxi, China to perform a preliminary study on the distribution of Se.

2 MATERIALS AND METHODS

2.1 *Study sites and sampling*

Pubei County and Guiping County are located in the south of Guangxi Zhuang Autonomous Region (GZAR). Pubei County was famous as the *Longevity Town in the World*. The population of the Pubei County was 93,000 and the longevity population ration could reach up to 12/100,000, which was far higher than the national average of 2.7/100000. Guiping County is a typical Chinese city, located in central Guangxi. A total of 55 soil samples and 39 rice samples were collected from 55 villages in Pubei County, and a total of 15 soil samples and 9 rice samples were randomly collected from 15 villages in Guiping County.

2.2 *Selenium concentration detection*

The total Se concentration was determined by HG-AFS (Beijing Titan Instrument Co. China) according to GB 5009.93-2010. Samples were digested by mixed acid (HNO_3-$HClO_4$ 4:1, v/v) and hydrogenated and then Se concentration was detected by AFS (Zhu et al. 2013).

3 RESULTS AND DISCUSSION

In Figure 1, the total Se concentrations of soils ranged from 349.45 to 1171.47 μg/kg in Pubei, with an average value of 659.53 ± 190.94 μg/kg. The total Se concentrations of rice ranged from 13.21 μg/kg to 164.30 μg/kg in Pubei, with an average value of 47.11 ± 27.42 μg/kg. Based on the classification of selenium-rich soil and rice samples conducted by Tan in 1996 (Tan 1996), almost 94.55% of soil samples (>400 μg/kg) were Se-rich soils.

In Figure 2, the range of Se concentration in the soils of 15 villages in Guiping was 183.26 to 2018.56 μg/kg, with an average value of 717.14 ± 567.86 μg/kg. About 53.33% of soil samples can be classified as Se-rich soils (Tan 1996). The total Se concentrations of rice seedlings ranged from 21.86 to 97.53 μg/kg in Guiping, with an average value of 47.11 ± 27.42 μg/kg.

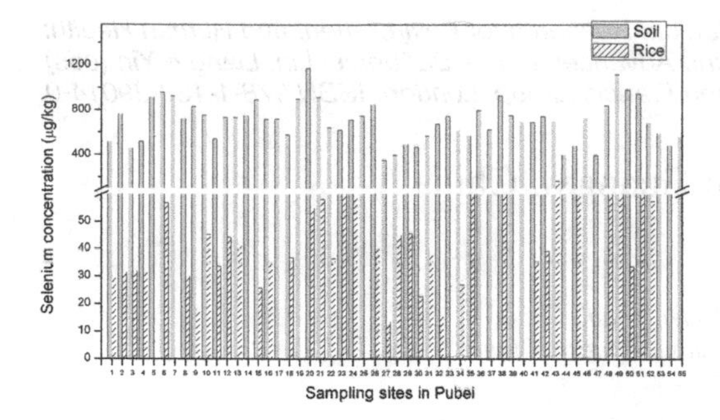

Figure 1. Selenium contents in soil and rice samples in Pubei.

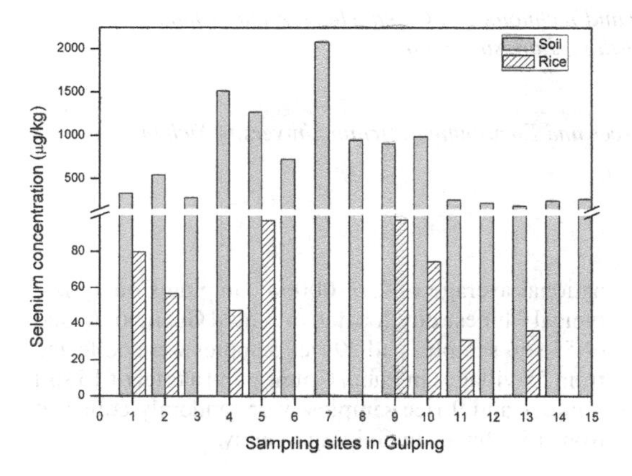

Figure 2. Selenium content in soils and rice seedling samples in Guiping.

4 CONCLUSIONS

Previous studies revealed that the concentration of Se in soil has a significant positive correlation with longevity index (the number of centenarians per 100,000 people) based on the sixth population census data of China (Liu et al. 2013). In the present study, more than 90% and about 50% of the soils could be identified as Se-rich soils in Pubei and Guiping, respectively, indicating that these two sites could be an ideal natural lab to study Se and longevity.

REFERENCES

Foster, H.D. & Zhang, L.P. 1995. Longevity and selenium deficiency evidence from the Peoples Republic of China. *Sci Total Environ* 170(1–2): 133–139.

Huang, B., Zhao Y.C., Sun, W.X. et al. 2009. Relationships between distributions of longevous population and trace elements in the agricultural ecosystem of Rugao County, Jiangsu, China. *Environ Geochem Health* 31(3): 379–190.

Lv, J.M., Wang, W.Y., Thomas, K., Li, Y., Zhang, F. & Yuan F. 2011. Effects of several environmental factors on longevity and health of the human population of Zhongxiang, Hubei, China. *Biol Trace Elem Res* 143(2): 702–716

Li, Y.H., Wang, W.Q., Luo, K.L. & Hairong, L. 2008. Environmental behaviors of selenium in soil of typical selenosis area, China. *J Environ Sci* 20(7): 859–864.

Liu, Y., Li, Y., Jiang, Y., Li, H., Wang, W. & Yang, L. 2013. Effects of soil trace elements on longevity population in China. *Biol Trace Elem Res* 153(1–3): 119–126.

Rayman, M.P. 2000. The importance of selenium to human health. *Lancet* 356(9225): 233–241.

Tan, J.A. 1996. *The Atlas of Endemic Disease and Their Environments in the People's Republic of China*. Beijing: Science Press.

Zhu, Y.Y., Yuan, L.X., Li, W., Lin, Z. Q., Banuelos, G. & Yin, X.B. 2013. A novel selenocystine-accumulating plant in selenium-mine drainage area in Enshi, China. *PLos One* 8(6): e65615.

Selenium Research for Environment and Human Health: Perspectives, Technologies and Advancements – Bañuelos, Lin, Liang & Yin (eds)
© 2020 Taylor and Francis Group, London, ISBN 978-1-138-39014-0

Selenium accumulation and speciation in soils along a climate gradient

J. Tolu, S. Bouchet, S.D. Chékifi & O. Hausheer
Swiss Federal Institute of Technology, Institute of Biogeochemistry and Pollutant Dynamics, Zürich, Switzerland
Institute for Atmospheric and Climate Science, Zürich, Switzerland
Swiss Federal Institute of Aquatic Science and Technology, Duebendorf, Switzerland

O.A. Chadwick
Department of Geography, University of California, Santa Barbara, California, USA

J. Helfenstein, E. Frossard & F. Tamburini
Institute of Agricultural Sciences, ETH Zürich, Switzerland

L.H.E. Winkel
Swiss Federal Institute of Technology, Institute of Biogeochemistry and Pollutant Dynamics, Zürich, Switzerland
Institute for Atmospheric and Climate Science, Zürich, Switzerland
Swiss Federal Institute of Aquatic Science and Technology, Duebendorf, Switzerland

1 INTRODUCTION

Selenium (Se) is an essential element for humans and animals. Main intake routes are via food consumption such as grains, cereals, and meat. Although Se is not known to be essential for plants, it can promote the growth and quality of crops (Haug et al. 2007). Selenium deficiency was estimated to be a major health problem for 0.5 to 1 billion people worldwide, and even more people probably consume less Se than is required for optimal protection against cancer, cardiovascular, and severe infectious diseases, including HIV (Haug et al. 2007).

A recent study used 33241 soil data points to model recent (1980–1999) global distributions of soil Se content (Jones et al. 2017). This study demonstrated that at large spatial scale, soil Se concentrations were dominated by climate–soil interactions, including pH, clay content, precipitation, aridity and soil organic matter (OM) content. Using moderate climate-change scenarios for 2080–2099, the authors predicted that changes in climate and soil organic carbon (OC) content will lead to overall decreased soil Se concentrations, which could increase the prevalence of Se deficiency. Although this study provides first indications of where crops may be or will become low in Se content, future prediction of Se bioavailability for plant requires a full understanding of Se cycling in soils.

The reasons why aridity, clay content and pH relate to soil Se content and influence its bioavailability for plants are relatively well understood (Fernandez-Martinez & Charlet 2009, Li et al. 2017). In arid soils where oxic conditions prevail, Se oxyanions, i.e. selenite (Se[IV]) and selenate, (Se[VI]), which are soluble and bioavailable, predominate over insoluble Se(0) or metal selenides (Se[-II]). Especially the most oxidized form, i.e. Se(VI), is present. Se(VI) is more mobile than Se(IV) and can be removed from the (top)soil by leaching. Selenium oxyanions can sorb to minerals, such as clay minerals and (oxy)hydroxides. This process is pH-dependent, explaining the importance of identifying clay content and pH in modeling soil Se concentrations (Fernandez-Martinez & Charlet 2009). In contrast, the effects of rainfall and OM on Se concentrations and bioavailability remain unclear. Rainfall may affect soil Se concentrations and bioavailability directly as a source of bioavailable Se(IV) and Se(VI), or indirectly by affecting soil properties, such as redox conditions, pH, or OM content (Bitterli et al. 2010). Organic matter can bind Se after it is reduced into selenide (Gustafsson and Johnsson, 1994) or via ternary complexation via Fe(III) (Martin et al., 2017). It can also indirectly influence Se retention and bioavailability by forming organo-mineral associations that protect Se adsorbed onto (oxy)hydroxides or by favoring reduction of Se oxyanions into Se(0) (Tolu et al. 2014a, Kaush & Pallud 2013). However, the prevailing mechanism of explaining the link between OM and Se is still unknown. One of the reasons for this knowledge gap, is the challenge to analyze the chemical form of Se, i.e., its speciation, in soils as most soils have relatively low Se content (Jones et al. 2017, Fernandez-Martinez & Charlet 2009).

The goal of our study was to better understand the role of rainfall and OM on Se concentrations, accumulation, and speciation in soils. For that purpose, we investigated Se concentrations and speciation in soils taken along the Kohala climate gradient in Hawaii. On the Kohala Mountain, soils developed on the same parent rock material with rainfall amounts varying from <300 to >3100 mm/y, which resulted in contrasting soil processes and properties (Vitousek & Chadwick 2013).

2 MATERIALS AND METHODS

Samples from soil depth profiles were taken at six sites along the Kohala Mountain, i.e. two dry sites, two sub-humid sites and two humid sites. Two parent rock materials were collected at the driest site.

Total carbon content was determined using a CNS analyzer (Euro EA 3000; Eurovectors SPA) and total inorganic carbon was measured with a Coulomat (CM 5015 Coulometer; UIC Inc). Total OC was calculated as the difference between total C and the inorganic carbon content. Selenium, as well as other trace elements, were quantified in the soils and parent rocks by inductively coupled plasma tandem mass spectrometry (Agilent 8900 ICP-QQQ) after microwave-assisted digestion. Extractions with ultrapure water, phosphate buffer, and NaOH were performed in parallel as described in Tolu et al. (2011). Total extracted Se was quantified by ICP-MS/MS, and the Se species were analyzed by high performance liquid chromatography (HPLC Agilent 1260) coupled to the ICP-MS/MS. Selenium oxyanions and seleno-amino acids (selenocystine, $SeCys_2$ and seleno-methionine, SeMet) were identified and quantified using anion exchange chromatography coupled to ICP-MS/MS as detailed in Tolu et al. (2011). This speciation analysis was complemented by a new method to characterize and quantify Se associated to OM (Tolu et al., unpubl.).

3 RESULTS AND DISCUSSION

Overall, Se content and Se accumulation with respect to parent rock were higher in wetter and sub-humid sites compared to driest sites. This result indicates that rainfall is a source of Se; however, other factors also play a role. Rainfall has an important influence on soil properties, such as soil moisture content and the accumulation of soil OM. We notably observed a significant positive correlation between soil OM and Se accumulation, suggesting that soil OM plays a role in the retention of Se, as was also found in previous studies (Li et al. 2017).

Regarding Se speciation, Se(VI) was found in water and phosphate extracts at the dry sites, which is in agreement with the general predominance of this Se species in oxic soils (Fernandez-Martinez & Charlet 2009). Se(IV) was found in all soil extracts, while the organic Se compounds $SeCys_2$ or SeMet were not identified, which is also in agreement with previous studies (Tolu et al. 2014a, b, Supriatin et al. 2015). This finding indicates that these species are either absent (or present in concentrations below the detection limit) or are degraded during soil drying process, as a previous study showed the stability of $SeCys_2$ and SeMet during water, phosphate, and NaOH extractions (Tolu et al. 2011). Mass balances between identified Se(IV), Se(VI), and total extracted Se indicate that only a small fraction of the extracted Se was identified (between 0 to 76% of Se). This small recovery is typical for studies on soil extracts using conventional LC-ICP-MS methods, as existing methods cannot identify more complex Se species, e.g. organic complexes. Therefore, we developed a new LC-ICP-MS/MS method to separate Se species in term of Se bound to OM, (organo)-mineral nanoparticles, and free ions. The recovery of Se species using this method ranged between 60 to 100% of Se extracted by ultrapure water and NaOH. The results obtained show that Se is bound to OM in soils containing from 1 to 26% of OM and that Se-OM associations occur in different size ranges, related to the content of soil OM.

Our study demonstrates that Se is associated to soil OM in a wide variety of size ranges, which is linked to the amount of soil OM.

REFERENCES

Bitterli, C., Bañuelos, G. & Schulin, R. 2010. Use of transfer factors to characterize uptake of selenium by plants. *J Geochem Explor* 107: 206–216.

Fernandez-Martınez, A. & Charlet, L. 2009. Selenium environmental cycling and bioavailability: a structural chemist point of view. *Rev Environ Sci Bio* 8: 81–110.

Gerrad, D.J., Droz, B., Greve, P. et al. 2017. Selenium deficiency risk predicted to increase under future climate change. *PNAS USA* 114(11): 2848–2853.

Gustafsson, J.P. & Johnsson, L. 1994. The association between selenium and humic substances in forested ecosystems — laboratory evidence. *Appl Organometallic Chem* 8: 141–7.

Haug, A., Graham, R.D., Christophersen, O.A. & Lyons, G.H. 2007. How to use the world's scarce selenium resources efficiently to increase the selenium concentration in food. *Microb Ecol Health Dis* 19: 209–228.

Kausch, M.F. & Pallud, C.E. 2013. Modeling the impact of soil aggregate size on selenium immobilization. *Biogeosciences* 10: 1323–1336.

Li, Z., Liang, D., Peng, Q., Cui, Z, Huang, J. & Lin, Z. 2017. Interaction between selenium and soil organic matter and its impact on soil selenium bioavailability: A review. *Geoderma* 295: 69–79.

Martin, P., Seiter, J. M., Lafferty, B.J. & Bednar, A.J. 2017. Exploring the ability of cations to facilitate binding between inorganic oxyanions and humic acid. *Chemosphere* 166: 192-196.

Supriatin, S., Weng, L. & Comans, R.N.J. 2015. Selenium speciation and extractability in Dutch agricultural soils. *Sci Total Environ* 532: 368–382.

Tolu, J., Le Hécho, I., Bueno, M., Thiry, Y. & Potin-Gautier, M. 2011. Selenium speciation analysis at trace level in soils. *Analytica Chimica Acta* 684: 126–133.

Tolu, J., Thiry, Y., Bueno, M, Jolivet, C., Potin-Gautier, M. & Le Hécho, I. 2014a. Distribution and speciation of ambient selenium in contrasted soils, from mineral to organic rich. *Sci Total Environ* 479–480: 93–101.

Tolu, J., Di Tullo, P., Le Hécho, I. et al. 2014b. A new methodology involving stable isotope tracer to compare simultaneously short- and long-term selenium mobility in soils. *Anal Bioanal Chem* 406: 1221–1231

Vitousek, P.M. & Chadwick, O.A. 2013. Pedogenic thresholds and soil process domains in Basalt-Derived soils. *Ecosystems* 16: 1379–1395.

Selenium Research for Environment and Human Health: Perspectives, Technologies and Advancements – Bañuelos, Lin, Liang & Yin (eds)
© 2020 Taylor and Francis Group, London, ISBN 978-1-138-39014-0

Accumulation and distribution of selenium in Brazil nut tree in relation to soil selenium availability

D.A. Castro, J.H.R. Souza & M.F. Moraes
Federal University of Mato Grosso, Cuiaba, Brazil

L. Wilson & M.R. Broadley
University of Nottingham, Leicestershire, UK

A.B.B. Tardin
Embrapa Agrosilvopastoral, Sinop, Brazil

R.M.B. Lima & K.E. Silva
Embrapa Western Amazon, Manaus, Brazil

P.J. White
James Hutton Institute, Dundee, UK

1 INTRODUCTION

Studies of the mineral composition of Brazil nuts (*Bertholletia excelsa*) have found that selenium (Se) is one of the most abundant elements, with concentrations reported in the literature ranging from 5 to 512 mg/kg Se in the seed (Dumont et al. 2006). These Se concentrations depend on Se bioavailability in the soil.

Geology determines the distribution of Se in the world. In general, areas with high levels of Se are less disseminated than those with low levels of Se (Fordyce 2010, Winkel et al. 2012). Information about the levels of Se in Brazilian soils is scarce. Nationwide studies of the total levels of Se both in plant products and soils are fundamental, particularly in light of the fact that half of the world's population suffers from a shortage of this essential nutrient (Zhao et al. 2009).

In this context, the aims of this study were: (1) to quantify Se in parts of Brazil nut trees grown in different locations; (2) to measure the levels of available and total Se in soil samples; and (3) to correlate soil Se present with the levels of Se found in the trees.

2 MATERIALS AND METHODS

The study was carried out in two forest areas planted with *Bertholletia excelsa*: One located in the state of Amazonas (AM) (3°0′30.63″S and 58°50′1.50″W), with 10-, 12-, and 14-year-old trees, and the other in the state of Mato Grosso (MT) (11° 35′20.3″S, 55° 17′34.7″ W), with 16- and 29-year-old trees.

Soil samples were collected for Se analysis at depths of 0–20, 20–40, 40–60, 60–80, and 80–100 cm, at six points uniformly distributed in a sample grid in each state (AM and MT). After collection, samples were prepared and sent to the School of Biosciences at the University of Nottingham, UK, where available and total levels of Se were determined by Inductively-coupled Plasma – Mass Spectrometry (ICP-MS, Agilent 7500ce Spectrometer, Stockport, UK) with sequential extraction using KNO_3 and KH_2PO_4. Total Se was determined in the 0-100-cm soil layer (TA) for correlation analyses with the levels in the trees.

Samples of about 200 g were collected from each tree part. For root samples, eight holes were bored to a depth of 5 cm, spaced 10 cm apart; for trunk samples eight cores were collected to a depth of 10 cm and spaced 50 cm apart (non-destructive sampling). Four branches were also collected from the lower base (LB) and four from the upper base (UB). In both cases samples were collected in four directions (N, S, W and E). For leaf samples, 50 g of new leaves (NL) and 50 g of old leaves (OL) were collected from the canopy in the N, S, W, and E directions.

All samples were prepared according to the method described by EMBRAPA (2000) to determine the dry matter. The milled samples were then sent to the School of Biosciences of the University of Nottingham for quantification of Se. Total Se content was also analyzed by ICP-MS in an Agilent 7500ce Spectrometer (Stockport, UK).

Data obtained in the two states were subjected to analysis of variance, with the qualitative means determined by the Scott-Knott test ($p \leq 0.05$), and the Spearman linear correlation coefficients (r) were calculated for application in the correlation matrix.

Table 1. Available and total Se (μg/kg) distributed at different depths in soils in the states of Amazonas (AM) and Mato Grosso (MT), in areas with Brazil nut trees.

Depth (cm)	Available Se			Total Se		
	AM	MT	CV (%)	AM	MT	CV (%)
0–20	23 a*	11 b	32	1363 a	120 b	12
20–40	26 a	14 b	68	1286 a	155 b	15
40–60	12 a	5,3 b	10	946 a	154 b	14
60–80	10 a	5,0 b	61	783 a	175 b	14
80–100	11 a	10 a	10	786 a	178 b	40

*Means with the same letter in each row are not statistically significant using the Scott-Knott test ($p > 0.05$).

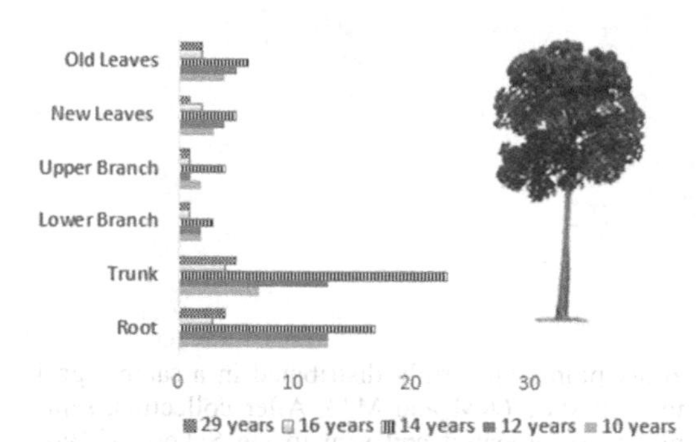

Figure 1. Distribution of Se levels (μg/kg) by parts of Brazil nut trees.

3 RESULTS AND DISCUSSION

Soil samples from the state of Mato Grosso had low concentrations of total and available Se, while those from the state of Amazonas had higher concentrations (Table 1). This variation between states can be attributed to the soil classes, concentration of Se, and quantity of organic matter present in these soils (Kabata-Pendias 2011).

The Se concentrations are presented in the different parts of the Brazil nut trees at different ages (Fig. 1). The highest Se concentrations were found in the trunk and roots, respectively, regardless of age. The Se values varied from 4 to 23 μg/kg in the trunk and 3 to 17 μg/kg of Se in the roots.

Correlation values are shown among the levels of Se available in the soil at different depths with parts of the Brazil nut trees at different ages (Fig. 2). These

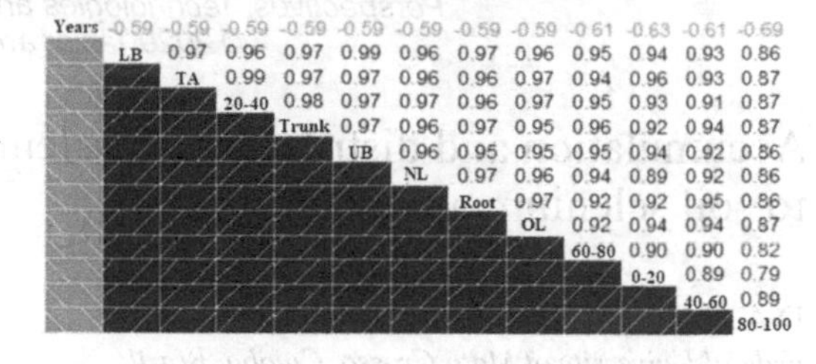

Figure 2. Result of the Spearman correlation matrix, corresponding to the levels of Se available in the soil at different depths with those found in parts of Brazil nut trees at different ages.

correlations were positive and did not depend on tree age. Therefore, the general pattern was the greater the Se content in soil, the higher the Se levels in the trees.

4 CONCLUSIONS

The trunk and roots of the Brazil nut trees were the parts containing the highest Se concentrations. The levels of Se found in this study were influenced by the quantity and availability of Se in the soil, irrespective of the age of the Brazil nut trees.

REFERENCES

Dumont, E., De Pauw, L., Vanhaecke, F. & Cornelis, R. 2006. Speciation of Se in *Bertholletia excelsa* (Brazil nut): A hard nut to crack? *Food Chem* 95(4): 684–692.

EMPRESA BRASILEIRA DE PESQUISA AGROPECUÁRIA – EMBRAPA. 1997. *Manual de métodos de análise de solo*, 2.ed. Rio de Janeiro: EMBRAPA.

Fordyce, F.M., Brereton, N., Hughes, J., Luo, W. & Lewis, F. 2010. An initial study to assess the use of geological parent materials to predict the Se concentration in overlying soils and in five staple foodstuffs produced on them in Scotland. *Sci Total Environ* 408(22): 5295–5305.

Kabata-Pendias, A. 2011. *Trace Elements in Soils and Plants*. Boca Raton: CRC Press.

Winkel, L.H.E., Johnson, C.A., Lenz, M. et al. 2012. Environmental selenium research: From microscopic processes to global understanding. *Environ Sci Tech* 46(2): 571–579.

Zhao, F.J., Su, Y.H. & Dunham, S.J. 2009. Variation in mineral micronutrient concentrations in grain of wheat lines of diverse origin. *J Cereal Sci* 49(2): 290–295.

Selenium Research for Environment and Human Health: Perspectives, Technologies and Advancements – Bañuelos, Lin, Liang & Yin (eds)
 ISBN 978-1-138-39014-0

Effect of straw derived DOM on selenite aging in Lou soil and mechanisms

D. Wang, W.X. Yang, M.K. Wang & D.L. Liang*
College of Natural Resources and Environment, Northwest A&F University, Shaanxi, China

1 INTRODUCTION

Selenium (Se) biofortification depends on Se content in the soil or exogenous applications of Se to the plant. The transformation of Se in soil determines its availability for plant absorption. Exogenous organic material will experience a series of biochemical transformations and stimulate fractionation and speciation transformations of Se in soil. Straw amendment will enhance soil-dissolved organic matter (DOM) and affect Se availability through the formation of DOM-Se, which involves hydrophilic acid-bound Se (Hy-Se), fulvic acid-bound Se (FA-Se), humic acid-bound Se (HA-Se) and hydrophobic organic matter bound-Se (HON-Se). The Hy-Se and FA-Se are the bioavailable forms for plant uptake in the environment (Ren et al. 2015). When selenite is applied into the soil, it is rapidly converted into soil solid-phase with much faster aging rate than that of selenate (Li et al. 2016). However, whether straw derived DOM affects selenite aging is still unknown. Hence, it is important to understand the dynamic transformation of DOM-Se during selenite aging to illustrate (1) the binding capacity of different DOM fractions with selenite; (2) the transformation of different DOM-bound Se species; and (3) mechanisms on the effect of straw-derived DOM on selenite availability.

2 MATERIALS AND METHODS

Lou soil (Eum-Orthic Anthrosol) samples were collected from Shaanxi Province in China. The basic physicochemical properties of the soil are as follows: soil pH was 8.14, cation exchange capacity (CEC) was 23.34 cmol/kg, electrical conductivity (EC) was 178.5 s/m, clay content was 39.50%, calcium carbonate was 55.0 g/kg, organic matter (OM) was 16.33 g/kg, amorphous iron was 1.2 g/kg, amorphous aluminum was 0.40 g/kg and total Se was 0.139 mg/kg. The soil was air-dried and sieved <2 mm. Two kilograms of dried soil were weighed into plastic bags and spiked to 1 mg Se/kg using Na_2SeO_3 in distilled water. The Se-spiked soil was mixed with maize straw (<2 mm) at the application rate of 0, 7500, or 15,000 kg/hm^2 and then kept under incubation at room temperature. The soil moisture was maintained at 70% of the water holding capacity by weighing and adding distilled water daily during the incubation period. At the different incubation times (weeks) of 0, 1, 2, 5, 9, 14, 26, and 52, after Se and straw were thoroughly mixed with soil and homogenized, a subsample of soil was taken at each designated week. The soil samples were air-dried, sieved, and then used for the DOM-Se fractionation analysis. The DOM-Se in soil solution was extracted and isolated to Hy-Se, FA-Se, and HA-Se (the HON-Se was ignored here because of the low content) in accordance with Wang et al. (2019). The functional groups of DOM were identified by ATR-FTIR. Data were analyzed by SPSS 21.0 and OMINIC 8.2.

3 RESULTS AND DISCUSSION

When straw and selenite were co-applied in soil, DOM-Se rapidly formed in soil solution. The dynamic changes of different DOM-Se fractions are shown in Figure 1. The DOM-Se was found pre-dominantly in the Hy-Se and FA-Se fractions, ac-counting for 62.7–89.8% and 18.9–22.1% of total Se in soil solution at initiation of aging. These two fractions declined with straw amounts amended during aging. In contrast, the HA-Se only accounts for 7.4–15.2% in after 52 w incubation, respectively. The hydrophobic component of DOM has stronger binding strength with soil than the hydrophilic component, whereas the hydrophilic component is easier to bond with metals since it is rich in carboxyl and hydroxyl groups (Li et al. 2011). Therefore, the Hy-Se dominated in soil solution during the entire aging period, while more HA-Se was transformed into soil solid phase.

The functional groups of DOM from soils collected at Weeks 1, 9, and 52 showed absorption bands occurring at 2994–3041 cm^{-1} (aliphatic hydroxyl group, -OH), 1682–1896 cm^{-1} (saturated and unsaturated aliphatic carboxyl groups, -C=O), 1361–1396 cm^{-1} (methyl group, -CH3) and 1059–1063 cm^{-1} (aromatic -C–O) (Table 1). The peak between 994–1006 cm^{-1} was assigned to inner sphere adsorption of Se(IV). No obvious transformation trends were found for each functional group during aging. The aromatization of straw derived DOM with 15,000 kg/hm^2 straw applied was more intensive than that with 7500 kg/hm^2 treatment when aliphatic hydroxyl groups, saturated and unsaturated aliphatic carboxyl groups were considered. These results indicated the aliphatic DOM-Se transformed into aromatic

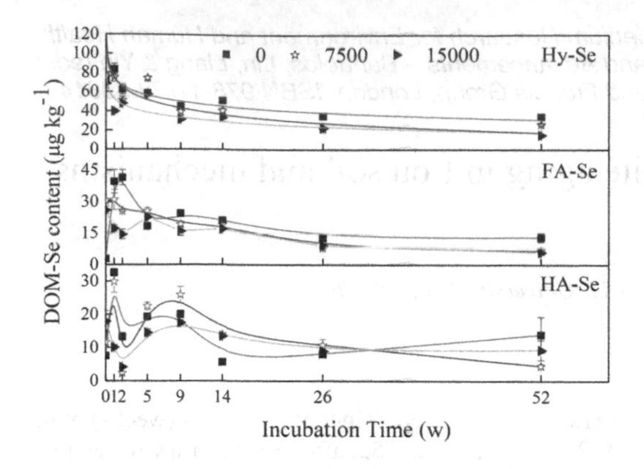

Figure 1. Concentration variations (μg/kg) of different DOM-Se fractions in soil solution from Lou soil under different straw application rates during 52 w incubation.

Table 1. ATR-FTIR results of different functional groups associated with selenite in Lou soil solution at Weeks 0, 9, and 52.

Straw Rate (kg/hm^2)	Time (W)	Bands Position (cm^{-1}) ν_{O-H}	$\nu_{C=O}$	δ_{CH3}	$\nu+\delta_{C-O}$	Se(IV)
0	0	–	–	–	–	1004
	9	–	–	–	–	1005
	52	–	–	–	–	1003
7500	0	3017	1689	1386	1060	1005
	9	3015	1703	1396	1059	997
	52	3018	1682	1387	1061	994
15,000	0	3041	1895	1364	1062	1005
	9	3032	1884	1384	1063	1006
	52	2994	1896	1361	1062	1005

DOM-Se during incubation. As more HA formed in the high straw applied soil, the availability of Se was reduced after more Se transformed into HA-Se fraction (Martin et al. 2017).

4 CONCLUSIONS

The straw derived DOM affected selenite availability in soil due to its composition and functional groups. Among the three DOM components, HA can easily incorporate with selenite and thus HA-Se became one of the important forms to reflect the availability of selenite. Additionally, the aromaticity of straw-derived DOM was enhanced with straw input and thus reduced Se availability. Although the effect of straw amendment on the aging process of selenite was studied, the mechanism for DOM derived from straw reacting with selenite needs further exploration in future studies.

REFERENCES

Li, J. Peng, Q. Liang, D.L. et al. 2016. Effects of aging on the fraction distribution and bioavailability of selenium in three different soils. *Chemosphere* 144: 2351–2359.

Li, T. Di, Z. Yang, X. & Sparks, D.L. 2011. Effects of dissolved organic matter from the rhizosphere of the hyperaccumulator *Sedum alfredii*, on sorption of zinc and cadmium by different soils. *J Hazar Mater* 192: 1616–1622.

Martin, D.P. Seiter, J.M. Lafferty, B.J. & Bednar, A.J. 2017. Exploring the ability of cations to facilitate binding between inorganic oxyanions and humic acid. *Chemosphere* 166: 192–196.

Ren, Z.L. Tella, M. Bravin, M.N. et al. 2015. Effect of dissolved organic matter composition on metal speciation in soil solutions. *Chem Geol* 398, 61–69.

Wang, D. Xue, M.Y. Wang, Y.K. et al. 2019. Effects of straw amendment on selenium aging in soils: Mechanism and influential factors. *Sci Total Environ* 657: 871–881.

Selenium Research for Environment and Human Health: Perspectives, Technologies and Advancements – Bañuelos, Lin, Liang & Yin (eds)
© 2020 Taylor and Francis Group, London, ISBN 978-1-138-39014-0

Physiological characteristics of selenate absorption into sunflower leaves

J.Y. Yang, Z.H. Fu, S.N. Liu, M.L. Chen, X. Wang, Q.Z. Sun, Z.Y. Zhao, M.Q. Gao, S.H. Xu, J.F. Wang, Z.X. Sun, H. Li, F.Y. Yu & L.H. Zhang*
Agricultural Faculty, Henan University of Science and Technology, Luoyang, Henan, China

1 INTRODUCTION

Selenium (Se) is an essential micronutrient for humans and other animals, and it plays an important role in antioxidant function, thyroid hormone metabolism, male fertility, and immune responses. Human Se is mainly acquired from plant foods in the diet. However, the average daily Se intake is insufficient in China to meet the requirements of human health because of the low Se concentrations in plant foods. Sunflower (*Helianthus annuus* L.) is an important oil crop for seed oil production. Sunflower oil contains the highest concentration of linoleic acid among the vegetable oils. People in many countries around the world love the seeds, and fried sunflower seeds are a favorite snack in China. Thus, increasing the Se concentration of sunflower seeds may be of great significance for improving Se intake. Production of Se-enriched sunflower is an efficient, inexpensive, and simple strategy to provide supplemental Se for the human diet.

Concentrations of Se in sunflower seeds are closely associated with Se levels in soils. Selenium levels are low in most parts of China, and it is difficult to reach high levels of Se in seeds when sunflower grows in low Se soils unless agronomic Se fortification is utilized. To increase Se concentration in sunflower seeds, it is necessary to apply Se to soil or plants. However, the efficiency of Se application into the soil is greatly reduced because of the negative effects exerted by soil type, redox potential, soil pH, microbial activity, and other factors, resulting in a higher cost of applying Se in soil. In comparison, foliar application of Se is postulated to be an alternative measure to increase Se concentration in sunflower seeds.

Although selenate uptake in plant roots has been investigated, the physiological characteristics of selenate absorption by sunflower leaf blades were not fully understood (Li et al. 2008, Zhang et al. 2006). In this study, we investigated the effects of Se concentration, Se application duration, light intensity, temperature, pH, respiration inhibitors, aquaporin inhibitors, and anions on selenate absorption in sunflower leaf blades to identify the path of selenate absorption into leaf blades.

2 MATERIALS AND METHODS

2.1 *Plant materials and growth conditions*

Selenate absorption was studied using sunflower. The sunflower seedlings were cultured in the full-strength Hoagland solution, which was renewed every 3 d. Three weeks later, the leaf blades of sunflower seedlings were excised and analyzed for selenate absorption (Zhang et al. 2010).

2.2 *Concentration-dependent kinetics of selenate absorption*

Leaf blades were transferred to an absorption solution containing 5.0 mM MES at different selenate levels (0, 1.0, 2.0, 3.0, 4.0, 5.0, 6.0, 7.0, 8.0, and 9.0 μM Na_2SeO_4, pH 5.0) for 3 h. After termination of selenate absorption, the leaf blades were rinsed for Se concentration analysis (Zhang et al. 2006, 2010, 2014).

2.3 *Selenate absorption experiments at different pH*

Leaf blades were excised and transferred to absorption solutions containing 100 μM $CaCl_2$, 5 mM MES and 2 μM Na_2SeO_4 of varying pH levels (3.0, 4.0, 5.0, 6.0, 7.0, and 8.0) for 3 h (Zhang et al. 2006, 2010, 2014).

2.4 *Selenate absorption experiments affected by light intensity*

Leaf blades were transferred to absorption solution containing 5 mM MES, 0.5 mM $Ca(NO_3)_2$ and 2 μM Na_2SeO_4 under conditions of dark or light (300 μmol/m^2/s), respectively, for 3 h.

2.5 *Selenate absorption experiments affected by CCCP and DNP*

Leaf blades were transferred to absorption solution containing 5 mM MES, 0.5 mM $Ca(NO_3)_2$ and 2 μM S Na_2SeO_4 with 1.0 μM CCCP or 20 μM DNP, respectively, for 3.0 h (Zhang et al. 2010, 2014).

2.6 *Selenate absorption experiments affected by $HgCl_2$ and $AgNO_3$*

Leaf blades were transferred to absorption absorption solutions containing 5 mM MES, 0.5 mM $Ca(NO_3)_2$, and 2.0μM Na_2SeO_4, with 20 μM $HgCl_2$ or 50 μM $AgNO_3$ (pH 3.0 or pH 5.0, respectively) for 3.0 h (Zhang et al. 2006, 2010).

2.7 *Competitive assay of selenate absorption with anions*

Leaf blades were transferred to absorption solution containing 5 mM MES, 0.5 mM $Ca(NO_3)_2$, and 2 μM Na_2SeO_4 with 5 mM anion, including 5 mM KNO_3, 5 mM K_2SO_4, 5 mM KH_2PO_4, and 5 mM K_2HPO_4, respectively, for 3 h.

3 RESULTS AND DISCUSSION

3.1 *Concentration-dependent selenate absorption*

Concentration-dependent kinetic experiment of selenate absorption in the leaf blades was investigated. Results showed that selenate absorption rate increased in proportion to the exogenous selenate levels. A linear regression equation and R^2 were $Y = 0.7456\,X + 0.5584$ and 0.91, respectively. The result indicated that selenate absorption was a passive process which depended upon concentration gradients as the driving forces.

3.2 *Selenate absorption at different pH values*

Selenate absorption of leaf blades varied with absorption solution pH values. Within the pH range of 3.0 to 8.0, leaf blades had the highest capacity to take up selenate at pH 3.0, followed by pH 4.0, with minimum absorption at pH 8.0. Selenate absorption decreased markedly with increasing pH from 3.0 to 5.0 and was not greatly affected by changes in pH from 6.0 to 8.0.

3.3 *Selenate absorption as affected by light intensity*

Light intensity was postulated to affect selenate diffusion through the cuticle due to its influence on stomata opening. The result revealed that selenate absorption rate was 1.21-fold higher under light than in dark, suggesting that selenate diffusion through the cuticle was partly attributed to stomata opening.

3.4 *Selenate absorption as affected by CCCP and DNP*

Effects of respiration inhibitors such as CCCP and DNP on selenate absorption were investigated in leaf blades. The results indicated that CCCP and DNP inhibited selenate absorption in leaf blades by 59% and 46% at pH 5.0, respectively, suggesting that selenate absorption was partly associated with energy metabolism at pH 5.0.

3.5 *Selenate absorption as affected by $HgCl_2$ and $AgNO_3$*

Effects of $HgCl_2$ and $AgNO_3$ on selenate absorption were investigated in leaf blades. The results revealed that $HgCl_2$ and $AgNO_3$ inhibited selenate absorption by 43% and 56% at pH 3.0, respectively, and by 23%, and 25% at pH 5.0, respectively, suggesting that selenate was partly taken up through aquaporins.

3.6 *Competitive absorption of selenate with anions*

It was postulated that absorption competition occurred between selenate and other anions. To assess this hypothesis, we investigated the effects of anions on selenate absorption in leaf blades. It was found that SO_4^{2-} inhibited selenate absorption by 60%. $H_2PO_4^-$, HPO_4^{2-}, and NO_3^- elicited relatively slight effects on selenate absorption. The results revealed that selenate shared common transporters with sulfate, suggesting selenate entered mesophyll cells via sulfate transporters.

4 CONCLUSIONS

Selenate mainly enters leaf blades by passive process, which is a rate-limiting step. H_2SeO_4 diffused across lipophilic path and stomata, while SeO_4^{2-} across aqueous pores and stomata in cuticle. H_2SeO_4 and SeO_4^{2-} entered mesophyll cells via aquaporins and sulfate transporters, respectively. Concentration gradient and selenate absorption in mesophyll cells provided a continual driving force for selenite penetration through the cuticle.

REFERENCES

Li, H.F., McGrath, S.P. & Zhao, F.J. 2008. Selenium uptake, translocation and speciation in supplied with selenate or selenite. *New Phytologist* 178: 92–102.

Zhang, L.H., Shi, W.M. & Wang, X.C. 2006. Difference in selenite absorption between high- and low-selenium rice cultivars and its mechanism. *Plant Soil* 282(1–2): 183–193.

Zhang, L.H., Yu, F.Y., Shi, W.M., Li, Y.J. & Miao, Y.F. 2010. Physiological characteristics of selenite uptake by maize roots in response to different pH levels. *J Plant Nutr Soil Sci* 173: 417–422.

Zhang, L.H., Hu, B., Li, W. et al. 2014. OsPT2, a phosphate transporter, is involved in active uptake of selenite in rice. *New Phytologist* 201: 1183–1191.

Selenium Research for Environment and Human Health: Perspectives, Technologies and Advancements – Bañuelos, Lin, Liang & Yin (eds)
© 2020 Taylor and Francis Group, London, ISBN 978-1-138-39014-0

Physiological characteristics of selenite absorption by soybean leaf blades

J.Y. Yang, Z.H. Fu, S.N. Liu, M.L. Chen, X. Wang, Q.Z. Sun, Z.Y. Zhao, M.Q. Gao, S.H. Xu, J.F. Wang, Z.X. Sun, H. Li, F.Y. Yu & L.H. Zhang*
Agricultural Faculty, Henan University of Science and Technology, Luoyang, Henan, China

1 INTRODUCTION

Selenium (Se) is an essential micronutrient for human health, and it plays an important role in selenoproteins by forming the active site as selenocysteine. Plant foods provide the main sources of Se for human beings. However, the average daily Se intake is oftentimes insufficient to meet the requirements of human health required for protection against cancer, cardiovascular diseases, and other severe infectious diseases. Soybean (*Glycine max* L.) is an important oil crop in the world, especially with a protein content as high as 35%–40%. Soybean seeds are often cultivated into bean sprouts or processed into soy milk, tofu, and beancurd sticks for people to eat. Therefore, increasing the Se content of soybean seeds is of great significance for improving daily Se intake. People can improve Se content in soybeans by improving cultivation techniques and applying Se to soil and plants. However, Se applied to the soil by mixing with chemical fertilizers is readily fixed and lost, resulting in a low utilization efficiency. In contrast, foliar spraying of Se was postulated to be more effective for increasing Se concentration in soybean seeds without experiencing negative effects within the soil profile.

Selenite absorption is well investigated in plants, but the physiological characteristics of selenite absorption by soybean leaf blades are not fully understood (Li et al. 2006, Zhang et al. 2006, 2010, 2014). In this study, we investigated the effects of Se concentration, pH, abscisic acid, polyethylene glycol, respiration inhibitors, aquaporin inhibitors, and anions on selenite absorption in soybean leaf blades to reveal the physiological characteristics of selenite absorption into leaf blades.

2 MATERIALS AND METHODS

2.1 *Plant materials and growth conditions*

Selenite absorption was studied in soybean. The soybean seedlings were initially cultured in full-strength Hoagland solution. The nutrient solutions were renewed every 3 d. Three weeks later, the leaf blades of soybean seedlings were excised and experiments were performed to examine selenite absorption (Zhang et al. 2010).

2.2 *Concentration-dependent kinetics assay of selenate absorption*

Leaf blades were transferred to an absorption solution containing 5.0 mM MES and different selenate levels (0, 1.0, 2.0, 3.0, 4.0, 5.0, 6.0, 7.0, 8.0, and 9.0 μM Na_2SeO_3, pH 5.0) for 3 h. After the termination of selenate absorption, the leaf blades were rinsed and prepared for Se analysis (Zhang et al. 2006, 2010, 2014).

2.3 *Selenite absorption experiments at different pH*

Leaf blades were transferred to absorption solutions containing 100 μM $CaCl_2$, 5 mM MES and 2 μM Na_2SeO_3 of varying pH levels (3.0, 4.0, 5.0, 6.0, 7.0, and 8.0) for 3 h. After the termination of selenite absorption, the leaf blades were rinsed, blotted, and oven-dried at 80°C for Se analysis (Zhang et al. 2006, 2010, 2014).

2.4 *Assay of selenite absorption affected by ABA and PEG*

Leaf blades were transferred to an absorption solution containing 5 mM MES, 0.5 mM $Ca(NO_3)_2$ and 2.0 μM Na_2SeO_3, with 15% PEG or 150 μM ABA (pH 3.0 or pH 5.0) for 3 h, respectively

2.5 *Selenite absorption experiments affected by CCCP and DNP*

Leaf blades were transferred to absorption solution containing 5 mM MES, 0.5 mM $Ca(NO_3)_2$, and 2 μM S Na_2SeO_3 with 1.0 μM CCCP or 20 μM DNP, respectively, for 3.0 h. After termination of selenite absorption, the leaf blades were rinsed, blotted, and oven-dried at 80°C in preparation for Se analysis (Zhang et al. 2014).

2.6 *Selenite absorption experiments affected by $HgCl_2$ and $AgNO_3$*

Leaf blades were transferred to absorption absorption solutions containing 5 mM MES, 0.5 mM $Ca(NO_3)_2$, and 2.0 μM Na_2SeO_3, with 20 μM $HgCl_2$ or 50 μM $AgNO_3$ (pH 3.0 or pH 5.0), respectively, for 3.0 h. After termination of selenite absorption, the leaf blades were rinsed, blotted, and oven-dried at 80°C in preparation for Se analysis (Zhang et al. 2006, 2010).

2.7 *Competitive assay of selenite absorption with anions*

Leaf blades were transferred to absorption solution containing 5 mM MES, 0.5 mM $Ca(NO_3)_2$, and 2 μM Na_2SeO_3 with 5 mM anion, including 5 mM KNO_3, 5 mM K_2SO_4, 5 mM KH_2PO_4, and 5 mM K_2HPO_4, respectively, for 3 h. After termination of selenite absorption, the leaf blades were rinsed, blotted, and oven-dried at 80°C in preparation for Se analysis.

3 RESULTS AND DISCUSSION

3.1 *Concentration-dependent selenite absorption*

Concentration-dependent kinetic experiment of selenite absorption was investigated in the soybean leaf blades. The results showed that selenite absorption rate increased in proportion to the exogenous selenite levels. A linear regression equation and R^2 were $Y = 1.7566\ X + 0.1763$ and 0.93, respectively, indicating that selenite absorption was a passive process which depended upon concentration gradients as the driving force.

3.2 *Selenite absorption in response to different pH values*

Selenite absorption in leaf blades varied with pH values in absorption solutions. Within the pH range of 3.0 to 8.0, leaf blades had the highest rate of selenite absorption at pH 3.0, followed by pH 4.0, with the minimum rate at pH 8.0. Selenite absorption decreased markedly with increasing pH from 3.0 to 5.0 and was not greatly affected by changes in pH from 6.0 to 8.0.

3.3 *ABA and PEG inhibited selenite absorption*

Effects of ABA and PEG on selenite absorption in wheat leaf blades were investigated to determine whether leaf blades take up selenite via stomata. The results indicated that ABA and PEG inhibited selenite absorption by 35%, and 26% at pH 3.0, respectively, and by 41%, and 33% at pH 5.0, respectively. These results revealed that the opening and closure of stomata affected selenite absorption.

3.4 *Selenite absorption as affected by CCCP and DNP*

Effects of CCCP and DNP on selenite absorption were investigated in leaf blades. The results indicated that CCCP and DNP inhibited selenite absorption in leaf blades by 63% and 54% at pH 5.0, respectively, suggesting that selenite absorption was partly associated with energy metabolism at pH 5.0.

3.5 *Selenite absorption as affected by $HgCl_2$ and $AgNO_3$*

Effects of $HgCl_2$ and $AgNO_3$ on selenite absorption were investigated in leaf blades. The results revealed that $HgCl_2$ and $AgNO_3$ inhibited selenite absorption by 56% and 66% at pH 3.0, respectively, and by 17% and 28% at pH 5.0, respectively, suggesting that selenite was partly taken up through aquaporins.

3.6 *Competitive absorption of selenite with anions*

Selenite was postulated to be competitively taken up with other anions. To assess this hypothesis, the effects of anions on selenite absorption was investigated in leaf blades. We found that $H_2PO_4^-$, HPO_4^{2-}, and NO_3^- can significantly inhibit selenite absorption. $H_2PO_4^-$ and HPO_4^{2-} elicited the strongest inhibitory effects on selenite absorption by 54% and 48%, followed by NO_3^-, which inhibited selenite absorption by 37%. SO_4^{2-} did not inhibit selenite absorption.

4 CONCLUSIONS

Selenite mainly enters leaf blades by passive process. H_2SeO_3 diffused across lipophilic path and stomata, while $HSeO_3^-$ across aqueous pores and stomata in cuticle. H_2SeO_3 and $HSeO_3^-$ entered mesophyll cells via aquaporins and Pi transporters, respectively. Concentration gradients and selenite absorption in mesophyll cells provided continual driving forces for selenite penetration through the cuticle.

REFERENCES

Li, H.F., McGrath, S.P. & Zhao, F.J. 2008. Selenium uptake, translocation and speciation in supplied with selenate or selenite. *New Phytologist* 178: 92–102.

Zhang, L.H., Shi, W.M., & Wang, X.C. 2006. Difference in selenite absorption between high- and low-selenium rice cultivars and its mechanism. *Plant Soil* 282(1–2): 183–193.

Zhang, L.H., Yu, F.Y., Shi, W.M., Li, Y.J. & Miao, Y.F. 2010. Physiological characteristics of selenite uptake by maize roots in response to different pH levels. *J Plant Nutr Soil Sci* 173: 417–422.

Zhang, L.H., Hu, B., Li, W. et al. 2014. OsPT2, a phosphate transporter, is involved in active uptake of selenite in rice. *New Phytol* 201: 1183–1191.

Biofortification

Selenium Research for Environment and Human Health: Perspectives, Technologies and Advancements – Bañuelos, Lin, Liang & Yin (eds)
© 2020 Taylor and Francis Group, London, ISBN 978-1-138-39014-0

Selenium biofortification: Accomplishments under field-growing conditions

G.S. Bañuelos
US Department of Agriculture, ARS, Parlier, California, USA

1 INTRODUCTION

Selenium (Se) content in the human diet is a topic of serious interest in public health systems around the world. Low dietary intake of Se can cause health disorders, including oxidative-related stress, epilepsy, fertility reduction, immune deficiency, and thyroid gland disorders (Rayman 2012, Zeng & Combs, 2008, Schomburg & Köhne 2008). Food products are the major source of Se for the general population and Se deficiencies can arise if dietary Se supply and intake is not adequate. The Se content of food is, however, highly dependent on the amount of bioavailable Se present in the soil and on the ability of plants to take up and accumulate Se. For this reason, it is vital to promote Se uptake by plants and to produce crops with higher Se concentrations and bioavailability in their edible tissues. In this regard, biofortification is an agronomy-based strategy that produces Se-enriched food products after growing crops in soils rich in Se or after Se has been applied to either plants, the soil, or seeds prior to planting. In the oral presentation, I will discuss real biofortification strategies that Bañuelos and other colleagues are practicing worldwide to produce Se-biofortified food products. Examples of producing biofortified food products in California will be shown below.

2 MATERIALS AND METHODS

2.1 *Field studies*

In central California, multi-year field biofortification studies were conducted on 1- to 5-ha field sites located at Five Points, Firebaugh, and Parlier, CA. The Se-rich soils were generally classified as Oxalis silty clay loam with a well-developed salinity profile. Soil salinity (EC = Electrical Conductivity) was <10 dS/m and water-extractable Se ranged from 0.12 to 0.25 mg/L. On Se-rich soils, selected crops were irrigated using Se-laden drainage water including (1) agretti (*Salsola soda*), (2) prickly-pear cactus (*Opuntia ficus-indica*), pistachio (*Pistacia vera* L.), and garlic (*Allium sativum*). On the Se-low soils (<0.01 mg Se/L), onions (*Allium cepa* L.) were treated with different forms of Se (selenite, selenate, selenomethionine) via foliar or soil application. Additionally, Se-enriched seed meals (approximately 2 mg Se/kg DW) from canola and Indian mustard were added as an organic source of Se (primarily as selenomethionine) to low-Se soils growing strawberries (*Fragaria x ananassa*).

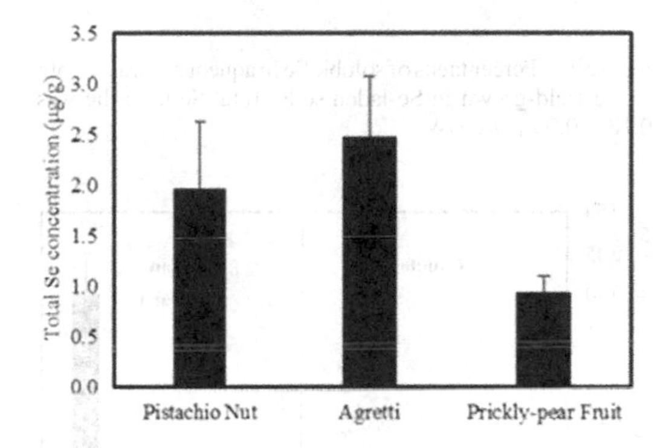

Figure 1. Total Se in different crops irrigated with Se-laden saline drainage water (75–100 µg Se/L).

2.2 *Chemical analysis*

Each year, agretti, prickly-pear fruit, strawberries, onion, and pistachio nuts were harvested. Plant samples were prepared and analyzed for Se and speciation analysis as described in detail by Bañuelos et al. (2012). Concentrations of total Se were measured by inductively-coupled plasma mass spectrometry (ICP-MS). The soluble Se forms in plant samples were extracted using aqueous proteolytic and non-proteolytic extractions and chemically speciated using high-performance liquid chromatography (HPLC) coupled to ICP-MS (Bañuelos et al. 2012).

3 RESULTS AND DISCUSSION

We have produced Se-enriched crops through biofortification strategies including irrigation with Se-laden water (Figs 1, 2), soil application of organic Se as plant material (Fig. 3), and foliar application of different forms of Se (Fig. 4). The uptake and absorption of Se readily occurred in all tested plant species. Selenium speciation analyses as soluble Se in aqueous extracts are also shown for garlic (Fig. 2) and onion (Fig. 4). In onion and garlic, Se speciation analyses showed the greatest percentages of Se-containing compounds were contained in organic Se forms.

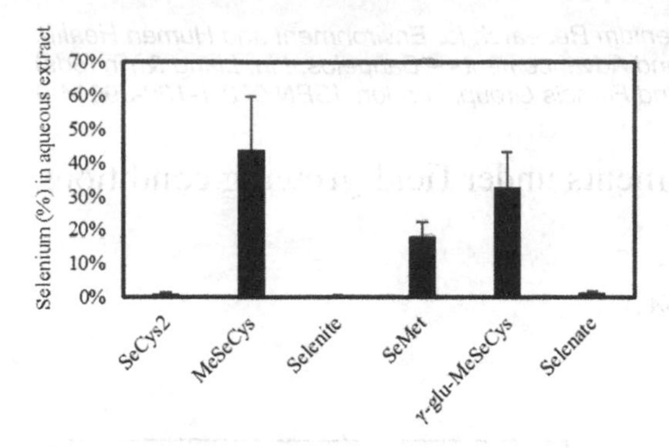

Figure 2. Percentages of soluble Se in aqueous extracts from garlic field-grown in Se-laden soils. Total Se in garlic was 0.18 ± 0.06 μg/g DW.

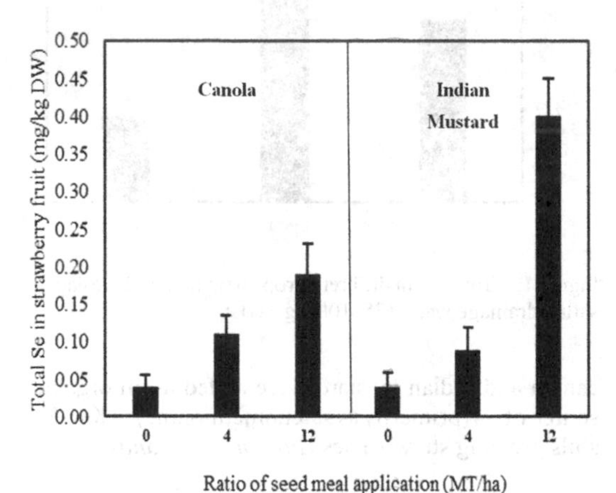

Figure 3. Total Se concentrations in strawberries grown in soils amended with Se-rich canola and Indian mustard seed meals. Mean total Se was 2.0 ± 0.2 mg/kg DM in seed meals.

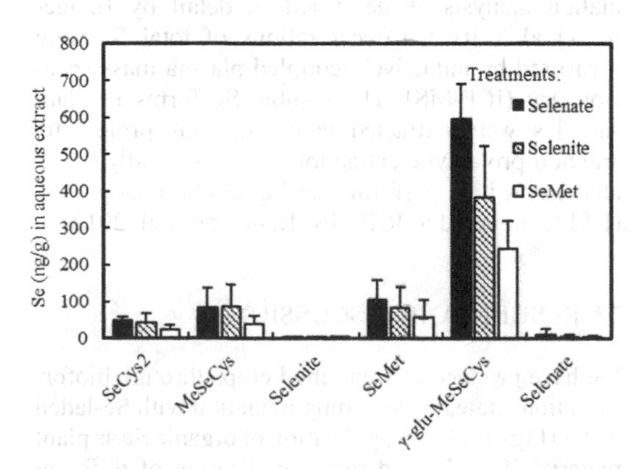

Figure 4. Soluble Se in aqueous extracts from field-grown onions treated with foliar (100g Se/ha) selenate, selenite, and SeMet, respectively. Total Se concentration (μg/g DW) in onions for each respective application of Se: 1.08 ± 0.31 selenate, 0.87 ± 0.23 selenite, and 0.48 ± 0.15 SeMet.

4 CONCLUSIONS

Natural biofortification is an environmentally friendly strategy for producing Se-enriched food products naturally in areas with high soil Se. Additionally, foliar and soil application of Se to regions low in Se is also effective for Se biofortification. Irrespective of biofortification strategy, consumption of Se-enriched vegetables containing different selenoamino acids can be a nutritionally sound strategy for effectively increasing Se dietary intake in Se-deficient areas throughout the world.

REFERENCES

Bañuelos, G.S., Walse, S.S., Yang, S.I. et al. 2012. Quantification, localization, and speciation of selenium in seeds of canola and two mustard species compared to seed meals produced by hydraulic press. *Anal Chem* 84(14): 6024-6030.

Rayman, M.P. 2012. Selenium and human health. *Lancet* 379(9822): 1256-1268.

Schomburg, L. & Köhrle, J. 2008. On the importance of selenium and iodine metabolism for thyroid hormone biosynthesis and human health. *Mol Nutr Food Res* 52(11): 1235-1246.

Zeng, H. & Combs Jr., G.F. 2008. Selenium as an anticancer nutrient: Roles in cell proliferation and tumor cell invasion. *J Nutr Biochem* 19(1): 1-7.

Selenium Research for Environment and Human Health: Perspectives, Technologies and Advancements – Bañuelos, Lin, Liang & Yin (eds)
© 2020 Taylor and Francis Group, London, ISBN 978-1-138-39014-0

Distribution characteristics of selenium in different soybean products during processing

X.F. Deng, Z.Q. Zhao, J.J. Zhou & X.W. Liu
Microelement Research Center, Huazhong Agricultural University, Wuhan, China
Hubei Provincial Engineering Laboratory for New-Type Fertilizer, Wuhan, China

1 INTRODUCTION

Selenium (Se) is an essential micronutrient for humans and animals. Soybean (*Glycine max* L.) has played an integral part in Asian foods for many centuries, and more recently, it has been integrated into Western diets (Lu et al. 2018). Compared to inorganic Se, organic Se is safer and more effective for absorption by the human body. Typically, more than 80% of the Se in crop kernels are organic (Eiche et al. 2015), and organic Se mainly exists in protein (Wang et al. 2013, Deng et al. 2017). Soybean protein content is relatively high, which accounts for roughly 40% of the dry weight of seeds (Hu et al. 2014). Soybean is widely planted and used for processing in China. They are harvested and processed into a variety of soybean products for human consumption. After a series of treatments and processing, Se concentration is bound to change. How Se is distributed in various Se-enriched soybean products is the primary concern in the development and utilization of Se-enriched soybean.

2 MATERIALS AND METHODS

2.1 *Experimental materials*

Soybean was obtained from field trials in Enshi, Hubei Province, China. The Se concentration of soybean is 11.47 mg/kg.

2.2 *Experimental design*

Selenium-enriched soybeans and Se-lowed soybeans were selected and soaked for 6 h according to the water-soybean ratio of 4:1, respectively. Soaking soybean seeds were germinated in incubator at 25°C. Sampling starts at Day 3 and stops at Day 6. Each treatment is repeated three times. The samples were divided into soybean sprouts and seed coats, and the Se concentration was determined.

The Se-rich soybean was soaked at room temperature for 10 h according to the water-soybean ratio of 4:1. According to the ratio of water-soybeans (dry soybeans) to 6:1, 8:1 and 10:1, beating with water for 4 minutes, respectively. Each treatment is repeated three times. With 80 mesh filter cloth filter, separation of soya-bean milk and bean dregs, soya-bean milk under 4°C saved for later use.

Tofu was produced with $CaSO_4$ (gypsum), $MgCl_2$ (bittern) or $C_6H_{10}O_6$ (Gluconolactone) as coagulants. Each treatment was repeated three times. Three tofu samples of the same quality were dried at 60°C and crushed to determine Se content.

2.3 *Determination of Se concentration*

The total Se concentration of samples were determined by adding HNO_3-$HClO_4$ (4:1, v/v) for digestion. The Se concentration in the filtrate was measured by hydride generation-atomic fluorescence spectrometry (HG-AFS-8220).

3 RESULTS AND DISCUSSION

3.1 *Selenium distribution characteristics of Se-enriched soybean sprouts during germination*

Selenium had no significant effect on the germination rate of soybean and sprout length. The Se concentration in soybean sprouts decreased significantly with the prolongation of germination time, while it remained stable without significant change in seed coat (Fig. 1). During the whole germination process, Se mainly accumulated in soybean sprouts, accounting for more than 95%. The recovery efficiency of total Se was over 90%, and that of Se in soybean sprouts was over 85%. Overall, the effective utilization rate of Se in soybean sprout production was very high.

3.2 *Effect of water-soybean ratio on Se distribution in Se-enriched soya-bean milk*

The yield of soya-bean milk increased from 48% to 64% with the increase of water-soybean ratio. The Se concentration of soya-bean milk and bean dregs decreased significantly with the increase of water-soybean ratio. The Se distribution ratio in soya-bean milk increased rapidly with the increase of the water-soybean ratio in the range of 6:1 – 10:1; however, the

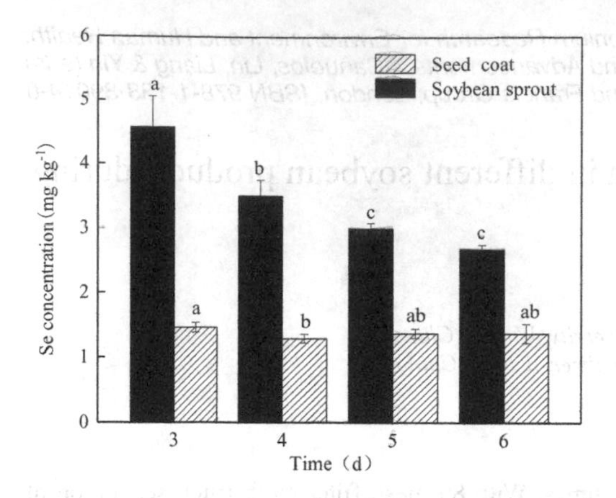

Figure 1. Effects of germination time on Se concentration in soybean sprouts and seed coats.

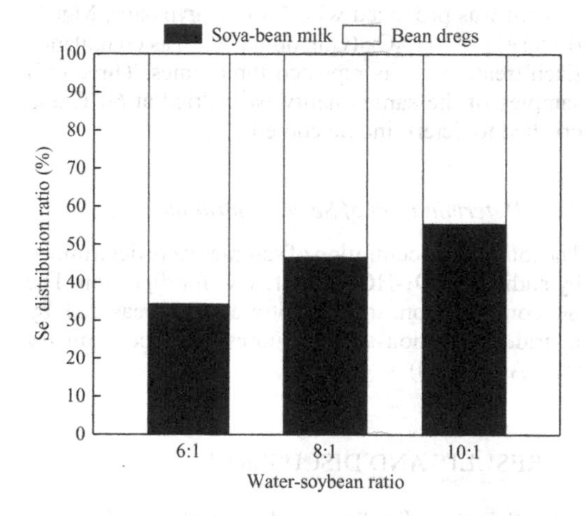

Figure 2. Effects of water-soybean ratio on Se distribution ratio in soybean milk and bean dregs.

opposite trend was observed in bean dregs (Fig. 2). From the viewpoint of Se retention of soya-bean milk, the optimum ratio of water-soybean is 10:1. Given that 44% to 65% of Se is distributed in bean dregs, the reuse of bean dregs is worth considering. The recovery of total Se was between 80% and 85%, while the recovery of Se in soya-bean milk increased from 29% to 44% with the increase of water-soybean ratio.

3.3 *Effect of coagulant on Se distribution in Se-enriched tofu*

Regardless of coagulants, the Se concentration in all parts followed the order of tofu > bean dregs > yellow serofluid (Fig. 3). Although the Se concentration of gluconolactone tofu is lower than that of gypsum and bittern tofu, the distribution ratio of Se in gluconolactone tofu is the highest at 57% in the process of tofu production. In general, gluconolactone tofu retains Se better.

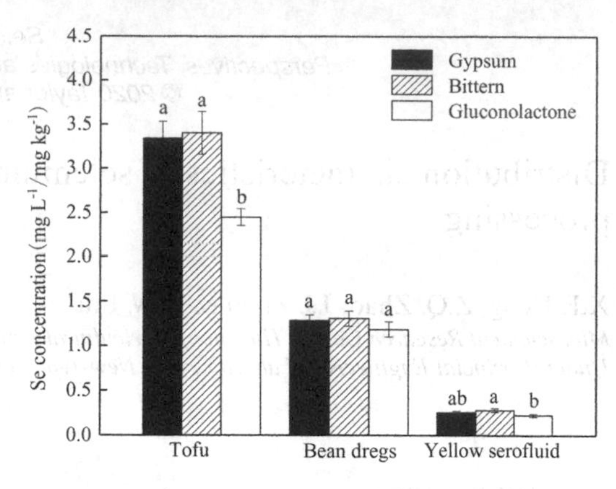

Figure 3. Effects of coagulants on Se concentration in tofu, bean dregs, and yellow serofluid.

4 CONCLUSIONS

For three Se-rich soybean products, the recovery efficiency of Se from soybean sprouts was the highest, above 85%, followed by tofu (43–53%), and soya-bean milk (29–45%). Therefore, Se-rich soybean can be processed into soybean sprouts for consumption, and Se supplementation has the best effect.

REFERENCES

Deng, X.F., Liu, K.Z., Li, M.F. et al. 2017. Difference of selenium uptake and distribution in the plant and selenium form in the grains of rice with foliar spray of selenite or selenate at different stages. *Field Crops Res* 211: 165–171.

Eiche, E., Bardelli, F., Nothstein, A.K. et al. 2015. Selenium distribution and speciation in plant parts of wheat (*Triticum aestivum*) and Indian mustard (*Brassica juncea*) from a seleniferous area of Punjab, India. *Sci Total Environ* 505: 952–961.

Hu, J.W., Zhao, Q., Cheng, X. et al. 2014. Antioxidant activities of Se-SPI produced from soybean as accumulation and biotransformation reactor of natural selenium. *Food Chem* 146: 531–537.

Lu, Y.L., Li, W.F. & Yang, X.B. 2018. Soybean soluble polysaccharide enhances absorption of soybean genistein in mice. *Food Res Intern* 103: 273–279.

Wang, Y.D., Wang, X. & Wong, Y.S. 2013. Generation of selenium-enriched rice with enhanced grain yield, selenium content and bioavailability through fertilisation with selenite. *Food Chem* 141: 2385–2393.

Selenium Research for Environment and Human Health: Perspectives, Technologies and Advancements – Bañuelos, Lin, Liang & Yin (eds)
© 2020 Taylor and Francis Group, London, ISBN 978-1-138-39014-0

Effects of soil-applied selenium on selenium content of Gala apple fruits

L. Liu, C.Q. Niu, J.T. Liu & L. Quan
Pomology Institute, Shanxi Academy of Agricultural Sciences, Shanxi Key Laboratory of Germplasm Improvement and Utilization in Pomology, Taiyuan, Shanxi, China

Z.M. Wang, L.X. Yuan & Z.K. Liu
Jiangsu Bio-Engineering Research Center for Selenium, Suzhou, Jiangsu, China

1 INTRODUCTION

Selenium (Se) is one of the essential trace elements for human and animals. It plays an important role in enhancing immune function and reducing carcinogenic substances (Nie et al. 2015). Developing selenium-rich agricultural products is of great significance to improve people's health. Presently, Se research on the application of exogenous Se is mostly concentrated on foliar application (Hu et al. 1998, Ning et al. 2013, He et al. 2015, Li et al. 2016). There are few reports on the application of selenium-rich organic fertilizer to the soil. In this study, different amounts of Se-enriched organic fertilizer were applied to soil at different growth stages of Gala apple fruit. Variation of Se content of selenium-enriched apple is discussed.

2 MATERIALS AND METHODS

The experiment was carried out in Taigu County Scientific Research Pilot Base of Pomology Institute, Shanxi Academy of Agricultural Sciences. Using 10-year-old *Malus* Gala/*Malus Octagonifolia* as experimental plant, three application stages for Se-fertilization treatments were selected: (A) flowering stage, (B) fruit expansion stage, and (C) both flowering and fruit expansion stages. For each application stage, there were four Se treatment levels: 0, 150, 225, and 300 g/hm^2. There were 10 groups, and only the 1st row of each group was treated. There were 20 trees in each row, and Se fertilizer was applied around the root area of each tree. Each treatment group was repeated three times. The field cultivation and management were carried out according to the field routine operation.

The Se-rich organic fertilizer (SETEK-BF-1709) was provided by Suzhou SETEK Co., Ltd., China, with Se content of 1000 mg/kg. Apple fruit samples were collected when harvested. All samples were washed with deionized water and then wiped dry. Fresh apple samples were ground into pulp and were acid-digested in mixed acids (HNO_3-$HClO_4$ at 4:1, v/v) and hydrochloric acid (Yuan et al. 2013). Total Se concentration was determined by HG-AFS (Beijing Titan Instrument, China), according to GB 5009.93-2010 hydride generation atomic fluorescence spectrometry. SPSS 19.0 was used for variance analysis of the data.

3 RESULTS AND DISCUSSION

The Se content in apples from the control group was 0.001 mg/kg in Taigu County, Jinzhong City, Shanxi Province. The Se accumulation in Gala apple fruit increased significantly by applying 150–300 g/hm^2 of Se in soil under all three Se-fertilization application stages (A, B, and C) (Table 1). There was a significant positive correlation between the amount of Se added to the soil and fruit Se content ($p < 0.01$). When Se was applied at 300 g/hm^2 at fruit expansion stage (B) and at both flowering and fruit expansion stages (C), the fruit Se content reached 0.020–0.021 mg/kg, which was 20 times higher than that of CK (Table 1). However, this finding is different from those reported via foliar application of Se fertilizer (Ning et al. 2013, Li et al. 2016, Hao et al. 2016). When the foliar Se treatment level increased, fruit Se content decreased. Li et al. (2003) indicated that Se was mostly bound to plant proteins in the form of seleno-amino acids and Se replaces sulfur (S) in sulfur-amino acids. Because apple tree is not a Se accumulator species, the S transporter could be saturated for Se uptake, and Se accumulation in apple fruit could thus be limited. In this study (compared with the control), increasing the rate of Se fertilizer application from 150 to 300 g/hm^2 increased the fruit Se content steadily at the rate of 0.003 mg/kg at flowering stage (A). However, with the application during the expansion stage (B) or during the flowering and fruit expansion stages (C), fruit Se concentration increased much more, compared to the flowering stage (A). The highest fruit Se concentration was 0.021 mg/kg which was observed with the 300 mg/kg Se fertilizer treatment at flowering and expansion application stages.

With an increase of Se application rate, Se accumulation in fruits did not show a linear increase but rather a gradual slowdown due to complex mechanistic processes of Se absorption, translocation, and distribution

Table 1. Effect of Se application on Se concentrations in Gala apple fruit.*

Se Fertilizer Treatment (g/hm^2)	Fruit Se Concentrations at Different Application Stages (mg/kg)		
	Flowering stage (A)	Expansion stage (B)	Flowering and expansion stages (C)
CK	0.001 ± 0.0004a	0.001 ± 0.0004a	0.001 ± 0.0004a
150	0.003 ± 0.0001b	0.003 ± 0.0001b	0.006 ± 0.0003b
225	0.006 ± 0.0002c	0.012 ± 0.0005c	0.014 ± 0.0006c
300	0.009 ± 0.0003d	0.020 ± 0.0006d	0.021 ± 0.0010d

* Values followed by different letters in a row are significantly ($p < 0.05$) different among different Se fertilization levels during the same growth stage.

in plant. Whether further increasing Se application in soil can significantly increase the Se content in fruits needs to be investigated.

Moreover, the fruit Se concentration with Se fertilizer applied at both flowering and expansion stages (C) was higher when application occurred at flowering stage (A) or at expansion stage (B), but it was much lower than the sum of fruit Se content fertilized with Se at flowering stage (A) and at fruit expansion stage (B) (Table 1). When Se fertilizer is applied to the soil, there are many factors affecting the bioavailability of Se for plant uptake. Chemical transformation of Se by trees or soil microbes can also play important roles in fruit Se accumulation (Wang et al. 2018). Thus, dividing Se fertilizer applications into multiple applications at different growth stages might be more useful for enhancing the accumulation of Se in apple fruits. In contrast, the effectiveness of Se was reduced when resulting in the reduction of Se content in fruit. Therefore, it is most effective for Se biofortification of apple to apply Se fertilizer to the soil at the fruit expansion stage.

4 CONCLUSIONS

The Se content in Gala apple fruit was significantly increased by applying Se fertilizer at rates of 150–300 g/hm^2 to soil at both flowering, expansion, or combined stages. Compared with control, the fruit Se content was significantly increased by 3–20 times. The Se application rate of 300 g/hm^2 at fruit expansion stage significantly increased the fruit Se concentration compared with that at the flowering stage.

REFERENCES

Hao, H.H., Zheng, T.T., Wang M. et al. 2016. Study on production technology of green Se-rich apple. *Modern Agr Sci Tech* (15): 66–71.

He, M.L., Gao, S.L., Liu X.Y. et al. 2015. Efficacy and production regulation for pollution-free Se-enriched apple. *Northern Fruits* 2: 25–27. (in Chinese)

Hu, S.B., Feng, G.Y., Zhao, X.N., Li, J.R. & Xue, C.Z. 1998. Researches on the Se absorption and accumulation characteristics of apple. *Acta Botanica Boreali – Occidentalia Sinica* 18(1): 110–115. (in Chinese)

Li, Q., Zheng, B., Liu, L. et al. 2016. Effects of spraying organic Se fertilizers on Se content and quality of apple fruits. *J Shanxi Agr Sci* 44(9): 1316–1319. (in Chinese)

Li, Y.S., Li, Y.N. & Chen, D.Q. 2003. Biological functions of selenium and the mechanism of selenium enrichment in plant. *J Hubei Agr College* 23(6): 476–480. (in Chinese)

Nie, J.Y., Kuang, L.X., Li, Z.X. et al. 2015. Se content of main deciduous fruits from China and its dietary exposure assessment. *Scientia Agricultura Sinica* 48(15): 3015–3026. (in Chinese)

Ning, C.J., Ding, N., Wu, G.L. et al. 2013. Effects of different Se spraying scheme on the fruit quality, total Se and organic Se contents in "Red Fuji" apple trees. *J Plant Nutr Fertilizer* 19(5): 1109–1117.

Wang, Z.M., Yuan, L.X., Zhu, Y.Y. et al. 2018. Study on standards of Se enriched agricultural products and Se-rich soil in China. *Soils* 50(6): 1080–1086.

Yuan, L., Zhu, Y., Lin, Z.Q. et al. 2013. A novel selenocystine-accumulating plant in Se-mine drainage area in Enshi, China. *PLoS ONE* 8(6): e65615.

Selenium Research for Environment and Human Health: Perspectives, Technologies and Advancements – Bañuelos, Lin, Liang & Yin (eds)
© 2020 Taylor and Francis Group, London, ISBN 978-1-138-39014-0

Reciprocal effects of soil selenium and nitrogen treatments on concentrations of Se and N in wheat

L. Yang, X.Y. Mao, F.Y. Sun, L. Li & T. Li*
Jiangsu Key Laboratory of Crop Genetics and Physiology; Key Laboratory of Plant Functional Genomics of the Ministry of Education; Jiangsu Key Laboratory of Crop Genomics and Molecular Breeding; Jiangsu Co-Innovation Center for Modern Production Technology of Grain Crops, Yangzhou University, Yangzhou, China

1 INTRODUCTION

Selenium (Se) is essential for human beings and animals and can be beneficial to plants. Wheat is a staple crop with the capacity to accumulate Se (Li et al. 2017). Nitrogen (N) is essential for wheat growth and development and is closely related to wheat yield and quality (Yuan et al. 2006, Feng et al. 2013). The objectives of this study were to understand the reciprocal effects of Se and N fertilization on the contents of these two elements and to propose strategy for Se biofortification under differential N nutrition.

2 MATERIALS AND METHODS

Two wheat varieties (7001 and Ning 7840) with spontaneous hypersensitive traits (SHR) are N dosage-dependent. These two wheat varieties were treated at jointing stage with nine different N and Se combinations, including three N levels (0, 3, and 6 g of carbamide per pot) and three Se levels (0, 15, and 30 mg of Na_2SeO_4 per pot) in 2016; and again with 16 different N and Se combinations, including four N levels (0, 1, 2, and 3 g of carbamide per pot) and four Se levels (0, 8, 16, and 24 mg of Na_2SeO_4 per pot) in 2017. The stems, leaves and grains were harvested separately at mature stage, and N and Se contents in different organs were assayed. Selenium concentrations were determined by inductively coupled plasma optical emission spectrometry (ICP-OES) (Thermo Scientific), and N concentration was determined by the Kjeldahl method.

3 RESULTS AND DISCUSSION

Nitrogen contents in different wheat organs increased with the increase of N application rate under the same Se treatment (Fig. 1). Similarly, wheat Se content increased with an increase of Se application rate in both years when exposed to the same N treatment. Interactions between Se and N applications on their respective contents were observed. With an increase of N application, the average Se content in organs initially increased and then decreased (Fig. 2), suggesting that Se accumulation was N-dependent. The moderate

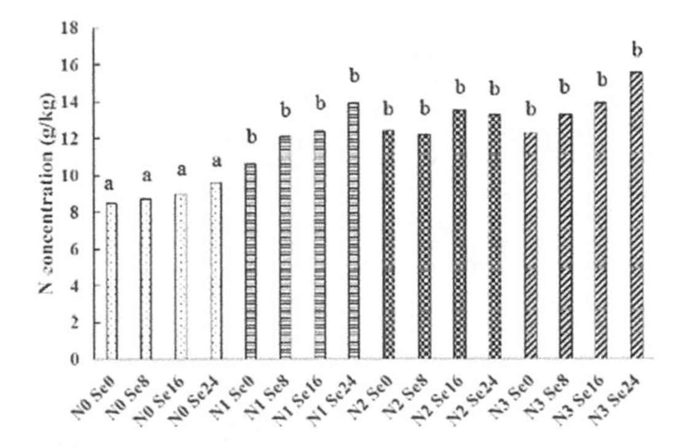

Figure 1. Concentrations of nitrogen in wheat organs. The same letter indicates that there is no significant difference between the treatments.

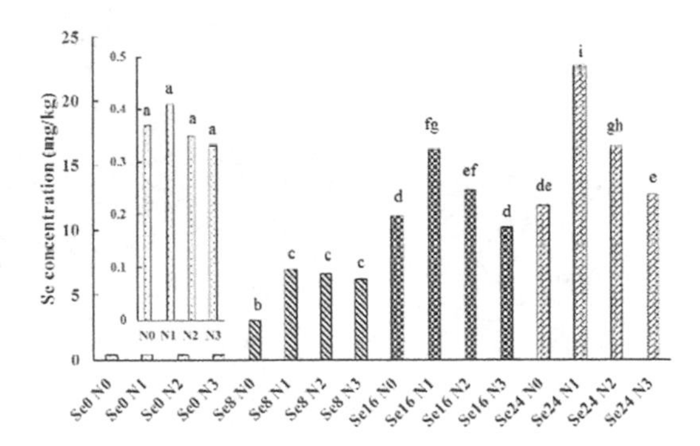

Figure 2. Concentrations of selenium in wheat organs. The same letter indicates that there is no significant difference between the treatments.

N application promoted the uptake of Se, whereas high N dosage application inhibited Se accumulation. However, N concentration increased with an increase of Se application, suggesting the Se treatment favors high N use efficiency.

4 CONCLUSIONS

The findings of this study can be useful for Se biofortification strategies under different soil conditions.

The data allow scientists, farmers, and food producers to balance Se biofortification and N application to meet the increasing requirements of both Se-enriched functional food and food security.

REFERENCES

Li, T. & Lan, G.F. 2012. Metabolism of selenium in plants and the strategies for selenium-biofortification using wheat. *Journal of Triticeae Crops* 32(1): 173–177. (in Chinese)

Yuan, J.C., Liu, C.J., E, S.Z. et al. 2006. Effect of nitrogen application rate and fertilizer ratio on nutrition quality and trace-elements contents of rice grain. *Plant Nutrition & Fertilizer Science* 12(2): 183–182. (in Chinese)

Feng, J.F., Zhao, G.C., Zhang, B.J. et al. 2013. Effect of topdressing nitrogen ratio on yield, protein components and physiological characteristics of winter wheat. *Journal of Plant Nutrition & Fertilizer* 19(4): 824–831. (in Chinese)

Tian, X.Y., Li, H.H. & Wang, Z.Y. 2009. Effect of Se application on contents of nitrogen, phosphorus and potassium Tartary buckwheat and content of available nutrients in soil. *J Soil Water Conserv* 23(3): 112–115.

Selenium Research for Environment and Human Health: Perspectives, Technologies and Advancements – Bañuelos, Lin, Liang & Yin (eds)
© 2020 Taylor and Francis Group, London, ISBN 978-1-138-39014-0

Selenium-enriched earthworm: A potential source of selenium supplement for animals

S.Z. Yue & Y.H. Qiao*
College of Resources and Environmental Sciences, China Agricultural University, Beijing, China

1 INTRODUCTION

Earthworm is rich in protein, accounting for about 54.6-71% of its dry weight (Sun et al. 2019). Therefore, it is often used to supplement animal or human nutrition. Earthworm has good bioaccumulation and conversion ability to selenium (Se) (Liu et al. 2001, Sun et al. 2014), which is also a good source of Se-enriched feed material for animals.

2 MATERIALS AND METHODS

Control and Se-treated earthworm *Eisenia fetida* samples were obtained in our previous experiment (Yue et al. 2019). The levels of Se treatment were: 0 (Se0), 20 (Se20), and 40 (Se40) mg Se/kg substrate dry weight. Soxhlet extraction was used to determine crude fat content, and crude polysaccharide was analyzed using the phenol-sulfuric acid method. Crude protein content was measured by the Kjeldahl method, and crude ash was quantified by burning the samples in a muffle furnace at 550°C.

Concentrations of Se in crude fat, crude polysaccharide, and crude ash were determined by ICP-MS, while crude protein Se content was calculated by the total Se concentration subtracted by the Se concentrations of crude fat, polysaccharide, and crude ash.

3 RESULTS AND DISCUSSION

Total Se concentrations in *E. fetida* body in the 0, 20, and 40 mg/kg Se treatments were 3.21, 97.78, and 151.56 mg/kg dry weight, respectively (Yue et al. 2019). The content of earthworm crude protein increased with increasing the Se concentration in substrate (Fig. 1a). The crude protein content (72.83%) was significantly higher with 40 mg/kg treatment than the crude protein content in control (64.26%). The Se addition helped to increase the content of crude polysaccharide (Fig. 1c), but decreased the contents of crude fat ($p < 0.05$, Fig. 1b) as well as crude ash (Fig. 1d), compared with control. Selenium concentrations in crude fat, crude polysaccharide, crude ash, and crude protein were all increased by Se addition in substrates (Fig. 1). Selenium was mostly distributed in crude protein with a maximum concentration of 138.56 mg/kg observed in the Se40 treatment.

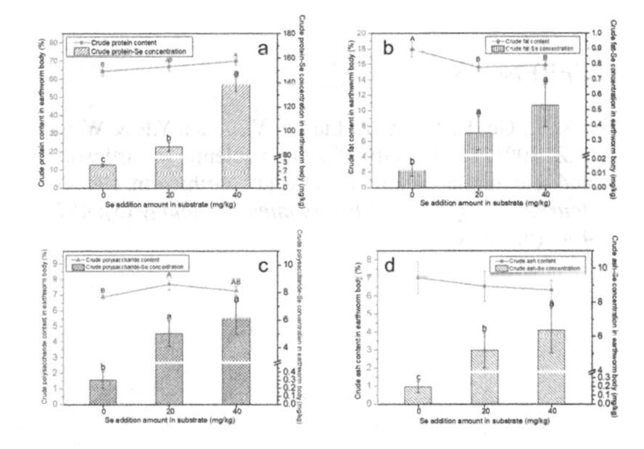

Figure 1. Effect of Se addition in substrate on the contents of earthworm crude protein (a), crude fat (b), crude polysaccharide (c), and crude ash (d), along with Se concentrations in those components.

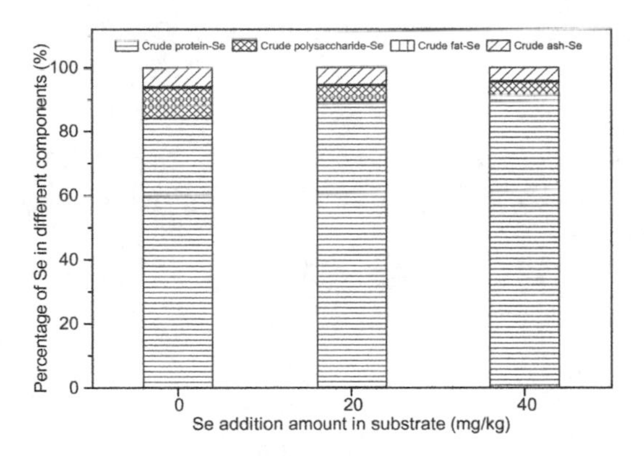

Figure 2. Selenium percentage in crude protein, crude fat, crude polysaccharide, and crude ash of earthworm body.

The distribution of Se among different components was shown as the percentages of the total Se in earthworm, and followed a descending order of crude protein (84.11–91.42%) > crude polysaccharide (4.04–9.45%) > crude ash (4.19–6.05%) > crude fat (0.35–0.39%) (Fig. 2).

4 CONCLUSIONS

Eisenia fetida has a great potential for Se bioaccumulation (Yue et al. 2019). The addition of Se in the substrate enhanced the synthesis of crude protein and crude polysaccharide in earthworm body, as well as inhibited the content of crude fat and crude ash. Consequently, quality of earthworms was improved to a certain extent. Crude protein in Se-enriched earthworm was 68.13–72.83%. Most of the Se is distributed in crude protein, and the organic Se was about 90% (Yue et al. 2019). Therefore, Se-enriched earthworm is a potential protein and Se supplement for animals or humans.

REFERENCES

Liu, X.H., Ge, F., Xu, Z.H., Liao, S.Y., Zhao, Y.F. & Wang, X.Z. 2001. The toxicity of sodium selenite to earthworm and selenium-accumulating effect of earthworm. *Chinese Journal of Applied and Environmental Biology* (5): 457–460. (in Chinese)

Sun, X.F., Qiao, Y.H., Sun, Z.J., Wang, C., Li, H.F. & Yue, S.Z. 2014. The cultivation and selenium enrichment of selenium enriched earthworm. *Journal of Agricultural Resources and Environment* 31(6): 570–574. (in Chinese)

Sun, Z.J., Liu, X.C., Sun, L.H. & Song, C.Y. 1997. Earthworm as a potential protein resource. *Ecol Food Nutr* 36(2–4): 221–236.

Yue, S.Z., Zhang, H.Q., Zhen, H.Y., Lin, Z.Q. & Qiao, Y.H. 2019. Selenium accumulation, speciation and bioaccessibility in selenium-enriched earthworm (*Eisenia fetida*). *Microchem J* 145: 1–8.

Selenium Research for Environment and Human Health: Perspectives, Technologies and Advancements – Bañuelos, Lin, Liang & Yin (eds)
© 2020 Taylor and Francis Group, London, ISBN 978-1-138-39014-0

Selenium biofortification and antioxidant activity in the medicinal mushroom *Cordyceps militaris*

T. Hu & Y.B. Guo*
College of Resources and Environmental Sciences, China Agricultural University, Beijing, China

1 INTRODUCTION

Selenium (Se) is a naturally-occurring non-metallic element that serves as an essential micronutrient for humans and animals. Insufficient intake of Se by mammals can contribute to the onset of "Keshan disease" and "white muscle disease", whereas excessive amounts can lead to toxicoses (Fordyce 2013). *Cordyceps militaris* is a highly nutritious ascomycetous fungus which is one of the most popular edible and medicinal mushrooms worldwide. The benefits of consuming *C. militaris* are due to its pharmacological functions affecting the circulatory system, as well as its antioxidant, anti-tumor, anti-metastatic, immunomodulatory, anti-inflammatory, and anti-bacterial activities (Das 2010). This study describes the development of Se-biofortified *C. militaris* fruiting bodies that combine the functions of the essential trace element and the medicinal properties of the mushroom in a potential dietary supplement. The presented results provide novel insights relevant to the cultivation of Se-biofortified *C. militaris* fruiting bodies.

2 MATERIALS AND METHODS

Sodium selenate [Se(VI)], sodium selenite [Se(IV)], and selenomethionine (SeMet) were respectively used in the experiment. The strain of *C. militaris* was cultured on potato dextrose ager. The mycelium was cultured on plates or by shake flask culture. The fruiting bodies were cultivated on rice medium. Selenium was supplied at the following concentrations: 5, 10, 20, and 40 μg per gram of medium.

The total Se and Se species were determined by hydride generation atomic fluorescence spectrometry (HG-AFS) and HPLC-HG-AFS, respectively.

The cordycepin, adenosine, and amino acid components were determined by HPLC and amino acid analyzer, respectively.

The antioxidant activity of *C. militaris* against 2,2-diphenyl-1-picryhydrazyl (DPPH) radicals by the soluble compounds were extracted from fruiting bodies of *C. militaris* with water and ethanol, respectively.

The data for the different treatments underwent analysis of variance ($p < 0.05$) with Duncan's multiple range test and t-test using SAS software. All data are provided as the mean ± standard deviation.

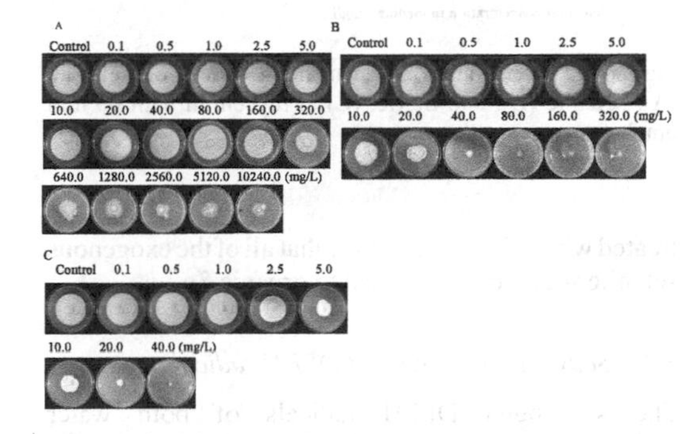

Figure 1. Morphology of mycelium under selenate (A), selenite (B), SeMet (C) treatments at different Se concentrations.

3 RESULTS AND DISCUSSION

3.1 *Mycelia growth and Se accumulation in mycelium in response to Se concentration*

The colony diameters of mycelium were inhibited with increasing Se concentration in the medium (Fig. 1). The mycelium of *C. militaris* was more sensitive to SeMet than selenite and selenate.

The accumulation of SeMet in mycelium was significantly higher than selenite and selenate treatments (Fig. 2).

3.2 *Selenium accumulation and transformation in fruiting bodies*

The total Se contents of fruiting bodies gradually increased with increasing Se concentrations in the medium. The Se concentrations in fruiting body were 130.9 μg/g and 128.1 μg/g in response to treatments with 40 μg/g selenate and selenite, respectively. These values were over 900-fold higher than the control values. In this study, selenate was mostly biotransformed to $SeCys_2$ and SeMet, as well as some Se(VI), but not Se(IV). Meanwhile, selenocystein ($SeCys_2$) and SeMet were identified only in fruiting bodies

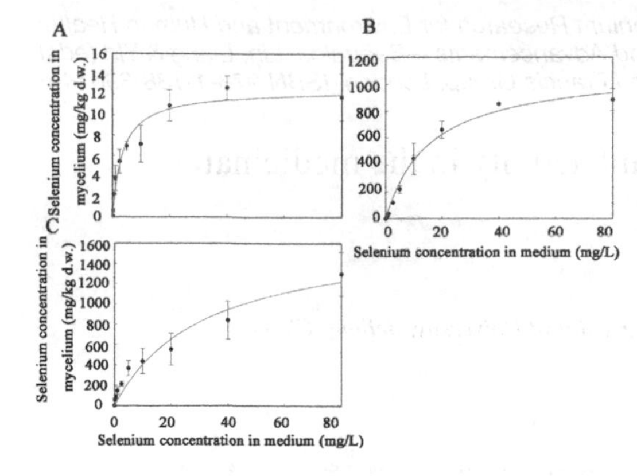

Figure 2. Concentration-dependent kinetics for selenate (A), selenite (B), and SeMet (C) in mycelium with 60 min uptake.

treated with selenite, implying that all of the exogenous selenite was biotransformed to organic Se.

3.3 *Scavenging activity of DPPH radicals*

The scavenged DPPH radicals of both water and ethanol extracts (12.5–100 mg/mL) from Se-biofortified *C. militaris* were significantly higher ($p < 0.05$) than that corresponding concentration from control *C. militaris*. The water crude extracts of *C. militaris* tested showed much stronger scavenging activities against DPPH compared to ethanol crude extracts. The water crude extract of *C. militaris* contained more active substances, including polysaccharid, adenosine, cordycepin, and cordycepic acid, than ethanol extract. *C. militaris* are usually used to make porridge, soup, and stew, and are soaked in wine. Based on these results, the cook method of Se-biofortified *C. militaris* with water has higher biological activity than ethanol.

4 CONCLUSIONS

We herein describe a potential method for Se-biofortified *C. militaris* fruiting bodies with potent antioxidant activity. The organic Se and cordycepin contents were especially increased in fruiting bodies treated with Se. The predominant Se species in Se-biofortified fruiting bodies were $SeCys_2$ and SeMet. Therefore, Se-biofortified *C. militaris* fruiting bodies could be used as Se supplements and potential natural antioxidants.

REFERENCES

Das, S.K. 2010. Medicinal uses of the mushroom *Cordyceps militaris*: Current state and prospects. *Fitoterapia* 81(8): 961–968.

Fordyce, F.M. 2013. Selenium deficiency and toxicity in the environment. In O. Selinus, J. Centeno, R. Finkelman et al. (eds.), *Essentials of Medical Geology*: 375–415. Dordrecht: Springer.

Selenium Research for Environment and Human Health: Perspectives, Technologies and Advancements – Bañuelos, Lin, Liang & Yin (eds)
© 2020 Taylor and Francis Group, London, ISBN 978-1-138-39014-0

Agronomic Se biofortification increases Se dietary intake of residents in Keshan disease area on Loess Plateau

D. Huang*, D. Yin, R. Yu, H. Liu, J. Wang & Z. Wang
College of Natural Resources and Environment, Northwest A&F University, Yangling, China

1 INTRODUCTION

Selenium (Se) is an essential element for animals and human beings. An adequate dietary intake of Se is necessary to keep humans, livestock and poultry healthy. Endemic Keshan-Beck disease on Loess Plateau was found to be related with low Se environments. In the 1970's, the prevalence of endemic Kashan-Beck disease was as high as 30% in two counties on southern part of Loess Plateau (Li et al. 1992). The soils in these areas are extremely low in Se, and the main crop – wheat – Se has less than 50 μg/kg (Liu et al. 2016). In recent years, there are still new cases of Keshan-Beck reported in this area. Whether the residents acquire enough Se via dietary intake to maintain their health is still unknown. A questionnaire survey was carried out in two counties to evaluate residents' dietary Se intake regarding endemic Keshan-Beck disease on Loess Plateau. Field experiments were carried on Loess Plateau at multiple wheat-growing field sites where we investigated the response of wheat and maize to Se fertilization.

2 MATERIALS AND METHODS

2.1 *Questionnaire survey*

A specifically designed questionnaire survey was conducted from July to August of 2017 to collect information on inhabitants' daily food intake. A total of 270 adult permanent residents were surveyed from nine natural villages, distributed in three counties, using the randomized stratified clustered sampling method. The Yongshou and the Linyou counties were the areas most affected by the Keshan-Beck disease in China during the 1970s–1980s. The Yangling District, a non-disease area in the Guanzhong Region of the Shaanxi Province, was chosen as control. Questions about food intake frequency and amount were used in the survey to calculate food diversity and daily food intake in 2016–2017.

2.2 *Field experiment*

Three separate field experiments were conducted to investigate response of maize and wheat to Se fertilization. Experiment 1 was a two-year experiment of soil and foliar application of selenite for maize in Yongshou County on Loess Plateau from 2009 and 2010. Treatments included soil application of 0, 150, 300, 450, and 600 g Se/ha and foliar application of 0, 11, 57, 114, 171, and 228 g Se/ha, respectively.

Experiment 2 was a two-year experiment (2013–2014) of soil-applied and foliar-applied selenite and selenate for growing wheat in the same area. The treatments were: selenate 15 g Se/ha soil-applied, selenate 18 g/ha foliar-applied, selenite 700 g Se/ha soil-applied, and selenite 45 g/ha foliar-applied. All Se fertilizers were only used in the first year.

Experiment 3 consisted of foliar selenite ($N_{a2}SeO_3$) at multiple locations in the main wheat production areas in China from 2010 to 2011. A total of 30 experimental stations from China Agricultural Research System working on wheat in 14 Provinces were involved in the study. Wheat received foliar application of 0.017% sodium selenite at middle and late jointing stages in a randomized complete block design with three replicates. The application rate was 116 g Se/ha.

3 RESULTS AND DISCUSSION

3.1 *The dietary Se intake of residents in Yongshou and Linyou*

The daily Se intake of the residents in Yongshou County and Linyou County were 25.0 μg/d and 35.0 μg/d, respectively, which was much lower than the residents in Yangling District at 61.8 μg/d. The dietary food of the Yongshou County and Linyou County residents consisted mainly of cereals. The daily intake of wheat (flour) corresponded to 46% and 37% of all food intake for the Yongshou County and Linyou County residents, respectively; these percentages were higher than those for the Yangling District, which were 24% of residents. However, due to the low Se content of wheat grains, cereals constituted only minor percentages of the total daily Se intake of 7%, 13%, and 12% for the Yangling District, Yongshou County, and Linyou County residents, respectively. If the Se content in wheat grain in Linyou and Yongshou counties could be enhanced to 100 μg/kg, then Se intake could be increased to 50.8 and 61.5 μg/d for the residents of the two counties. There is an urgent need to improve Se content in staple food in the low-selenium soil areas such as Linyou and Yongshou counties.

3.2 *Response of maize to Se fertilization*

Soil Se application increased the Se content in maize grain (Fig. 1). When the Se rate was increased from 0 to 600 g/ha, the grain Se content increased by 55-fold (from 3.7 μg/kg to 206 μg/ kg) and 118-fold (from 0.6 μg/kg to 71.5 μg/kg) in 2009 and 2010, respectively. Correlation analysis showed that 1 g of Se/ha increased the grain by 0.33 μg/kg in 2009 but only 0.12 μg/kg in 2010 (Fig. 1). Foliar application of Se also remarkably increased the Se content in maize grain. When foliar-applied Se was increased from 0 to 228 g/ha, the Se content in grains increased by 168-fold (from 11 μg/kg to 1863 μg/kg) and 329-fold (from 7 μg/kg to 2312 μg/kg) successively during the two-year period. Foliar Se application showed a significantly greater potential for increasing the Se level in maize grains, with 1 g of Se/ ha increasing the grains by 8.67 μg/kg and 8.23 μg/kg in 2009 and 2010, respectively.

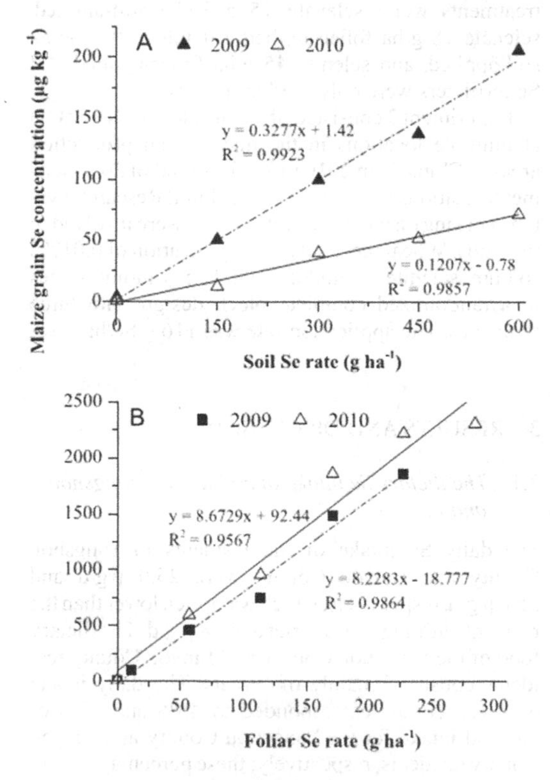

Figure 1. Relationship between the Se content in maize grain and soil (A) and foliar (B) Se application rates.

3.3 *Response of wheat to Se fertilization*

Fertilization with selenate (SeO_4) and selenite (SeO_3) can both significantly increase wheat grain Se above 100 μg/kg continuously for two years with four treatments. The wheat response to Se(VI) was better than Se(IV) with higher Se biofortification index of 4.7 and 15.6 μg/kg (or g/ha) compared to 0.32 and 8.0 μg/kg (or g/ha), respectively. In the second year, only treatment with soil application of Se(IV) had residual effect on wheat grain Se, and it was still greater than 100 μg/kg. The total use efficiency of Se fertilizer is higher with foliar application than soil application, and the Se use efficiency of soil-applied Se(IV) is the lowest.

The results of multiple location field experiment showed that Se application significantly increased grain Se content. When no Se was applied, grain Se content averaged 31.0 μg/kg, and applying Se at 116 g/hm increased the grain Se content to an average of 647.8 μg/kg, which reached selenium-enriched level, but was still below toxic level. For 1.0 g Se/ha applied, grain Se content was increased by an average of 5.3 μg/kg and applying 51 g Se/hm could increase the Se content of wheat grain from the average of 31.0 μg/kg to above 300 μg/kg.

4 CONCLUSIONS

It is not difficult to increase Se concentration in the main staple crop maize and wheat grain with Se fertilization methods. The dietary Se intake of residents in the area of Keshan disease can be increased to 51~62 μg/d when wheat grain Se concentration were enhanced above 100 μg/kg. However, the application methods and the Se fertilizer rates need to be considered with different crops and biofortification goals.

REFERENCES

Li, J., Chen, D. & Ren, S. 1992. Environmental factors affecting human body's low selenium: Investigation of the Kashin-Beck Disease Area in Weibei, Shaanxi Province. *Environ Sci* 13: 16–22.

Liu, H., Yang, Y., Wang, Z. et al. 2016. Selenium content of wheat grain and its regulation in different wheat production regions of China. *Scientia Agricultura Sinica* 49: 1715–1728. (in Chinese)

Selenium Research for Environment and Human Health: Perspectives, Technologies and Advancements – Bañuelos, Lin, Liang & Yin (eds)
© 2020 Taylor and Francis Group, London, ISBN 978-1-138-39014-0

Selenium speciation and content of oat kernels in response to foliar and soil selenium fertilization

J.H. Li, A.N. Guo, Q. Xia, X.J. Dai, Zh.Q. Gao & Z.H.P. Yang*
College of Agricultural, Shanxi Agricultural University, Taigu, Shanxi, China

CH.Y. Wang*
Crop Science Research Institute, Shanxi Academy of Agricultural Sciences, Taiyuan, Shanxi, China

1 INTRODUCTION

Selenium (Se) is an essential trace element in animals, plants, and humans (Schwarz et al. 1999). Many studies have shown that appropriate intake of Se can help reduce the risk of cancer and improve the body's immune system (Combs et al. 2004). Among several common cereal crops (wheat, rice, corn, barley, oats), wheat has the strongest Se accumulation capacity, while oats have the lowest Se accumulation capacity (Lyons et al. 2003). Studies on the absorption and utilization of Se in crops such as wheat, rice, and barley have been reported extensively; however, there are few studies on oats (Li et al. 2009, Zhou et al. 2008).

This experiment was designed to study the Se enrichment of oats by foliar and soil applications of Se and to measure the effects of each application treatment on Se speciation and content in oat kernels.

2 MATERIALS AND METHODS

The main plots were planted with oat cultivars Jinyan 21 and Jinyan 19, while the sub-plots were treated with four types of Se: Non-selenium fertilizer as control, soil Se fertilizer (SA), foliar Se fertilizer (FA), and soil + foliar spraying (SA + FA). The kernels were harvested at the mature stage of the oats, and their total Se content was determined after microwave digestion by inductively-coupled plasma mass spectroscopy (ICP-MS). Selenium speciation in kernel tissues was determined after enzymatic hydrolysis by High Performance Liquid Chromatography and Inductively-Coupled Plasma Mass Spectrometry (HPLC-ICP-MS).

3 RESULTS AND DISCUSSION

Selenium content was between 40 and 125 μg/kg in Se-enriched oats. This level was an appropriate concentration of Se in oat kernels according to the National Standard (DB/T 22499-2008) of 100–300 μg/kg and recommended human intake of 50 μg per day. Results showed that foliar and soil application of Se strongly influenced the total Se contents in kernels (Table 1). The lowest total Se content in kernels was found in control, and the highest Se accumulation was recorded in the SA+FA treatment. The FA treatment had higher grain total Se contents than the SA treatment. There were four identified peaks in the chromatograph of the sample, including selenocystine (SeCys2), methyl-selenocysteine (MeSeCys), Se^{4+}, and selenomethionine (SeMet). The peak at retention time of 350,000 s was not identified, because it did not match with any one of the Se standard compounds (Fig. 1). Inorganic Se species, such as selenate (SeO_4^{2-}) and selenite (SeO_3^{2-}), are generally

Table 1. Effects of Se fertilization on total Se content (μg/kg) in oat.*

Varieties	Control	FA	SA	SA+FA
JinYan21	41.72 d	94.44 b	52.27 c	125.7 a
JinYan19	40.22 d	90.78 b	69.38 c	121.9 a

*Control=without Se; SA=Se-fertilizer topdressing in soil; FA = Foliar spraying of Se-fertilizer; SA + FA = Topdressing in soil and foliar spraying of Se-fertilizer. Means with different letters are not statistically different ($p < 0.05$).

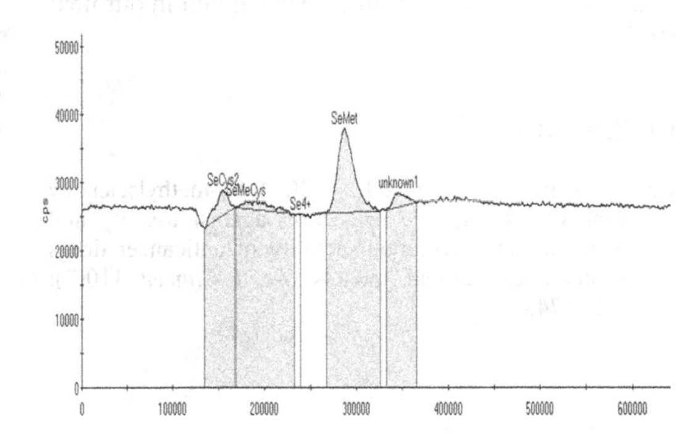

Figure 1. Selenium speciation in oat kernels treated with Se fertilizer through soil surface and foliar spray application. X-axis is retention time (s).

considered to be the most bioavailable and potentially toxic Se species (Fordyce et al. 2005). Organic Se species such as SeCys2, SeMet, and MeSeCys are generally considered less toxic (Rayman et al. 2008, Zhu et al. 2009). Furthermore, Se bioaccessibility greatly depends on Se chemical species (Fairweathertait et al. 2010). It is well known that selenocysteine methyltransferase (SMT) catalyzes the methylation reaction of selenocysteine (SeCys) and Se-methylselenocysteine. Se-methylselenocysteine is a natural Se-containing amino acid, and a preferred source of Se intake for human health according to recommendation by CFDA (GB 14880-2012, National Standard for Nutritional Fortification Substances by Industries, Ministry of Health of P.R. China). Previous studies also suggested that Se-methyl-selenocysteine has higher anti-cancer activity, because it can induce apoptosis of cancer cells, inhibit tumor vascular development, and prevent early cancer cell expansion (Chen et al. 2012, Cao et al. 2014). Therefore, future research on Se-methyl-selenocysteine accumulation in oat kernels will be beneficial.

4 CONCLUSIONS

We conclude that both soil application and foliar application (SA + FA) treatments will enhance the total Se in oat kernels. The application's effect on Se speciation in oat kernels needs to be determined in our next study.

REFERENCES

Cao, S., Durrani, F.A. & Tóth, K. 2014. Se-methylselenocy steine offers selective protection against toxicity and potentiates the antitumour activity of anticancer drugs in preclinical animal models. *Br J Cancer* 110(7): 1733–1743.

Chen, T. & Wong, Y.S. 2008. Selenocystine induces apoptosis of A375 human melanoma cells by activating ROS-mediated mitochondrial pathway and p53 phosphorylation. *Cell Mol Life Sci* 65(17): 2763–2775.

Combs, G.F. 2004. Status of selenium in prostate cancer prevention. *Br J Cancer* 91(2): 195–199.

Fairweathertait, S.J., Collings, R. & Hurst, R. 2010. Selenium bioavailability: current knowledge and future research requirements. *Am J Clin Nutr* 91(5): 1484S.

Fordyce, F.M. 2005. Selenium deficiency and toxicity in the environment. In O. Selinus (ed), *Essentials of Medical Geology*: 373–415. New York: Springer.

Li, Y.L. & Gao, H.M. 2009. Effects of spraying Na2SeO3 on yield of wheat. *Chinese Agricultural Science Bulletin* 25(18): 253–255.

Lyons, G., Stangoulis, J. & Graham, R. 2003. High-selenium wheat: biofortification for better health. *Nutr Res Rev* 16(1): 45–60.

Rayman, M.P., Infante, H.G. & Sargent, M. 2008 Food-chain selenium and human health: spotlight on speciation. *Br J Nutr* 100(2): 254–268.

Schwarz, K. & Foltz, C.M. 1999. Selenium as an integral part of factor 3 against dietary necrotic liver degeneration. *Nutrition* 15(3): 255.

Yao, Z., Zhang, X.-P., Deng, Z.-Y. & Li, H-Y. 2012. Advances in the chemical synthesis methods, metabolic pathways and main bioactivities of L-Se-methyselenocysteine. *Academic Periodical of Farm Products Processing* 11: 122–125. (in Chinese)

Zhou, X.B., Shi, W.M. & Yang, L.Z. 2008. Study on mechanisms of selenium accumulation in rice grains. *Plant Nutrition and Fertilizer Science* 14(3): 503–507. (in Chinese)

Zhu, Y.G., Pilon-Smits, E.A.H., Zhao, F.J., Williams, P.N. & Meharg, A.A. 2009. Selenium in higher plants: understanding mechanisms for biofortification and phytoremediation. *Trends Plant Sci* 14 (8): 436–442.

Selenium Research for Environment and Human Health: Perspectives, Technologies and Advancements – Bañuelos, Lin, Liang & Yin (eds)
© 2020 Taylor and Francis Group, London, ISBN 978-1-138-39014-0

Selenium-biofortified herbs as antivirals?

G.H. Lyons
School of Agriculture, Food and Wine, The University of Adelaide, Waite Campus, Urrbrae, South Australia, Australia

1 INTRODUCTION

Viruses comprising RNA cause a range of conditions in humans, from relatively mild (e.g. rhinoviruses causing the common cold) to very severe (e.g. Ebola, with a case fatality of around 60%). Influenza, polio, measles, hepatitis A and C and AIDS are all caused by RNA viruses. The Spanish Flu pandemic of 1918-19 and the Black Death of the 14th century in Europe have been the most devastating disease events in history, each causing an estimated 50 million fatalities (Scott & Duncan 2001, Benedictow 2005, Chandra & Kassens-Noor 2014). Some researchers regard the rapid spread and symptoms of the Black Death to be more indicative of an Ebola-type hemorrhagic RNA viral disease than the plague bacillus (Scott & Duncan 2001). In livestock, RNA virus diseases include bird flu, equine flu, swine fever and foot and mouth disease.

RNA viruses are more unstable and subject to higher mutation rates than DNA viruses. This makes control of the diseases they cause more difficult. Pharmaceutical antiviral drugs, e.g. Relenza, Tamiflu, are quite ineffective against most RNA viruses (Jefferson et al. 2014). This, combined with the large number of incident cases of RNA viral diseases globally, along with their frequent severity in terms of morbidity and mortality, underline the imperative to identify effective alternative antiviral agents. Even a modest therapeutic benefit would translate to a large improvement globally.

2 SELFHEAL AND SELENIUM v RNA VIRUSES

Numerous plants and their active components exhibit antiviral activity (Mishra et al. 2013), and one of the most effective is Selfheal, *Prunella vulgaris*, a mint-family herb. Evidence includes activity against HIV (Kageyama et al. 2000), lentivirus (Brindley et al. 2009) and Ebola (Zhang et al. 2016). Key therapeutic components of *Prunella* include betulinic acid, prunellin, delphinidin, oleanolic acid, rosmarinic acid, and ursolic acid (Meng et al. 2014).

Low selenium (Se) (i.e. less than 70 μg/L in plasma, a level common in Sub-Saharan Africa and parts of China) reduces immunocompetence and thus increases the susceptibility of the host to infection. However, and perhaps more importantly, low Se also influences the genetic make-up of the viral genome. Under Se deficiency, inherently unstable RNA viruses tend to destabilise and mutate to more virulent forms (Beck et al. 2004).

Keshan disease killed many people in Se-deficient regions of China, but it has been largely controlled with selenized salt. In most cases, a malignant cardiotropic variant of the Coxsackie B3 virus was implicated (Christophersen et al. 2013). Laboratory studies found that Coxsackie B3-resistant mice become susceptible under Se and vitamin E deficiency (Beck et al. 2003). Other RNA viruses exhibit increased virulence in Se deficient regions. Hemorrhagic fever with renal syndrome (HFRS) caused by hantaviruses and transmitted by rodents is a public health issue in China. A study found the incidence of HFRS in humans was around six times higher in severely Se-deficient areas and twice as high in moderately deficient areas compared to non-deficient areas (Fang et al. 2015).

A plausible explanation for the association (shown in several studies) of low Se status and severity of HIV disease may be reduced effectiveness of cellular systems of antioxidant defence and enhanced transcription of "the AIDS gene" HIV-1 nef. These changesurther deplete Se, leading to immunodeficiency (Christophersen et al. 2013, Taylor et al. 2016). A similar hypothesis has been advanced for Ebola, which appears, in common with a number of other RNA viruses, to encode selenocysteine. Biosynthesis of this protein could impose a high Se demand on the host, leading to lipid peroxidation, cell membrane breakdown and hemorrhagic symptoms (Ramanathan & Taylor 1997). A role for Se in Ebola treatment (Lyons 2014) is supported by Chinese researchers who treated patients in an outbreak of viral hemorrhagic fever with oral sodium selenite, which resulted in a rapid drop in mortality. After nine days of Se dosage, the death rate fell from 100% (untreated) to 37% (treated) in the very severe cases, and from 22% to zero in the less severe cases (Hou 1997).

3 CONCLUSIONS

Evidence suggests a useful role for both Se and *Prunella vulgaris* against RNA viruses. Why not combine them to increase the potency of the intervention? Research is needed to determine the most effective way to biofortify Prunella with Se, whether foliar or soil-applied, and at what dose. Of course, Prunella could be naturally Se-biofortified if grown on high-selenium soils in places like Enshi, China. In vitro trials with human and animal cells infected with RNA viruses and in vivo animal studies are needed initially, where biofortified Prunella could be compared with the unbiofortified herb plus inorganic Se. Judicious application of this surprising element, e.g. in combination with proven antiviral herbs, may prove to be a useful weapon against RNA viral diseases that threaten humanity.

REFERENCES

Beck, M., Williams-Toone, D. & Levander, O. 2003. Coxsackievirus B3-resistant mice become susceptible in Se/vitamin E deficiency. *Free Radic Biol Med* 34: 1263–1270.

Beck, M., Handy, J. & Levander, O. 2004. Host nutritional status: the neglected virulence factor. *Trends Microbiol* 12: 417–423.

Benedictow, O. 2005. The Black Death: the greatest catastrophe ever. *History Today* 55: 42–49.

Brindley, M., Widrlechner, M., McCoy, J. et al 2009. Inhibition of lentivirus replication by aqueous extracts of *Prunella vulgaris*. *Virology* 6:8.

Chandra, S., & Kassens-Noor, E. 2014. The evolution of pandemic influenza: evidence from India, 1918-19. *BMC Infect Dis* 14: 501.

Christophersen, O.A., Lyons, G., Haug, A. & Steinnes, E. 2013. Selenium (Ch 16). In B.J. Alloway (ed), *Heavy Metals in Soils: Trace Metals and Metalloids in Soils and their Bioavailability*: 429–463. Dordrecht: Springer.

Fang, L.-Q., Goeijenbier, Zuo, S.-Q., Wang, L.-P. et al. 2015. The association between hantavirus infection and selenium deficiency in mainland China. *Viruses* 7: 333–351.

Hou, J-C. 1997. Inhibitory effect of selenite and other antioxidants on complement-mediated tissue injury in patients with epidemic hemorrhagic fever. *Biol Trace Elem Res* 56: 125–130.

Jefferson, T., Jones, M., Doshi, P., Del Mar, C. et al. 2014. Neuraminidase inhibitors for preventing and treating influenza in healthy adults and children. *Cochrane Database Syst Rev* 4: CD008965.

Kageyama S., Kurokawa, M. & Shiraki, K. 2000. Extract of Prunella vulgaris spikes inhibits HIV replication at reverse transcription in vitro and can be absorbed from intestine in vivo. *Antivir Chem Chemother* 11: 157–164.

Meng, G., Wang, M., Zhang, K., Guo, Z., & Shi, J. 2014. Resesarch progress on the chemistry and pharmacology of *Prunella vulgaris* species. *OAlib* 1: 1–19.

Mishra, K., Sharma, N., Drishya, D. et al 2013. Plant derived antivirals: a potential source of drug development. *J Virol Antivir Res* 2: 2.

Ramanathan, C. & Taylor, E.W. 1997. Computational genomic analysis of hemorrhagic fever viruses. *Biol Trace Elem Res* 56: 93–106.

Scott, S. & Duncan, C. 2001. *Biology of Plagues: Evidence from Historical Populations*. Cambridge, England: Cambridge University Press.

Taylor, E.W., Ruzicka, J.A., Premadasa, L. & Zhou, L. 2016. Cellular selenoprotein mRNA tethering via antisense interaction with Ebola and HIV-1 mRNAs may impact host Se biochemistry. *Curr Top Med Chem* 16: 1530–1535.

Zhang, X., Ao, Z., Bello, A. et al. 2016. Characterization of the inhibitory effect of an extract of *Prunella vulgaris* on Ebola virus glycoprotein (GP)-mediated virus entry and infection. *Antivir Res* 127: 20–31.

Selenium Research for Environment and Human Health: Perspectives, Technologies and Advancements – Bañuelos, Lin, Liang & Yin (eds)
© 2020 Taylor and Francis Group, London, ISBN 978-1-138-39014-0

Effect of selenium biofortification on yield and quality of sweet potato

H.F. Li, Y.M. Huang, Y.Q. Li, J.F. Hua, T.Y. Chen & C.R. Wu
Upland Crops Research Institute, Guangxi Academy of Agricultural Sciences, Nanning, China

1 INTRODUCTION

Selenium (Se) is a trace element with multiple biological functions. Low Se dietary intakes can result in health disorders and some diseases, such as reduced fertility and immune function, increased risk of cancers, and thyroid and cardiovascular diseases. Overt Se deficiency has caused endemic Keshan or Kaschin-Beck disease in areas of China with low Se diets (Tan et al. 2002). Unfortunately, the uneven distribution of Se over the Earth's surface causes wide Se-deficient areas, leading to increased risk of the already mentioned health issues. Biofortification with Se is considered an effective approach for addressing the human Se deficiency status (Alfthan et al. 2015). Sweetpotato is an excellent candidate for the health-centered strategy of producing functional foods with high levels of Se. The present research aimed to assess evaluate the effects of soil selenium content on agronomic characters, quality, and yield of sweet potato grown in soil enriched with different forms of Se.

2 MATERIALS AND METHODS

2.1 *General methods*

The greenhouse-grown experiment was conducted on sweetpotato (*Ipomoea batatas* L. *Lam*) cultivar Guishu10 at the Guangxi Academy of Agricultural Sciences Mingyang Base, Nanning region, China in 2018. The trial was carried out on a latosolic red soil with pH 7.1, 8.7% organic matter, 449 mg/kg N, 430 mg/kg P_2O_5, 3928 mg/kg K_2O, and 0.23 mg/kg total soil Se content.

Pot experiment was carried out with five treatments of sodium selenite: 0 mg/kg (control), 4 mg/kg (Se4), 8 mg/kg (Se8), 12 mg/kg (Se12), and 16 mg/kg (Se16). After sodium selenite was added, each pot was incubated for two weeks prior to planting. A randomized block design was selected with fifteen replicates per treatment, and each pot was planted with two sweet potato seedlings.

2.2 *Sample preparation and chemical analysis*

Plants were harvested 125 d after planting. Plants were separated into roots, stems, leaves, and tubers. Plant samples were cleaned, weighed, and cut into small pieces. An aliquot of fresh tissues was dried at 60°C for 3 days to constant weight, and prepared for the determination of Se and heavy metal concentrations. Selenium contents and speciation were determined using the methods described by Li et al. (2017). Soluble and reduced sugars, starch, and protein analyses were also performed (Li, unpubl.).

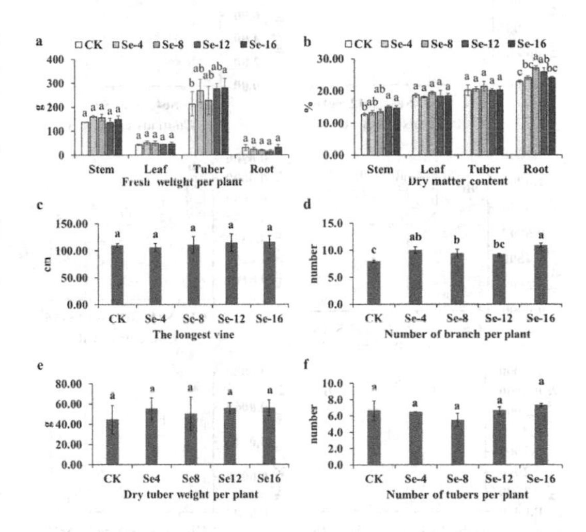

Figure 1. Effects of soil Se content on agronomic characters and yield on harvest time.

2.3 *Statistical analysis*

Data were processed by two-way analysis of variance, and mean separations were performed through Duncan's multiple range test with reference to a 0.05 probability level, using SPSS software version 14.0.

3 RESULTS AND DISCUSSION

3.1 *Agronomic traits*

The different experimental treatments showed significant interactions on tuber weight per plant, stem dry matter, root dry matter, and number branches per plant (Figs 1a, b, d).

3.2 *Quality and heavy metal*

Contents of soluble sugars, reducing sugars, starches and proteins of the Se treatments was statistically significant from those of the control (Fig. 2a). In comparison with control, there was no significant difference in the content of lead (Pb) and cadmium (Cd) in tubers. The content of Cd in tubers was reduced by

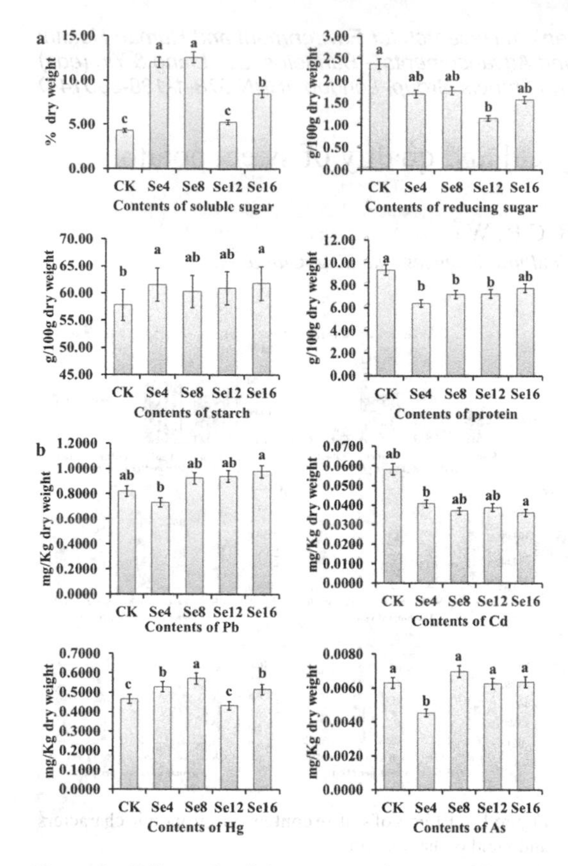

Figure 2. Effects of soil Se concentration on quality and heavy metal content in tubers on harvest time.

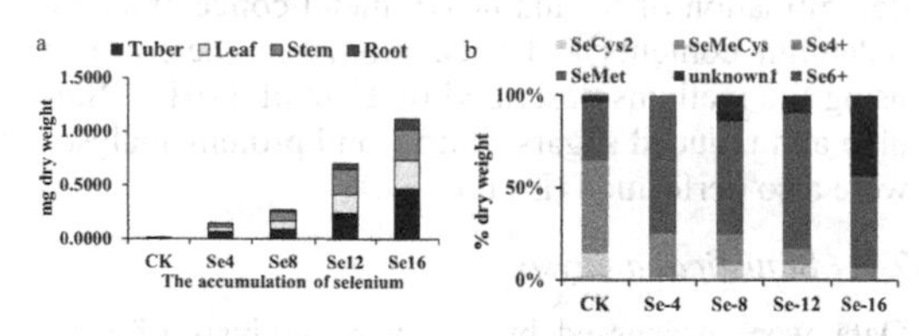

Figure 3. Effects of soil Se concentration on plant Se content and components in tubers on harvest time.

Se treatments. However, the contents of arsenic (As) and mercury (Hg) were affected (Fig. 2b).

3.3 *Selenium content and components*

The Se content in the roots, stems, leaves and tubers of sweet potato increased significantly with an increase of soil Se concentration (Fig. 3a). Irrespective of adding selenite or selenate to the soil, there were no detection of Se differences in the tubers with different treatments. However, the percentages of $SeCys_2$, SeMeCys, SeMet, and unknowns were different with the experimental treatments (Fig. 3b). With an increase of soil Se concentration, the percentage of SeMeCys in tubers gradually decreased. In the treatment of Se16, the proportion of unknown Se compounds increased significantly in tubes.

4 CONCLUSIONS

The results showed that soil Se content had different effects on plant characteristics, quality indexes, Se, and heavy metal contents of sweet potato plants. Low concentration of Se significantly increased the soluble sugar content and starch content but reduced the protein content. Selenium had obvious antagonistic effect on the absorption of Cd, and a low concentration of Se exerted an antagonistic effect on As and Pb. With an increase of soil Se concentration, the Se content in stems, leaves, tubers and roots were significantly increased, and the selenium components in tubers was significantly changed. We recommend that increasing Se supply in soil is an effective way to improve soil quality and Se content in sweet potatoes tubes.

REFERENCES

Alfthan, G., Eurola, M., Ekholm, P. et al. 2015. Effects of nationwide addition of selenium to fertilizers on foods, and animal and human health in Finland: From deficiency to optimal selenium status of the population. *J Trace Elem Med Biol* 31: 142–147.

Li, T., Sun, F.Y., Gong, P. et al. 2017. Effects of nano-selenium fertilization on selenium concentration of wheat grains and quality-related traits. *Journal of Plant Nutrition and Fertilizer* 23(2): 427–433. (in Chinese)

Tan, J.A., Zhu, W.Y., Wang, W.Y. et al. 2002. Selenium in soil and endemic diseases in China. *Sci Total Environ* 284: 227–235.

Selenium Research for Environment and Human Health: Perspectives, Technologies and Advancements – Bañuelos, Lin, Liang & Yin (eds)
© 2020 Taylor and Francis Group, London, ISBN 978-1-138-39014-0

Selenium biofortification of green tea in Se-rich areas in Guangxi, China

W.X. Jia
Jiangsu Bio-Engineering Research Center for Selenium, Suzhou, Jiangsu, China
School of Resources and Environment, Shanxi Agricultural University, Taigu, Shanxi, China

Z.M. Wang, L.X. Yuan, X.B. Yin, Z.K. Liu & J.P. Song
Jiangsu Bio-Engineering Research Center for Selenium, Suzhou, Jiangsu, China

P.X. Liang & Z.P. Jiang
Agricultural Resource and Environment Research Institute, Guangxi Academy of Agricultural Science, Guangxi, China
Selenium-rich Agricultural Research Center, Nanning, Guangxi, China
Development and Quality Control, Ministry of Agriculture and Rural Affairs, Ankang, Shaanxi, China

1 INTRODUCTION

Selenium (Se) is an important trace element in the ecological and biological environment (Dai et al. 1995). The abundance of Se in the diet is closely related to human health characteristics (Zhao et al. 2005). The diet is the main source of Se intake for humans (Zhang 2017). While foliar application of Se fertilizer is not a common practice for production of Se-enriched tea, it is important to determine whether the application of Se fertilizer to the soil can effectively increase the Se content in tea leaves, and, consequently, in tea beverage.

In this study field trials to produce Se-enriched tea were carried out in Guangxi, China. Selenium-rich organic fertilizer was applied to soil growing tea plants. Concentrations of water extractable Se in tea leaves were analyzed, and the extraction rate of Se in tea leaves was calculated.

2 MATERIALS AND METHODS

2.1 *Experimental design*

The two tea varieties were "Fuyun 6" in Heng County (A) and "Hua 26" in Lingyun County (B). The Se-rich fertilizer (SETEK-BF-002) was provided by Suzhou SETEK Co, Ltd, China, with a total Se content of 1000 mg/kg. Two levels of Se fertilizer were applied in each of two study areas: 0 (CK) and 225 kg/hm^2. Each treatment had three replicates. The Se-rich fertilizer was applied to the A location with "Fuyun 6" in April 2018 and to the B location with "Hua 26" in January 2018.

2.2 *Sampling and sample preparation for Se measurement*

In December 2017, soil samples from A and B areas were collected before applying Se fertilizer. Soil samples were naturally air dried (about a week), ground, and passed through 0.15-mm sieves. After continuous extraction, total Se and available Se were measured.

In July 2018, tea leaf samples in both treatment and control were collected from the two study areas (i.e. A-CK, A+Se, B-CK, B+Se). Tea leaf samples were rinsed with deionized water, and half of the samples were oven-dried at 60°C for 24 hours for total Se analysis. The other half of the tea leaf samples were prepared for tea. According to GB/T 22776-2018 entitled *Methodology for Sensory Evaluation of Tea*, 3-g tea were measured and brewed in 150 mL boiling water for 4 min, and then the tea soup was quickly passed through 0.45-μm filter and kept at 4°C in a refrigerator for dissolved Se measurement. The amount of extractable Se in tea leaves and the Se extraction rate were further calculated. Determination of total Se content in tea and soil samples was performed according to GB 5009.93-2010.

Table 1. Selenium content and bioavailable Se ratio in soil.

	Soil A	Soil B
Total Se content (μg/kg)	1545.1 ± 194.2	1690.6 ± 163.6
Bioavailable Se ratio (%)	10.4 ± 3.2	15.8 ± 2.7

3 RESULTS AND DISCUSSION

3.1 *Selenium contents in soil*

The total and bioavailable Se contents in soils are shown in Table 1. Total Se contents in soils at A and B study areas were 1.55 mg/kg and 1.69 mg/kg, respectively, which can be classified as Se-rich soils. The bioavailable Se ratio in soils from both A and B regions was not at a high level, and accounted for 10.4% and 15.8%, respectively. Studies have shown that bioavailable Se in Se-rich red soil in southern China is generally low (Xie et al. 2017), indicating

Table 2. Total Se in tea leaves and dissolved Se in tea soup.

	Tea Samples from Heng County (A)		Tea Samples from Lingyun County (B)	
	Control	Se Treatment	Control	Se Treatment
Total Se Concentration (μg/kg)	163.8 ± 47.2	1386.6 ± 27.0	139.2 ± 24	2923.1 ± 288.9
Tea soup content (μg/L)	2.2 ± 0.9	6.6 ± 0.9	3.6 ± 1.1	15.3 ± 3.4
Se dissolution rate (%)	66.5	23.8	100.0*	26.1
Se dissolve amount (μg/kg)	108.9	330.2	139.2**	763.2

* The value was calculated as 130.3%, which exceeds 100%, and the dissolution rate is considered to be 100%; ** The value was calculated by the 100% Se dissolution rate from tea. references.

that the bioavailable Se ratio in areas with high total Se contents is not necessarily high.

3.2 *Selenium contents in tea*

Concentrations of total Se in tea are shown in Table 2. Compared with control, Se concentrations in tea of both A+Se and B+Se samples increased by 8 times and 21 times, respectively. The Se fertilizer applied to soil effectively increased the Se content in tea. Meanwhile, the tea product also meets the national industry standard in terms of total Se content (0.25-4.0 mg/kg) according to "Selenium-enriched Agricultural Products" (GH/T 1135-2017). However, the Se content was only 0.164 mg/kg in control sample at the A location and 0.139 mg/kg at the B location, which is lower than the lower limit of Se-enriched tea (GH/T 1135-2017). This indicates that the Se contents in tea cannot be guaranteed by the total Se content alone (Wang et al. 2018).

3.3 *Selenium dissolution in tea*

The Se content of tea soup, Se dissolution rate from tea and amount of 3 g tea were listed in Table 2. The Se dissolution rate of tea was reduced compared with CK group in both regions, but the Se dissolution amount of tea after Se biofortification was higher than that in the CK group. In our research, the increased Se dissolution amount in tea soup indicates that Se biofortification by soil application methods is a good solution for people to supplement Se by diet.

4 CONCLUSIONS

The results showed that tea grown in Se-rich soil cannot be guaranteed to be Se-enriched tea. However, the application of Se fertilizer to soil can improve Se accumulation in tea and meet the industry standard of Se-enriched tea in China. Selenium-enriched tea after biofortification shows lower extraction rate of Se but higher dissolved amount compared with the non-Se-biofortified tea. These results provide a theoretical basis for the cultivation of Se-rich tea by soil application of Se fertilizer.

REFERENCES

Dai, W. & Geng, Z.C. 1995. Study of soil selenium. *Northwestern Forestry* 10(3): 93–97.

Wang, Z.M., Yuan, L.X., Zhu, Y.Y. et al. 2018. On standards of Se-enriched agricultural products and Se-rich soil in China. *Soil* 50(06): 1080–1086.

Xie, B.T., He, L., Jiang, G.J. et al. 2017. Regulation and evaluation of selenium availability in Se-rich soils in southern China. *Rock Mine Test* 36(03): 273–281.

Zhao, N.M., He, F. & Zhang, Q.A. 2014. Dissolution of some compounds from Ziyang Se-rich tea during the process of brewing. *Agri Process* (19): 45–48.

Zhang, F.S. 2017. Comparative study of organic Se content and Se dissolution of artificial Se and natural Se rich tea. *Agri Technol* 37(23): 16–17.

Selenium Research for Environment and Human Health: Perspectives, Technologies and Advancements – Bañuelos, Lin, Liang & Yin (eds)
© 2020 Taylor and Francis Group, London, ISBN 978-1-138-39014-0

Functional agriculture in China: 11 years of research and practices

X.B. Yin*
Key Lab for Functional Agriculture, USTC (Suzhou), Jiangsu, China
School of Earth and Space Sciences, University of Science and Technology of China, Hefei, China
Engineering Center for Functional Agriculture, Nanjing University (Suzhou), Jiangsu, China
Shanxi Institute for Functional Agriculture, Taigu, Shanxi, China
Key Laboratory of Se-enriched Products Development and Quality Control, Ministry of Agriculture and Rural Affairs and National-local Joint Engineering Laboratory of Se-enriched Food Development, Ankang, Shaanxi, China
Guangxi Selenium-rich Agricultural Research Center, Nanning, Guangxi, China

Z.M. Wang, L.X. Yuan, X.Q. Lu, F. Li & Z.K. Liu
Jiangsu Bio-Engineering Research Center for Selenium, Suzhou, Jiangsu, China

Q.Q. Chen
College of Life Science and Resources and Environment, Yichun University, Yichun, Jiangxi, China

1 INTRODUCTION

In 2008, Qiguo Zhao, academician of Chinese Academy of Sciences (CAS), first developed a new conception of functional agriculture in "*Agricultural Science and Technology in China: A Road Map to 2050*" (Zhao 2011). In cooperation, Dr. Yin's team at University of Science and Technology of China (USTC) has been conducting systematic studies and industrialization practices of functional agriculture since 2008.

Until now, functional agriculture has been widely applied in 10 provinces in China, including Jiangsu, Guangxi, Shanxi, Ningxia, Jiangxi, Anhui, Hebei, Heilongjiang, Hubei, and Shaanxi provinces along with Thailand, which was defined as "10+1" (Fig. 1). Thailand Institute of Scientific and Technological Research (TISTR) and the USTC Team are jointly studying the standards and the certification for selenium (Se)-enriched agricultural products such as rice, banana, and pineapple in Thailand.

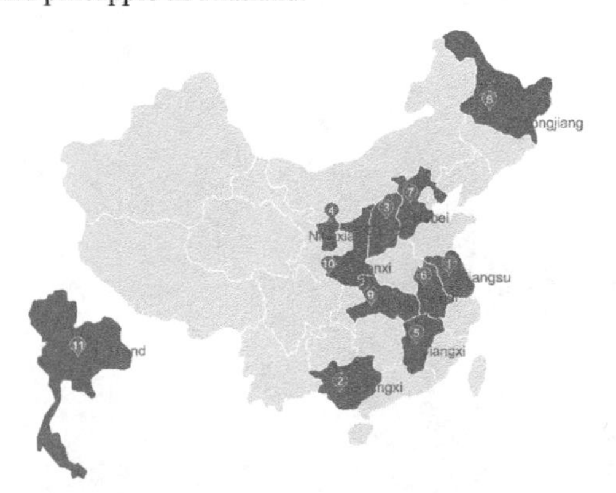

Figure 1. The "10+1" demonstration in the world, including 10 Chinese Provinces and one country – Thailand.

Table 1. The Se-enriched agriculture products in China.

Se-rich products	Province	Area (ha)	Se content (μg/kg)
Watermelon	Ningxia	6000	10–50
Rice	Jiangsu	6000	100–500
Tea	Guangxi	5000	500–2000
Vegetables	Hebei & Shanxi	8000	10–100

2 PROGRESS ON FUNCTIONAL AGRICULTURE IN CHINA

2.1 *The standards for Se-rich agriculture products*

The USTC research laboratory and the Setek company have jointly conducted demonstration projects to produce Se-enriched rice and watermelon on 6000 hectares, Se-enriched tea on 5000 hectares, and Se-enriched vegetables on 8000 hectares (Table 1).

Based on these projects, the first Chinese industrial standard of "Se-enriched Agricultural Products" was issued by Chinese Supply & Marketing Industry, which identified the range of total Se contents and the percentages of Se species (i.e. Se amino acids) in different Se-biofortified products (Table 2).

To ensure the quality of the functional agricultural products, a "nine-step procedure" of functional agriculture was established (Fig. 2).

2.2 *The basic standard substances*

Since the composition of biofortified fertilizer generally depends on the soil properties and climatic conditions, worldwide cooperation is urgently needed to conduct experiments. The USTC research laboratory and Setek have developed a series of basic

Table 2. The content and speciation of Se in products.

Se-enriched products	Total Se (mg/kg)	Se amino acid** (%)
Cereals	0.10~0.50	>65
Beans	0.10~1.00	>65
Tubers (DW*)	0.10~1.00	>65
Vegetables (DW)	0.10~1.00	>65
Edible mushrooms (DW)	0.10~5.00	>65
Meat	0.10~0.50	>70
Eggs	0.10~0.50	>70
Tea	0.10~4.00	>60

*DW = dry weight; **Selenium amino acid contains selenomethionine, selenocystine, and methyl-selenocysteine.

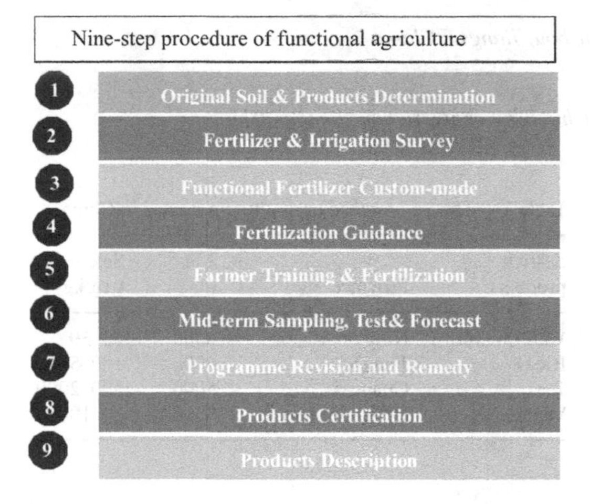

Figure 2. The nine-step procedure of functional agriculture (Setek, unpubl.).

Table 3. The content of Se and the speciation in standard substances.

	Broccoli	Corn
Total Se (mg/kg)	300	50
$SeCys_2$(%)	10.1	21.7
MeSeCys (%)	22.5	–
Se^{4+}(%)	–	–
SeMet (%)	57.5	74.9
Se^{6+}(%)	–	–
Others (%)	9.9	3.4

standard substances, including Se fertilizers and Se-enriched food materials (e.g. broccoli, corn and wheat). Their Se concentrations and speciation values are shown in Table 3.

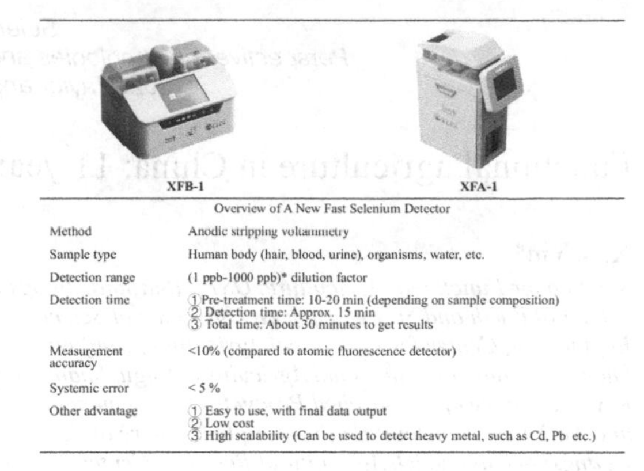

Overview of A New Fast Selenium Detector	
Method	Anodic stripping voltammetry
Sample type	Human body (hair, blood, urine), organisms, water, etc.
Detection range	(1 ppb-1000 ppb)* dilution factor
Detection time	① Pre-treatment time: 10-20 min (depending on sample composition) ② Detection time: Approx. 15 min ③ Total time: About 30 minutes to get results
Measurement accuracy	<10% (compared to atomic fluorescence detector)
Systemic error	< 5 %
Other advantage	① Easy to use, with final data output ② Low cost ③ High scalability (Can be used to detect heavy metal, such as Cd, Pb etc.)

Figure 3. The Fast Selenium Analyzer and major features.

2.3 *The Fast Se Analyzer*

To quickly determine concentrations of Se in human samples and agricultural products, a Fast Selenium Analyzer (Fast Se-1) has been developed by Setek and Yaozhi Co., Ltd. In general, the whole procedure takes only 30 minutes. The parameters and performance of the equipment are shown in Figure 3.

2.4 *The FAST Network*

To promote the development of science and technology of functional agriculture, we strongly suggest establishing a Functional Agriculture Science and Technology Network (FAST Network), and innovation chain for global products. The functional agriculture laboratory at USTC in Suzhou, China can function as the FAST Network hub.

REFERENCE

Zhao, Q.G. 2011. *Agricultural Science and Technology in China: A Roadmap to 2050*: 100-126. Beijing and Berlin: Science Press and Springer.

Selenium Research for Environment and Human Health: Perspectives, Technologies and Advancements – Bañuelos, Lin, Liang & Yin (eds)
 ISBN 978-1-138-39014-0

Exploration of Se-rich bioproducts generated from (waste)water as fertilizers

J. Li, L. Otero-Gonzaleza & G. Du Laing
Laboratory of Analytical Chemistry and Applied Ecochemistry, Faculty of Bioscience Engineering, Ghent University, Ghent, Belgium

P.N.L. Lens
UNESCO-IHE Institute for Water Education, Delft, The Netherlands

I. Ferrer
Group of Environmental Engineering and Microbiology, Department of Civil and Environmental Engineering, Universitat Politècnica de Catalunya, Barcelona, Spain

1 INTRODUCTION

Selenium (Se) is an essential trace element for humans and animals with a narrow window between deficiency and toxicity levels (Boyd 2011, Fordyce 2013, Rayman 2000). Application of conventional chemical Se fertilizers to increase the Se content in crops can result in secondary soil and water contamination due to the low utilization rate of Se. Therefore, it may be beneficial to produce slow-release Se-enriched organic fertilizers locally from Se-containing (waste)water. This process may contribute to the worldwide drive for resource recovery. In this study, we aim to evaluate the bioavailability of Se released from two Se-enriched biomaterials (duckweed and anaerobic sludge) added to soil, and to assess the potential of these two biomaterials as Se fertilizers.

2 MATERIALS AND METHODS

2.1 *Crops (Se bioaccumulator) screening*

Five crops (green beans, wheat, maize, pakchoi, and lettuce) were planted on sandy soil supplemented with 0, 0.5, 1.0, 3.0, and 5.0 mg/kg of Se added as sodium selenate (Na_2SeO_4). The concentration of Se in different crops was measured by ICP-MS after harvest, drying, grinding, and digesting with acids.

2.2 *Fertilization with Se-enriched biomaterials*

Green beans were screened by the crops screening experiment in the experiment of 2.1 as a Se accumulator and grown on sandy and loamy soils fertilized with Se-enriched duckweed and anaerobic sludge at rates equivalent to 1.0 and 5.0 mg Se/kg soil. Non-planted treatments served as controls. The mobility and solubility of Se were assessed by regularly measuring the Se concentration in the soil pore water at a depth of 5.0 cm and in the different bean tissues. Meanwhile, the carbon content in soil pore water was determined. The Se composition in the Se-enriched duckweed and anaerobic sludge was analyzed by HPLC-ICP-MS and XANES (X-ray Absorption Near Edge Structure), respectively.

3 RESULTS AND DISCUSSION

Green beans had relatively higher Se accumulation compared to the other four crops, while wheat had the lowest Se accumulation among the five tested crops. Considering the high Se accumulation and high protein content of bean seeds, green beans were selected and designated in this study as a Se accumulator for the current experiments.

The results of Se-enriched biomaterials supplement showed that Se was mainly present as the hexavalent selenate form in duckweed and as the zerovalent form in anaerobic sludge. The different Se species in the two biomaterials resulted in distinct Se release patterns. After 3 d of incubation, the application of 1.0 and 5.0 mg Se/kg soil of duckweed increased the Se concentration in pore water to 537 and 4375 µg/L for sandy soil, and 413 and 1238 µg/L for loamy soil, respectively; whereas the sludge amendment of 1.0 and 5.0 mg Se/kg soil led to an increment of Se content in pore water to 65 and 322 µg/L for sandy soil, and 72 and 387 µg/L for loamy soil, respectively. However, increasing the incubation time from 3 to 42 d reduced the Se content in soil pore water by 92% for sandy soil and 89% for loamy soil in non-planted treatments amended with 5 mg Se/kg soil of duckweed. The decrease of bioavailable Se followed a second-order equation. By contrast, the Se concentration in the pore water of soils supplied with sludge remained stable during the entire incubation period.

The application of Se-enriched duckweed and anaerobic sludge significantly increased the Se concentration in the different tissues of beans (seeds, leaves, stems, and roots), which was supported by the positive correlation between Se in the soil pore water and Se in the beans' tissues. In addition, the Se concentration in beans fertilized with sludge was 1 to 3 times higher than in those amended with duckweed. This result is attributed to the higher carbon content (TC and TOC) in soil pore water with duckweed application and the different Se species released into soil solution between Se-enriched duckweed and sludge.

The estimation of daily Se intake through bean seeds fertilized with Se-enriched duckweed and sludge showed that the seeds of beans produced in this study can contribute to achieving the recommended daily Se intake for human diets. Overall, Se-enriched green beans were successfully produced after amending the soil with Se-enriched duckweed and anaerobic sludge, demonstrating their potential as organic Se-rich fertilizers for Se deficient regions.

4 CONCLUSIONS

Both Se-enriched biomaterials can be used as slow-release Se fertilizers to improve Se uptake by beans. Selenium contained in duckweed is released to the soil faster than Se released from sludge, while Se-enriched sludge is more efficient than Se-enriched duckweed for improving the Se concentration in beans.

REFERENCES

Boyd, R. 2011. Selenium stories. *Nat Chem* 3: 570.

Fordyce, F.M. 2013. Selenium deficiency and toxicity in the environment. In O. Selinus, B.J. Alloway, J.A. Centeno et al. (eds), *Essentials of Medical Geology (Revised Edition)*: 375–416. Berlin: Springer.

Rayman, M.P. 2000. The importance of selenium to human health. *Lancet* 356(9225): 233–41.

Selenium Research for Environment and Human Health: Perspectives, Technologies and Advancements – Bañuelos, Lin, Liang & Yin (eds)

Establishment and optimization of yeast selenium fermentation system

Y.L. Xu, B. He, L. Meng, M. Qi & D.J. Tang*
Ankang Se-enriched Product Research and Development Center, Ankang, China
Key Laboratory of Se-enriched Products Development and Quality Control, Ministry of Agriculture and Rural Affairs, Ankang, China

S.Y. Yang*
Xi'an Jiaotong University, Xi'an, China

1 INTRODUCTION

Selenium (Se) is an essential trace element for the human body (Thomson 2004). However, the distribution of Se is uneven, and over two-thirds of the world is low in Se (El-Ramady et al. 2014). People can supplement Se by eating Se-enriched foods. Compared with inorganic Se, Se-enriched yeast has better bioactivity and bioavailability (Zeng et al. 2011, Alzate et al. 2010). In this study, Se-enriched yeast was evaluated, and a fermentation system was established and optimized to increase the yield of Se-enriched yeast. This work will be a foundation for the industrial production of Se-enriched yeast.

2 MATERIALS AND METHODS

2.1 *Strains*

Yeast PJ-1, PJ-2, and PJ-3 stains were obtained from School of Life Sciences in Xi'an Jiaotong University.

2.2 *Major instruments*

The following equipment were used for the study: Sw-cj-1fd clean bench from Sujing Antai, 5-L fermentation tank from Shanghai Boxing, THA-3560c vertical sterilizer from Tsao Hsin, and a fixed oscillation incubator.

2.3 *Determination of total selenium*

The content of total Se was determined by catalytic kinetic spectrophotometry of phenylhydrazine hydrochloride. After acid digestion, organic Se was transformed into selenite ions (Se^{4+}). Se^{4+} can catalyze the oxidation of phenylhydrazine hydrochloride by $KClO_3$ to form azo ion and then generate red azo dye with chromotropic acid. The absorbance of the red azo dye is proportional to the concentration of Se in a certain range. The absorbance can be determined by UV spectrophotometer at 520 nm, and the Se concentration in Se-enriched yeast can be obtained by the standard curve method.

Zero ml (blank), 1 ml, 2 ml, 3 ml, 4 ml, and 5 ml of selenite standard solution (0.5 μg Se/ml) were placed in 50-ml conical flasks, respectively. We added 0.4 ml EDTA (0.02 M), 2 ml HCl-KCl buffer (0.2 M), 2 ml $KClO_3$ solution (0.6 M), 2 ml chromotropic acid (0.03 M), 1 ml phenylhydrazine hydrochloride (0.03 M), and water bring up to volume of 15 ml. After heating in boiling water bath for 20 min and cooled for 3 minutes, the absorbance was measured at 520 nm.

2.4 *Screening Se-enriched yeast*

The Se concentrations for 100 ml Se-enriched YEPS medium treatments were 40 μg/ml and 80 μg/ml. The culture temperature was 30°C. After cultivation for 72 h, the OD_{600} and yeast culture dry weights, yield, and Se concentrations were determined. This Se-enriched yeast was chosen as the starting strain for other measurements.

2.5 *Optimization of Se accumulation in YEPS medium*

The Se concentrations for 100 ml YEPS medium treatments were 30μg/ml, 45μg/ml and 60 μg/ml and the culture temperature was 30°C. After 72h of culture, the OD_{600} and dry weights were determined, and the obtained amount of Se-enriched yeast was recorded. The supernatant was discarded by centrifugation for 20 minutes, and Se-enriched yeast powder was prepared by freeze-drying. The Se concentration in Se-enriched yeast was determined, and the yields under different levels of Se treatments were calculated.

2.6 *Establishment and optimization of 5 L and 10 L batch/fed-batch fermentation systems*

To reduce production costs, the medium used for the fermentation system was optimized, including the following: 365 g/L sucrose, 5 g/L yeast powder, 5 g/L peptone, 13 mg/L sodium selenite, 15 g/L $MgSO_4 \cdot 7H_2O$, 0.9 g/L $CaSO_4 \cdot 2H_2O$, 20 g/L K_2HPO_4, and pH 3.5.

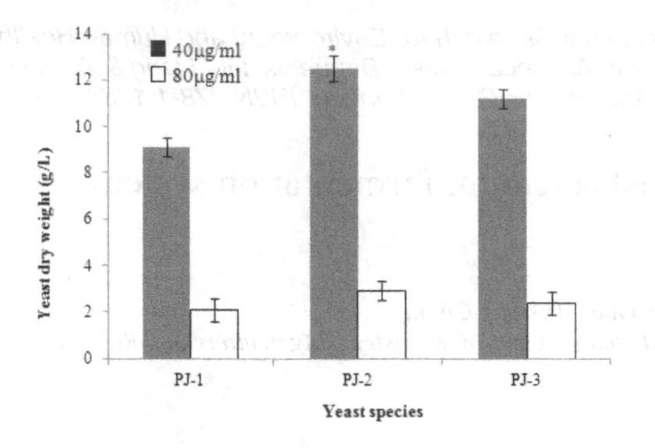

Figure 1. The yeast biomass production at different levels of Se treatments.

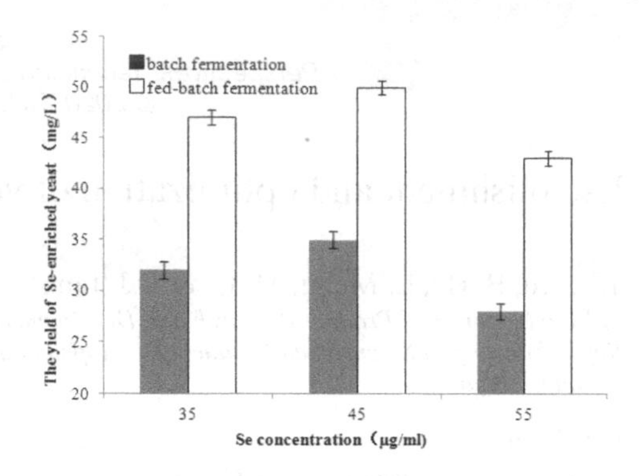

Figure 2. Comparison of the yield of Se-enriched yeast between batch and fed-batch fermentation system levels of Se treatments.

To verify and further determine whether the optimal Se concentration obtained under medium fermentation conditions can be applied for industrial scaled fermentation tank conditions, different Se concentrations of 35 μg/ml, 45 μg/ml, and 55 μg/ml were selected, and the yield of Se-enriched yeast in 5L and 10-L batch/fed-batch fermentation tanks under different Se concentration were also compared.

3 RESULTS AND DISCUSSION

3.1 *The standard curve of Se determination*

The linear regression between the measured absorption value (X) and the Se concentration (Y) was $Y = 0.1869X$ ($R^2 = 0.9968$). The linear range was 0 μg/ml to 2.5 μg/ml.

3.2 *Screening of Se-enriched yeast*

The culture temperature of Se-enriched YEPS medium treated with Se of 40 μg/ml or 80 μg/ml was 30°C for a cultivating time period of 72 h. The determination results of OD600 and yeast culture dry weight were shown in Figure 1, showing that PJ-2 yeast has relatively high Se resistance. If the yeast biomass is used as the standard to select Se-enriched yeast, PJ-2 will be used as the selected strain to carry out other subsequent optimization experiments.

3.3 *The optimization results of 5 L and 10 L batch/fed-batch fermentation systems*

The yields of Se-enriched yeast with different Se concentrations of 5-L and 10-L batch/fed-batch fermentation systems were compared, and the results showed that the yield of Se-enriched yeast reach the maximum at a Se concentration of 45 μg/ml, which is consistent with the YEPS medium condition. Furthermore, the yield of Se-enriched yeast under fed-batch fermentation was significantly increased compared with the batch fermentation (Fig. 2).

With the enlargement of the fermentation tank volume from 5 L to 10 L, the dry weight and the yield of Se-enriched yeast were not significantly affected, indicating that the fermentation system established and optimized in the experiment was stable and reliable. The yields of 5-L and 10-L fed-batch fermentation systems were 243 mg and 550 mg, respectively.

4 CONCLUSIONS

Based on the strain screened and improved YEPS medium, a stable and reliable fed-batch fermentation system has been established, which significantly increased the yield of Se-enriched yeast and could meet the requirements of large-scale industrial production.

REFERENCES

Alzate, A., Pérez-Conde, M., Gutiérrez, A. et al. 2010. Selenium-enriched fermented milk: a suitable dairy product to improve selenium intake in human. *Int Dairy J* 20(11): 761–769.

El-Ramady, H., Abdalla, N., Alshaal, T. et al. 2015. Selenium in soils under climate change, implication for human health. *Environ Chem Lett* 1: 1–19.

Thomson, C.D. 2004. Assessment of requirements for selenium and adequacy of selenium status: a review. *Eur J Clinic Nutr* (58): 391–402.

Zeng, H., Jackson, M.I., Cheng, W.H. et al. 2011. Chemical form of selenium affects its uptake, transport, and glutathione peroxidase activity in the human intestinal Caco-2 cell model. *Biol Trace Elem Res* 143(2): 1209–1218.

Selenium Research for Environment and Human Health: Perspectives, Technologies and Advancements – Bañuelos, Lin, Liang & Yin (eds)
 ISBN 978-1-138-39014-0

Selenium biofortification of a wheat crop in northeastern Pakistan using ^{77}Se as a tracer

S. Ahmad, S.D. Young & E.H. Bailey
School of Biosciences, University of Nottingham, Nottingham, UK

M.J. Watts
British Geological Survey, Nottingham, UK

M. Arshad & S. Ahmed
Mountain Agriculture Research Centre Gilgit, Pakistan Agriculture Research Council, Gilgit, Pakistan

1 INTRODUCTION

Selenium (Se) is an important dietary micronutrient for human and animal health and it is a vital component of various selenoproteins (Zarczynska et al. 2012) that play an essential role in regulating multiple body functions, including metabolism, thyroid hormone synthesis, and normal functioning of the thyroid gland (Rayman 2000). Its deficiency can result in various health disorders, such as cardiovascular diseases, cancer, and reduced fertility (Broadley et al. 2007). Previous research in northeast (NE) Pakistan has demonstrated that Se content in locally produced food is inadequate and, consequently, daily intake may fall below the WHO recommended level of 50–70 μg/d. Our study aimed to assess the viability of Se biofortification in NE Pakistan, a region considered to be selenium-deficient. A further aim was to quantify the effect of residual soil Se on subsequent crops and to assess the fate of residual soil Se. Wheat was selected for the study because it accounts for almost 75% of daily energy intake for the local population in this region (Zia et al. 2014). An enriched stable isotope (^{77}Se) was used to distinguish between fertilizer- and soil-derived Se.

2 MATERIALS AND METHODS

2.1 *Site selection*

An agricultural field at the Mountain Agriculture Research Centre (MARC) Gilgit station was selected. A total of 20 plots (2 m × 2 m) was established with 0.4 m borders between consecutive plots. Local agriculture practices were followed; for example, wheat was hand-sown, and flood irrigation of field site was performed when required.

2.2 *Preparation and application of ^{77}Se (Se_{Fert})*

Stock solutions of fertilizer selenite ($^{77}Se^4$) and selenate ($^{77}Se^6$) were prepared from elemental Se according to the method described by Mathers et al. (2017). Both Se species (designated as Se_{Fert}), were applied at three levels (0, 10, and 20 g) in a randomized block design with four replicates.

2.3 *Plant and soil sampling*

Plants were sampled by harvesting the entire central 1 m^2 of each plot and, from that, 10% of plants were subsampled. The plant samples were divided into grain, chaff and straw for each plot. A five-point composite soil sample was also collected from the central 1 m^2 of each plot, air-dried, and sieved to <2 mm. After H1, a maize crop was planted, grown on the site, and subsequently sampled (H2) and analyzed to investigate the effect of residual Se.

3 RESULTS AND DISCUSSION

Both species of Se_{Fert} (Se^4_{Fert} and Se^6_{Fert}) enhanced Se concentration in the wheat plants (grain, chaff, and straw) at harvest (Fig. 1), but Se^6_{Fert} was more effective compared to Se^4_{Fert}. A single application of 10 and 20 g/ha Se^4_{Fert} resulted in a 14- and 32-fold increase in grain Se compared to an extremely low grain Se concentration of 1.42 μg/kg in control plots. The same application rates of Se^6_{Fert} produced a 35- and 95-fold increase in grain Se over control plots. The concentration of Se in chaff was similar to that in grain and substantially greater than Se in the straw in all cases. The application of different levels and species of Se had no effect on straw, chaff, or grain yield. The mean soil-to-grain transfer factors for Se derived from soil (Se_{Soil}) and fertilizer (Se_{Fert}) were 0.32 and 9.9, respectively, which demonstrated that Se_{Fert} was more bioavailable at the first wheat harvest (H1). As expected, the transfer factor of Se^6_{Fert} was higher than that of Se^4_{Fert}. Selenium analysis of soil samples showed that a considerable proportion of Se_{Fert} was lost from the system after wheat and maize

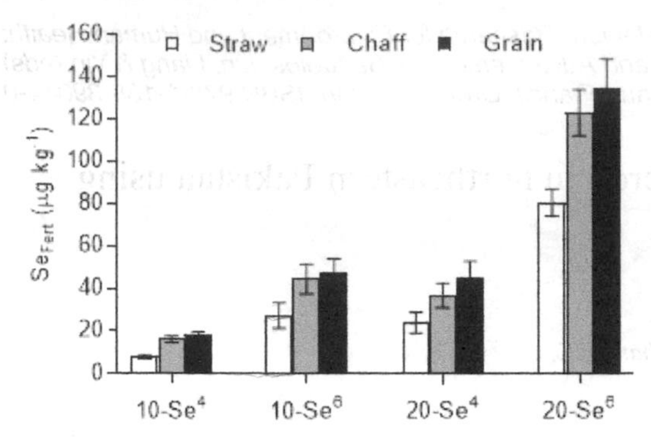

Figure 1. Selenium concentration in plant tissues originating from fertilizer Se (Se_{Fert}). Treatments indicate the Se application (10 and 20 g/ha) and the Se species (selenite and selenate). Error bars represent standard error of means (n = 4).

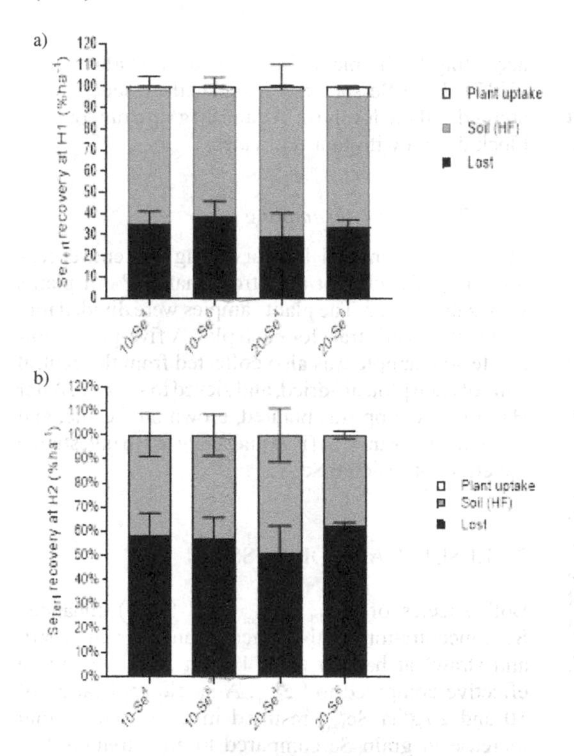

Figure 2. Status of fertilizer Se (Se_{Fert}) after (a) first wheat harvest (H1) and (b) a second 'residual' maize crop (H2). Treatments indicate the Se application (10 and 20 g/ha) and the Se species (selenite and selenate).

harvests (H1 and H2) (Figs 2a, b). The soil texture was sandy with a low organic carbon concentration (1.6%), which could contribute to the loss of Se_{Fert} from the topsoil.

Further data including fractionation of Se_{Fert} in the soil and the Se_{Fert} concentration in residual crops (maize and wheat) will be discussed in the presentation.

4 CONCLUSIONS

The application of selenate was more effective in increasing grain Se content compared to selenite. Application of 20 g/ha selenate was sufficient to raise grain Se concentration to the required level in the study area. However, before implementing it on a large scale we recommend to conducting further research to investigate the fate of residual soil Se, because large proportion of added Se was lost from soil in this experiment, which may have consequences for water quality.

REFERENCES

Broadley, M.R., White, P.J., Bryson, R.J. et al. 2007. Biofortification of UK food crops with selenium. *Proc Nutr Soc* 65: 169–181.

Mathers, A.W., Young, S.D., McGrath, S.P., Zhao, F.J., Crout, N.M.J. & Bailey, E.H. 2017. Determining the fate of selenium in wheat biofortification: an isotopically labelled field trial study. *Plant Soil* 420: 61–77.

Rayman, M.P. 2000. The importance of selenium to human health. *Lancet* 356: 233–241.

Zarczynska, K., Sobiech, P., Radwinska, J. & Rękawek, W. 2012. Effects of selenium on animal health. *J Elementol* 18(2): 329–340.

Zia, M.H., Watts, M.J., Gardner, A. & Chenery, S.R. 2014. Iodine status of soils, grain crops, and irrigation waters in Pakistan. *Environ Earth Sci* 73: 7995–8008.

Selenium Research for Environment and Human Health: Perspectives, Technologies and Advancements – Bañuelos, Lin, Liang & Yin (eds)
© 2020 Taylor and Francis Group, London, ISBN 978-1-138-39014-0

Selenium biofortification in grain crops in Brazil

L.R.G. Guilherme, C. Oliveira, A.P. Corguinha, M.A. Silva, G.F. Sousa, F.A. Namorato, P.E. Cipriano, T.S. Lara, J.H. Lessa & G. Lopes
Department of Soil Science, Federal University of Lavras, Lavras, Minas Gerais, Brazil

F.A.D. Martins
State of Minas Gerais Agricultural Research Corporation, Lavras, Minas Gerais, Brazil

1 INTRODUCTION

Biofortification is used to enhance the contents of minerals/vitamins in agricultural crops, providing enriched food to the population. While genetic biofortification comprises the selection of varieties with a greater potential for nutrient uptake, agronomic biofortification relies on fertilization strategies to provide extra nutrients to plants. This approach is relevant in tropical regions of the world, were most soils are nutrient-depleted (Lopes & Guilherme 2016). Selenium (Se) has many benefits in plant metabolism, yet it is not essential for crops. Still, due to its importance for humans and animals, Se is one of the target elements of biofortification programs.

In Brazil, recent advancements of research on the benefits of Se to cropping systems and human health (Lopes et al. 2017) have led to regulations on selenium-carrying fertilizers. The association of Se with nitrogen or phosphate fertilizers is an alternative strategy that may be feasible to increase the concentration of this element in food crops. Foliar applications of Se is sometimes most effective for biofortification strategies, as it provides the food crops, i.e. grains, with Se using smaller doses.

Selenate has been the preferred form of Se in biofortification programs, since it is easily translocated to the shoot and slowly forms selenoaminoacids following its absorption (Li et al. 2008).

Food consumed by Brazilians has low Se content (Ferreira et al. 2002). Many studies on Se biofortification of food crops have been conducted in Brazil, yet field data is still limited. This work reports recent studies conducted by our research group on Se biofortification potential of grain crops in Brazil (i.e. rice, beans, soybeans, wheat) under field conditions.

2 MATERIALS AND METHODS

Rice: Sodium selenate (Na_2SeO_4) was soil-applied as a solution (38 g Se/ha) or with monoammonium phosphate – MAP (40 g Se/ha) (Table 1). As a foliar spray, Se was applied as sodium selenite (Na_2SeO_3) and sodium selenate (10 and 40 g Se/ha), 40 days after sowing.

Common beans: Na_2SeO_4 was applied at two rates (0 and 48 g/ha) (Table 1) via top-dressing normal prilled urea (without Se) or selenium-enriched urea (480 mg Se/kg) 30 days after emergence of the seedlings (100 kg/ha of urea).

Soybeans: Na_2SeO_4 was applied (80 g Se/ha) at planting with selenium-enriched conventional MAP (C-MAP) and selenium-enriched enhanced-efficiency MAP (E-MAP) to the soil, as well as applied via foliar sprays (10, 40, and 80 g Se/ha) (Table 1).

Wheat: Selenium was applied (Na_2SeO_4) via foliar sprays (0, 12, 21, 38, 68, and 120 g Se/ha) at 36 and 41 days after sowing (Table 1).

Selenium content in the grains was determined in acid extracts following the 3051A method (USEPA 1998), and Se measurements were made either by graphite furnace atomic absorption spectrometry or inductive coupled plasma emission spectrometry. All analyses followed a rigid QA/QC protocol.

3 RESULTS AND DISCUSSION

Rice and beans are staple crops consumed in large scale in Brazil. Compared with rice (up to 0.43 mg Se/kg grain DW; Table 1), common beans have shown to be a promising grain crop to be biofortified with Se (up to 1.71 mg Se/kg grain DW) for similar Se rates). Soybean – up to 5.21 mg Se/kg grain DW with foliar Se spray – is also a crop with a great potential for Se biofortification, since it is a protein-rich food and is largely used in the food and feed industry. Lastly, wheat Se biofortification via foliar sprays – up to 2.86 mg Se/kg grain DW also represents a great strategy for providing Se to humans, as wheat is the second most widely cultivated cereal in the world and also a highly consumed food crop.

Among all grain crops evaluated, wheat appears to be the one that can significantly contribute Se to daily diets. Rice can still provide >50% of the recommended Se daily intake for adults (Table 2).

4 CONCLUSIONS

Our work has shown that Se biofortification of grain crops in Brazil can be achieved with great success by

Table 1. Selenium biofortification in grain crops in Brazil, using selenate as the selenium source.

Grain crop	Soil Characteristics pH	Clay (%)	Soil (S) or foliar (F) application	Se rate (g/ha)	Yield (t/ha)	Se content in grain (mg/kg DW)	Reference
Rice	5.7	67	S (Solution)*	38	3.651	0.29	Unpublished data
Rice	5.8	67	S (Se-enriched MAP)**	40	1.942	0.43	Unpublished data
Rice	5.8	67	F (Solution)†	10	2.173	0.31	Unpublished data
Rice	5.8	67	F (Solution)†	10	1.825	0.31	Unpublished data
Rice	6.1	45	S (MAP)**	40	5.012	0.22	Unpublished data
Rice	6.1	45	F (Solution)†	10	4.085	0.15	Unpublished data
Rice	6.1	45	F (Solution)†	10	4.338	0.13	Unpublished data
Beans	5.2	33	S (Se-enriched urea)‡	48	4.556	1.15	Unpublished data
Beans	5.2	59	S (Se-enriched urea)‡	48	3.301	1.26	Unpublished data
Beans	6.1	42	S (Se-enriched urea)‡	48	2.512	1.65	Unpublished data
Beans	5.9	15	S (Se-enriched urea)‡	48	3.861	1.71	Unpublished data
Soybeans	6	51	S (Se-enriched C-MAP)§	80	5.515	0.45	Unpublished data
Soybeans	6	51	S (Se-enriched E-MAP)§	80	5.481	0.40	Unpublished data
Soybeans	6	51	F (Solution)†	10	4.895	0.69	Unpublished data
Soybeans	6	51	F (Solution)†	40	4.667	2.69	Unpublished data
Soybeans	6	51	F (Solution)†	80	5.042	5.21	Unpublished data
Wheat	5.1	42	F (Solution)†	12	0.752	0.38	Lara et al. (2019)
Wheat	5.1	42	F (Solution)†	21	1.027	0.71	Lara et al. (2019)
Wheat	5.1	42	F (Solution)†	38	0.730	1.16	Lara et al. (2019)
Wheat	5.1	42	F (Solution)†	68	0.585	1.83	Lara et al. (2019)
Wheat	5.1	42	F (Solution)†	120	0.957	2.86	Lara et al. (2019)

* Selenium supplied via soil as sodium selenate solution; **Selenium supplied via soil as Na_2SeO_4 coating MAP granules (Se-enriched MAP 120 mg Se/kg); †Selenium supplied via foliar spray as Na_2SeO_4 or Na_2SeO_3; ‡Selenium supplied via soil as Na_2SeO_4 coating urea (Se-enriched urea 480 mg Se/kg); §Selenium supplied via soil as Na_2SeO_4 coating MAP granules (Se-enriched MAP 490 mg Se/kg).

Table 2. Percentage of recommended daily intake (RDI) eating Se-enriched grains.

Crop	Grain Se (mg/kg)*	Grain ingestion (g/person/day)** Brazil	World	Se intake (mg/person/day) Brazil	World	% RDI‡ Brazil	World
Rice	0.43	91	148	39	64	56	91
Wheat	2.86	147	184	420	526	601	752
Soybeans	5.21	NA	50†	NA	261	NA	372
Common beans	1.71	50	200	86	342	122	489

* Highest value found in our studies; ** FAO 2019, Brazil 2018, Bouis et al. 2011); †Based on recommended intake of soybeans in China; ‡RDI of Se for an adult of 0.07 mg/day (Kipp et al. 2015); NA = not available.

many different agronomic strategies. This success is relevant not only for assuring food security (i.e. better food quality) for the Brazilian population and for those countries that import agricultural products from Brazil, but also beneficial for using best management practices for enriching crops with Se for human health.

REFERENCES

Bouiss, H.E. 2011. Biofortification: a new tool to reduce micronutrient malnutrition. *Food Nutri Bull* 32(S): 31–S40.

Brazil. Ministério da Agricultura, Pecuária e Abastecimento (Ministry of Agriculture, Livestock and Food Supply). 2018. ANUÁRIO DA AGRICULTURA BRASILEIRA. Agrianual 2018. São Paulo: FNP Consultoria e Comércio.

FAO. 2019. *AMIS Market Database*. http://statistics.amis-outlook.org/data/index.html#COMPARE.

Ferreira, K.S., Gomes, J.C., Bellato, C.R. & Jordão, C.P. 2002. Concentrações de selênio em alimentos consumidos no Brasil. *Revista Panamericana de Salud Publica/Pan American Journal of Public Health* 11 (3): 172–177.

Kipp, A.P. 2015. Revised reference values for selenium intake. *J Trace Elem Med Biol* 32(1): 195–199.

Lara, T.S., Lessa, J.H.L., Souza, K.R.D. et al. 2019. Selenium biofortification of wheat grain via foliar application and its effect on plant metabolism. *J Food Composit Analy* 81(10): 18–26.

Li, H.F., McGrath, S.P. & Zhao, F.J. 2008. Selenium uptake, translocation and speciation in wheat supplied with selenate or selenite. *New Phytologist*, 178(1): 92–102.

Lopes, A.S. & Guilherme, L.R.G. 2016. A career perspective on soil management in the Cerrado region of Brazil. *Adv Agron* 137:1–72.

Lopes, G., Ávila, F.W. & Guilherme, L.R.G. 2017. Selenium behavior in the soil environment and its implication for human health. *Ciência e Agrotecnologia*, 41(6): 605–615.

United States Environmental Protection Agency (USEPA). 2007. Microwave assisted acid digestion of sediments sludge, soils, and oils. EPA SW 846 3051A:30. Washington: US EPA.

Effects of Se on plant, animal, and fish health

Selenium Research for Environment and Human Health: Perspectives, Technologies and Advancements – Bañuelos, Lin, Liang & Yin (eds)
© 2020 Taylor and Francis Group, London, ISBN 978-1-138-39014-0

Impact of selenium application on selenium and phytic acid content in cowpea seeds

V.M. Silva
UNESP – Univ Estadual Paulista, Jaboticabal-SP, Brazil

A.R. Reis*
UNESP – Univ Estadual Paulista, Tupã-SP, Brazil
UNESP – Univ Estadual Paulista, Jaboticabal-SP, Brazil

1 INTRODUCTION

Selenium (Se) is essential element for humans and animals (White 2016). Selenium can promote growth, act in oxidative stress combat, and may be considered as a beneficial element for higher plants (Reis et al. 2017). Combs Jr. (2001) estimated that more than one billion people may suffer from Se deficiency. Selenium deficiency presents the third highest deficiency risk of any mineral micronutrient in Africa (Joy et al. 2014).

The availability of Se in the soil determines its concentration in edible parts of agricultural products. Increasing concentrations of soil Se through the application of Se fertilizers increases Se uptake by plants and improves nutritional quality of food (Reis et al. 2018, Silva et al. 2018). The addition of Se together with NPK fertilizers in agricultural areas seems to be an effective and safe way to reduce Se deficiency in human and animal nutrition, as evident from decades of monitoring programs carried out in Finland (Hartikainen et al. 2005).

Phytic acid (PA) (or myo-inositol hexaphosphate, InsP6 or IP6) is the main compound of phosphorus (P) storage in seeds, comprising between 60 and 90% of all P in cereals. The bioavailability Fe, Zn, Cu, and Mg in human diets is often reduced by the presence of PA, which is considered an antinutrient (Raboy 2003). For this reason, there are commonly severe mineral micronutrient deficiencies in populations with diets composed primarily of cereals, particularly in vulnerable groups such as children and pregnant women from poor regions that commonly ingest low amounts of bioavailable Ca, Fe, Mg, and Zn (Silva et al. 2019). Thus, strategies to reduce PA concentrations in edible parts of plants are also needed in order to enhance the nutritional quality of foods.

The hypothesis of this study is that the biofortification of cowpea (*Vigna unguiculata* L. Walp.) with Se may affect the concentrations of other essential mineral elements, including macronutrients, micronutrients, and antinutrients such as PA in seed. In addition, Se biofortification using Se chemical forms, such as selenate or selenite, could have contrasting effects on seed nutritional quality. The objective of this study is to evaluate the effects of Se biofortification using selenate or selenite on the elemental composition and PA concentrations of cowpea seeds. A better understanding of factors that influence Se concentration in cowpea seeds will inform strategies for increasing the dietary intake of Se safely using this legume.

2 MATERIALS AND METHODS

The experiment was performed at São Paulo State University (UNESP) in the municipality of Selviria, Mato Grosso do Sul State, Brazil (20°20'43"S; 51°24'7"W). Soil was prepared with subsoiling, heavy disking, medium disking (twice), and leveling. Sowing was carried out on October 18, 2016, with spacing of 0.45 m between rows and a sowing density of 11.2 seeds/m^2. Fertilization of the planting furrow consisted of 20 kg/ha K_2O applied as KCl (33 kg/ha) and 20 kg/ha P_2O_5 applied as single superphosphate (110 kg/ha).

Cowpea seeds of the BRS- Tumucumaque variety were treated with pyraclostrobin (25 g/L commercial product), thiophanate-methyl (225 g/L commercial product), and fipronil (250 g/L commercial product) at 2 mL product per kg of seeds. After the seeds were dried, they were inoculated with a premium peat inoculum for cowpea (strain SEMIA 6462, product registration number SP 00581-10030-1, 2.0 × 10^9 colony forming units/g, BIOMAX, São Joaquim da Barra city, Brazil), at 8 g/kg of seed.

The seed material was collected at 77 DAS. 0.20 g dry weight (DW) was digested and analyzed according to the methodology described by Thomas et al. (2016). PA analyses were performed using an enzymatic dephosphorylation reaction to estimate PA concentration from the difference between total P and free P assayed using the phytic acid assay kit (Megazyme) described in Silva et al. (2019).

3 RESULTS AND DISCUSSIONS

The application of Se resulted in a substantial increase in Se concentrations in seed of cowpea plants. Seed Se

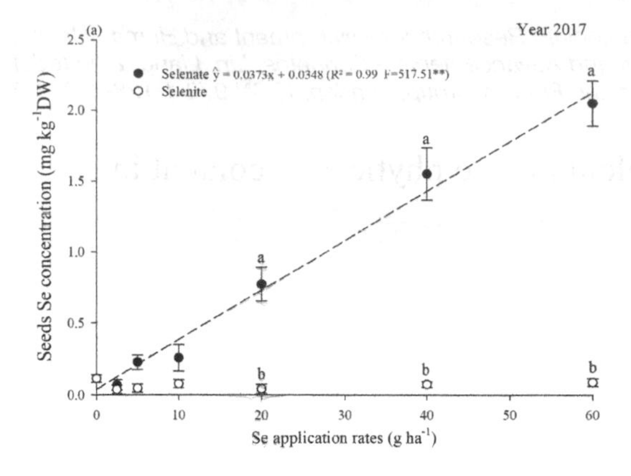

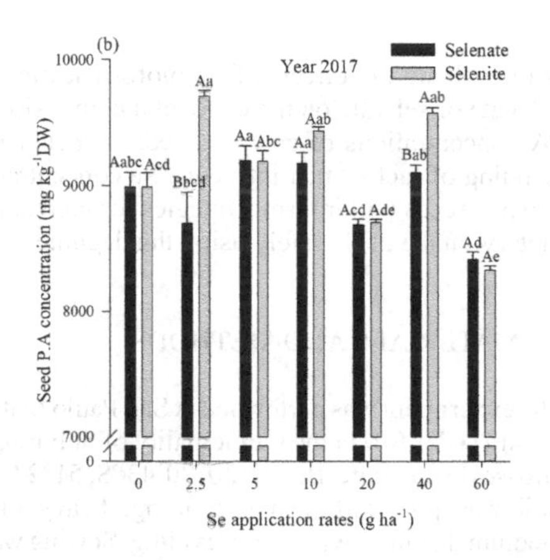

Figure 1. The effect of applying selenate or selenite on seed Se concentration (a) and phytic acid concentration (b). The standard error of the mean ($n = 4$) is indicated by error bars. Different letters indicate difference between means according to a Tukey test ($p \leq 0.05$). CV (%) = 42.68 (a); 1.74 (b).

concentration was affected by an interaction between chemical form and concentration of Se applied. The application of selenate resulted in greater seed Se concentration than the application of selenite (Fig. 1a). Seed Se concentration increased with increasing selenate application ranging from 0.04 to 2 mg/kg DW. In plants, the uptake of selenate and its translocation from roots to shoots was faster than that of selenite. Only a minor fraction of the selenite taken up by roots was translocated from roots to shoots, since it is readily converted into organic forms (White 2016). The conversion of selenate into organic Se forms is slower, and this allows greater mobility in plants (Silva et al. 2019).

Seed PA concentration was affected by both the chemical form and concentration of Se applied (Fig. 1b). Phosphate and selenite can compete for uptake by plant roots (White 2016), but the application of neither selenate nor selenite influenced P concentration in seeds (data not shown). Similarly, there was no obvious relationship between the PA concentration in seeds and the application of Se as either selenate or selenite (Fig. 1b).

4 CONCLUSION

Selenium uptake and accumulation in seeds of cowpea plants depends upon the Se source (selenate or selenite). The application of 10 g Se/ha as selenate to cowpea plants provides sufficient seed Se to increase daily human Se intake by 13 to 14 μg/d without affecting plant biomass or crop yield. Furthermore, the application of 10 g Se/ha as selenate to cowpea plants had no effect on the concentrations of PA in seed. Thus, the present study demonstrates an agronomic strategy to biofortify cowpea seeds with Se for benefiting Se nutrition in humans.

REFERENCES

Combs, G.F. 2001. Selenium in global food systems. *Br J Nutr* 85: 517–547.

Hartikainen, H. 2005. Biogeochemistry of selenium and its impact on food chain quality and human health. *J Trace Elem Med Bio* 18: 309–318.

Joy, E.J.M., Ander, E.L., Young, S.D. et al. 2014. Dietary mineral supplies in Africa. *Physiol Plant* 151: 208–229.

Raboy, V. 2003. *myo*-Inositol-1,2,3,4,5,6-hexakisphosphate. *Phytochem* 64: 1033–1043.

Reis, A., El-Ramady, H., Santos, E.F., Gratão, P.L. & Schomburg, L. 2017. Overview of selenium deficiency and toxicity worldwide: affected areas, selenium-related health issues, and case studies. In L.J. De Kok & M.J. Hakesford (eds), *Plant Ecophysiology* (11): 209–230. Cham: Springer International Publishing.

Reis, H.P.G., Barcelos, J.P.Q., Furlani Junior, E. et al. 2018.Agronomic biofortification of upland rice with selenium and nitrogen and its relation to grain quality. *J Cereal Sci* 79: 508–515.

Silva, V.M., Boleta, E.H.M., Lanza, M.G.D.B. et al. 2018. Physiological, biochemical, and ultrastructural characterization of selenium toxicity in cowpea plants. *Environ Exp Bot* 150: 172–182

Silva, V.M., Boleta, E.H.M., Martins, J.T. et al. 2019. Agronomic biofortification of cowpea with selenium: effects of selenate and selenite applications on selenium and phytate concentrations in seeds. *J Sci Food Agric* doi.org/10.1002/jsfa.9872

Thomas, C.L., Alcock, T.D., Graham, N.S. et al. 2016. Root morphology and seed and leaf ionomic traits in a *Brassica napus* L. diversity panel show wide phenotypic variation and are characteristic of crop habit. *BMC Plant Biol* 16: 214.

White, P.J. 2016. Selenium accumulation by plants. *Ann Bot* 117: 217–235.

Selenium Research for Environment and Human Health: Perspectives, Technologies and Advancements – Bañuelos, Lin, Liang & Yin (eds)
© 2020 Taylor and Francis Group, London, ISBN 978-1-138-39014-0

Selenium supplementation to salmon feeds: Speciation, toxic mode of action, and safe limits

M.H.G. Berntssen, P.A. Olsvik, J.D. Rasinger, V. Sele, K. Hamre & R. Ørnsrud
Institute for Marine Research (IMR), Bergen, Norway

T.K. Sundal & L. Buttle
Cargill Innovation Centre, Dirdal, Norway

M. Hillestad
BioMar AS, Trondheim, Norway

M. Betancor
Institute of Aquaculture, Faculty of Natural Sciences, University of Stirling, Stirling, UK

1 INTRODUCTION

1.1 *Selenium supplementation to aquafeeds*

Due to a rapid growth in aquaculture and limited access to marine resources, marine fish oil and fish meal in feeds to carnivorous marine fish species such as Atlantic salmon (*Salmo salar*) have been replaced with plant ingredients in the last decade. The change from marine to plant feed ingredients has altered the nutritional composition of salmon feeds, reducing the levels of essential micro-nutrients that are naturally high in fish meal such as selenium (Se). Several studies have indicated the need for Se supplementation in plant-based feed to marine carnivorous fish to maintain the natural high levels of Se in marine food products and to cover Se requirements in farmed fish. Selenium has a narrow range between its toxic and its beneficial effects. Supplementation of aquafeeds with organic Se-methionine (SeMet) forms or inorganic selenite hence requires toxicological assessment to set safe upper limits that protect fish health and consumer safety (Berntssen et al. 2017, Berntssen et al. 2018, Berntssen et al. 2019).

In Atlantic salmon, little is known about the toxic mode of action of Se. Thus, the first objective was to assess the underlying toxic mechanisms and sensitivity of both dietary selenite and SeMet-yeast in Atlantic salmon by using classic endpoints of Se toxicity, as well as overall metabolomics profiling approaches to assess non-target end-points of toxicity. Secondly, based on the established toxic endpoints, safe limits for Se supplementation regarding fish health were assessed using benchmark dose modelling. Finally, the feed-to-fillet transfer of Se supplementation to Atlantic salmon feeds throughout the seafood production chain was assessed, thus expanding on earlier food safety risk assessments setting the upper limit for feed Se supplementation for food-producing animals.

2 MATERIALS AND METHODS

2.1 *Feeding trials*

Atlantic salmon (147 g) were fed a low natural background organic Se diet (0.45 mg Se/kg, wet weight [ww]) fortified with 5 graded levels of inorganic selenite (0.45, 5.4, 11.0, 29.4, or 60.0 mg/kg ww) or organic SeMet (0.45, 6.2, 16.2, 21, or 39 mg/kg ww) in triplicate tanks for 3 months. For the second lowest exposure levels (5.4 and 6.2 mg/kg ww selenite and SeMet, respectively), uptake and elimination kinetics were assessed by feeding the 3-month-exposed fish control diets for a following depuration period of 3 months. Metabolomic profiles of the exposure groups were determined, as described by Berntssen et al. (2017). Briefly, extracted samples were split into equal parts for analysis on gas chromatography coupled to mass spectrometry (GC/MS) and liquid chromatography coupled to tandem mass spectrometry (LC/MS/MS) platforms. Water-soluble metabolites were separated on a LC coupled to an LTQ MS with an electrospray ionisation (ESI) source. Biological effects of excess Se supplementation was assessed by targeted biomarkers of Se toxicity pathways (markers identified by metabolomics), as well as general adverse effect parameters (plasma biochemistry, hematology, liver histopathology, and growth). Safe limits were set by model-fitting the effect data in a dose-response (lower bound) bench mark dose (BMDL) regression evaluation (Berntssen et al. 2018). The elimination and uptake rates were used in a simple one-compartmental kinetic model to predict levels in fillet based on long-term (whole production cycle) feeding with given dietary Se levels. Selenium speciation analysis in feed and muscle was assessed by anion, cation, and reversed phase exchange high-performance liquid chromatography (HPLC)-ICP-MS for selenite/selenite, SeMet,

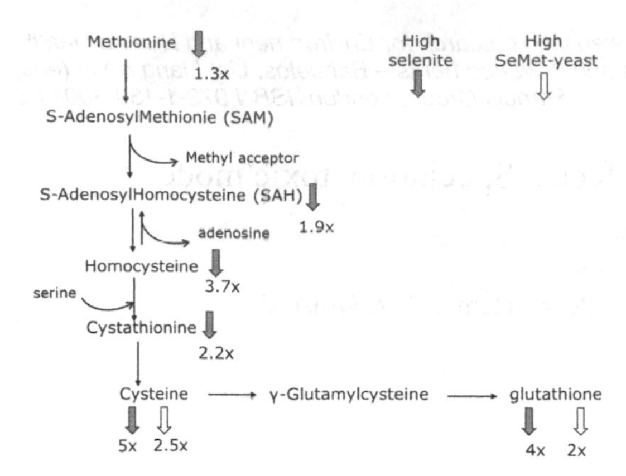

Figure 1. Overview of liver metabolites in the predicted s-Adenosylmethionine (SAM) pathways in Atlantic salmon that are significantly ($p < 0.05$) affected by selenite (grey arrow) or selenomethionine (SeMet) (open arrow) fortified diets for 3 months. Arrow indicates significant increase or decrease of metabolite compared to control.

and selenocystein (SeCys), respectively (Berntssen et al. 2019).

3 RESULTS AND DISCUSSION

3.1 *Metabolomics*

Main metabolic pathways significantly affected by 15 mg/kg selenite, and to a lesser extent 15 mg/kg SeMet, were lipid catabolism, endocannabinoids synthesis, and oxidant/glutathione metabolism. Disturbance in lipid metabolism was reflected by depressed levels of free fatty acids, monoacylglycerols, and diacylglycerols as well as endocannabinoids. Specific for selenite was the significant reduction of metabolites in the S-Adenosylmethionine (SAM) pathway, indicating a use of methyl donors that could be allied with excess Se excretion (Fig. 1).

3.2 *Safe limits fish health*

Fish fed dietary SeMet had no mortality in any of the dietary groups. In contrast, fish fed selenite showed mortality when fed 29 and 60 mg/kg. Reduced glutathione was significantly reduced compared to control group in fish fed 5.4 and 11 mg/kg selenite and 16, 21, and 39 mg/kg SeMet. For plasma parameters as a marker for liver injury, fish fed 11 mg/kg selenite had only elevated alkaline phosphatase (ALP), and fish fed 5.4 mg/kg reduced alanine aminotransferase (ALAT). Plasma liver injury was supported by histopathological alterations, such as hepatocyte degeneration and focal necrosis, in fish fed both 5.4 and 11 mg/kg selenite and fish fed the two highest SeMet level (21 and 39 mg/kg). Plasma markers of kidney function and protein and energy metabolism, such as creatinine and total protein, were also significantly reduced in fish fed the highest SeMet level (39 mg/kg).

Table 1. Uptake (α) and depuration rates ($t_{1/2}$ and k_2) in salmon fed selenite or selenomethionine (SeMet).

	Feed Se (mg/kg)	α	$t_{1/2}$ (day)	k_2 (10^{-3}/day)
SeMet	5.4	0.148 ± 0.016^a	779 ± 188^a	0.98 ± 0.39^a
Selenite	6.2	0.012 ± 0.001^b	339 ± 103^b	1.80 ± 0.45^b

3.3 *Feed-to-fillet kinetics and transfer model*

Table 1 gives the uptake (α) and depuration rates ($t_{1/2}$ and k_2) for SeMet and selenite, which are the model equation parameters. Based on these parameters, model predictions for Atlantic salmon fed plant-based feeds low in natural Se and supplemented with either 0.2 mg selenite or SeMet per kg gave fillet levels of 0.042 and 0.058 mg Se/kg (ww).

4 CONCLUSIONS

With regards to fish health, for selenite-exposed fish, a safe feed limit (as assessed BMDL regression analyses) was set at 1–2 mg/kg ww feed, based on liver SAM pathway depletion, plasma ALAT, and liver histopathology. For SeMet-fed fish, the safe feed limit was higher than for selenite with a BMDL of 2.8 mg/kg ww, based on plasma creatinine and liver histopathology and lipid metabolism.

Based on model-feed-to-fillet transfer predictions of an entire salmon food production cycle and the European Food Safety Authority (EFSA) risk assessment of Se feed supplementation for food producing terrestrial farm animals, the supplementation with 0.2 mg selenite per kg would likely be safe for the the most sensitive group of consumers (toddlers). However, supplementing feed to farm animals, including salmon, with 0.2 mg SeMet per kg would give a higher (114%) Se intake than the safe upper intake limit for toddlers.

REFERENCES

Berntssen, M.H.G., Betancor, M., Caballero, M.J. et al. 2018. Safe limits of selenomethionine and selenite supplementation to plant-based Atlantic salmon feeds. *Aquaculture* 495: 617–630.

Berntssen, M.H.G., Lundebye, A.K., Amlund, H., Sele, V. & Ørnsrud, R. 2019. Feed-to-fillet transfer of selenite and selenomethionine additives to plant-based feeds to farmed Atlantic salmon fillet. *J Food Prot* (in press).

Berntssen, M.H.G., Sundal, T.K., Olsvik, P.A. et al. 2017. Sensitivity and toxic mode of action of dietary organic and inorganic selenium in Atlantic salmon (*Salmo salar*). *Aquat Toxicol* 192: 116–126.

Selenium Research for Environment and Human Health: Perspectives, Technologies and Advancements – Bañuelos, Lin, Liang & Yin (eds)
© 2020 Taylor and Francis Group, London, ISBN 978-1-138-39014-0

Selenium protects oilseed rape against *Sclerotinia sclerotiorum* attack

Q. Cheng, C.X. Hu, W. Jia, J.J. Ming & X.H. Zhao
College of Resources and Environment, Huazhong Agricultural University, Wuhan, China

1 INTRODUCTION

Sclerotinia stem rot (SSR), caused by the necrotrophic fungal pathogen *Sclerotinia sclerotiorum* (Lib.) de Bary, poses a significant threat to yield and seed quality worldwide. Applying excessive fungicides to control *S. sclerotiorum* causes a lot of potential environmental problems. Thus, eco-friendly control methods are necessary to be developed. Selenium (Se) can sometimes serve as a beneficial element for plant growth and development. Selenium, as well as dissolved organic matter derived from rape straw pretreated with Se in soil, can inhibit mycelial growth of *S. sclerotiorum* by damaging the membrane system and interfering with its metabolism (Jia et al. 2018, 2019).

The aims of the study were to (1) test the protective effect of Se on *S. sclerotiorum* infection in oilseed rape leaves and (2) reveal the underlying physiological mechanisms of Se improving the resistance of oilseed rape against *S. sclerotiorum*. Results of this study will provide a new direction not only for the control of *S. sclerotiorum*, but also for promoting the industrial development of the oilseed rape.

2 MATERIALS AND METHODS

Experimental soil was collected from eco-agriculture base, College of Resources and Environment (N 30° 28' 43", E 114° 21' 10"), Huazhong Agricultural University, Wuhan, China. Four levels of Se were established in the pot experiment (0, 0.1, 0.5, and 1.0 mg/kg soil treatment was denoted as control, $Se_{0.1}$, $Se_{0.5}$, and $Se_{1.0}$, respectively). Na_2SeO_3 was used for the Se treatments. We used seeds of rape (*Brassica napus* var Zhongshuang No.9).

S. sclerotiorum (JZJL-13) was selected as the fungus for this study. At the flowering stage of oilseed rape, a 2-day-old mycelial agar plug (4 mm in diameter) was placed face down on the leaf, adjacent to the midrib. Each plant was inoculated for three leaves. The diameter of each lesion was measured 5 days post inoculation (dpi) using the crossing method. Disease incidence (%) = (number of completely wilted leaves/number of inoculated leaves) × 100%.

Both a low Se concentration ($Se_{0.1}$) and a high Se concentration ($Se_{1.0}$) were selected to further investigate ultrastructure of rape leaves post inoculation using optical microscopic observations and transmission electron microscopy.

Gaseous exchange parameters were also measured including: net photosynthetic rate (Pn), stomatal conductance (Gs), intercellular CO_2 (Ci), and transpiration rate (Tr) were determined using a LI-6400XT portable photosynthesis system (LI-COR, Lincoln, USA).

For biochemical measurements, rape leaves (0.5 g) were ground in liquid nitrogen, and then homogenized with different extracting buffers to determine catalase (CAT), superoxide dismutase (SOD), peroxidase (POD), hydrogen peroxide (H_2O_2), and superoxide radicals (O_2^-). Frozen leaves were homogenized in 0.8 mL methyl alcohol and then centrifuged at 12000 × *g* for 15 min at 4°C. Next, 200 μL supernatant were transferred into a glass vial to conduct further analysis using LC-MS. Rape leaves were also collected, and the gene expressions of four defense gene (*CHI, EDS1, NPR1,* and *PDF 1.2*) in rape leaves were measured by real-time qPCR.

Metabonomics analysis was performed by SIEVE software, and t-test was used to compare the two means using SPSS 20.0.

3 RESULTS AND DISCUSSION

3.1 *Disease assessment*

Selenium applied to soil significantly shortened spot diameter of rape leaves. Especially, $Se_{0.5}$ caused the minimum of spot diameter and decreased it by 32.6% compared to control. $Se_{0.1}$ and $Se_{0.5}$ also reduced incidence of SSR caused by *S. sclerotiorum*, and decreased spot diameter by 36.4% and 45.5%, respectively.

3.2 *Selenium maintained internal fine structure and improved the gas exchange of rape leaves*

The leaf is a specialized organ evolved for photosynthesis. Low Se concentrations helped maintain chloroplast structure, while high Se level caused more serious damage to rape leaves once infected with *S. sclerotiorum*. $Se_{0.5}$ increased net photosynthetic rate (Pn), stomatal conductance (Gs) and transpiration rate (Tr) after inoculation and alleviated damage of rape leaves induced by *S. sclerotiorum*. Low Se levels helped to maintain leaf internal structure.

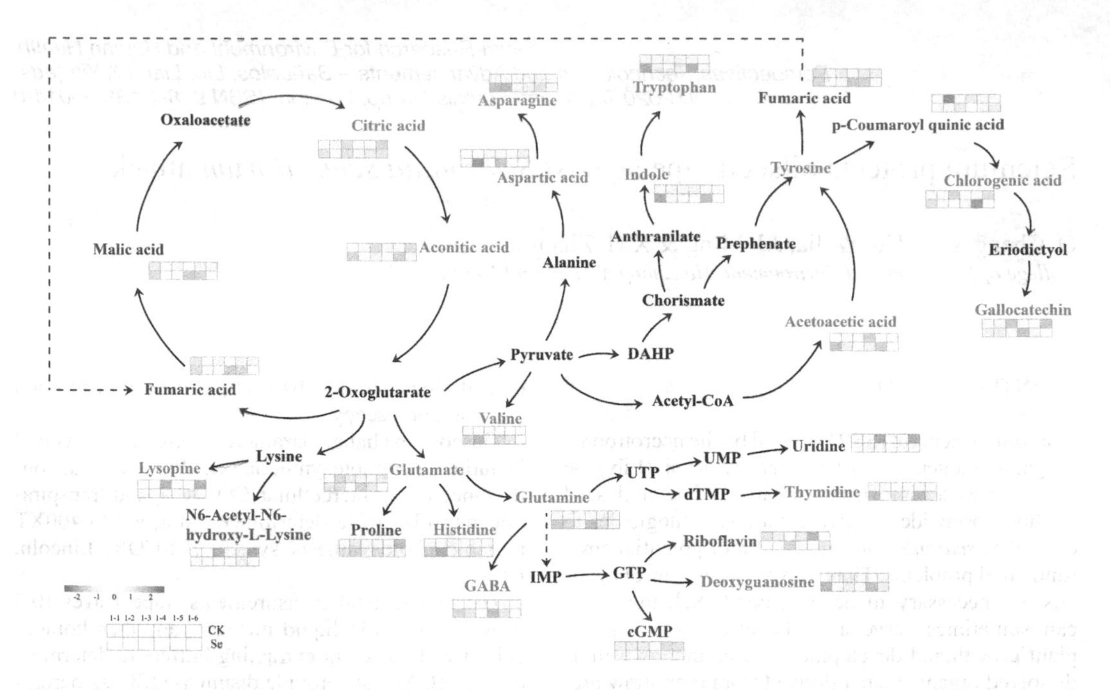

Figure 1. Pathway analysis of metabolites in *S. sclerotiorum* infecting leaves of rape in selenium-treated soil.

3.3 *Selenium improved antioxidant enzyme activities under S. sclerotiorum invasion*

Low Se concentrations (0.1 or 0.5 mg/kg) improved antioxidative enzyme activities (CAT, PPO, SOD, and POD) and reduced H_2O_2 and O_2^- when suffering an attack from *S. sclerotiorum*.

3.4 *Selenium upregulated defense gene levels*

Transcriptional levels of defense genes (*CHI, EDS1, NPR1 and PDF 1.2*) responded differently to Se application. However, all treatments showed similar variation tendency over time compared to control.

3.5 *Selenium supplement in soil changed metabolism in rape leaves*

When Se was applied, 21 metabolites were upregulated, while 13 metabolites downregulated in rape leaves. The metabolic network structure of rape leaves was shown in Figure 1. The addition of Se into soil promoted amino acid metabolism, including tryptophan, tyrosine, lysopine, histidine, L-glutamic acid and proline. Furthermore, compounds involved in TCA cycle changed remarkably, such as upregulated citrate and aconitic acid as well as downregulated malic acid and fumaric acid.

4 CONCLUSIONS

Proper Se concentration applied to soil improved oilseed rape resistance against *S. sclerotiorum*, thus contributing to small lesion diameter and lower incidence. In addition, Se led to a combined action of enhanced antioxidant system, upregulated defense gene expression and amino acid metabolism of oilseed rape.

REFERENCES

Jia, W., Hu, C.X., Ming, J.J. et al. 2018. Action of selenium against *Sclerotinia sclerotiorum*: damaging membrane system and interfering with metabolism. *Pest Biochem Physiol* 150: 10–16.

Jia, W., Hu, C.X., Xu, J.Y. et al. 2019. Dissolved organic matter derived from rape straw pretreated with selenium in soil improves the inhibition of *Sclerotinia sclerotiorum* growth. *J Hazard Mater* 369: 601–610.

Selenium Research for Environment and Human Health: Perspectives, Technologies and Advancements – Bañuelos, Lin, Liang & Yin (eds)
© 2020 Taylor and Francis Group, London, ISBN 978-1-138-39014-0

Selenium negatively regulates *Fusarium graminearum* growth and mycotoxin production

X.Y. Mao, L. Yang, Y.H. Zhang & T. Li*
Jiangsu Key Laboratory of Crop Genetics and Physiology/ Key Laboratory of Plant Functional Genomics of the Ministry of Education/ Jiangsu Key Laboratory of Crop Genomics and Molecular Breeding/Jiangsu Co-Innovation Center for Modern Production Technology of Grain Crops, Yangzhou University, Yangzhou, China

Q. Hu
College of Food Science and Engineering, Yangzhou University, Yangzhou, Jiangsu, China

1 INTRODUCTION

Fusarium head blight (FHB) of wheat caused by *Fusarium graminearum* not only hinders grain development, but it leads to significant reduction in yield, and also produces trichothecene mycotoxins such as deoxynivalenol (DON) and zearalenone, which are a serious concern for human and animal health (Paranidharan et al. 2008). Selenium (Se) is essential for human and animals, and can be beneficial for plants (Sun *et al.* 2010). The objective of this study is to study the effect of Se on *F. graminearum* growth and its DON production under *in vitro* conditions.

2 MATERIALS AND METHODS

2.1 *Experimental design*

Fusarium graminearum isolate PH-1 was cultured on PDA medium for 72 h and was then cleaved into round fungal blocks of 6 mm in diameter with a puncher. One block of fungal inoculum was placed onto the center of TBI solid medium with prior supplementation with sodium selenite (Na_2SeO_3), sodium selenate (Na_2SeO_4), selenomethionine (SeMet), or selenocystine (SeCys2), with three treatment levels of 0, 150, and 300 mg/L for each Se chemotype, and three replicates per treatment. The medium for each was then incubated at 25°C in an incubator.

2.2 *Effect of Se on the growth of* F. graminearum

The diameter of the colony was measured every 24 h. The size of the colony is determined by calipers. The growth inhibition rate of the hyphae was calculated by [(Treatment – Control) / Control] *100.

2.3 *Effect of Se on DON production*

The medium was incubated for 96 h and then crushed with an 80-micron nylon filter mesh. The smashed medium was subjected to a 1:1 (vol/vol) dilution with water and then transferred into a conical flask. Afterwards, the diluted smashed medium was incubated at 25°C at 200 rpm on a rotary shaker. After filtration, the toxin content was determined by enzyme-linked immunosorbent assay kit (Casco Biotech, Zhuozhou, China).

2.4 *Selenium inhibited* F. graminearum *growth*

The fungal block was transferred from the TBI solid medium containing 300 mg/L Se (on which the growth of *F. graminearum* completely stopped), and then to a new solid PDA medium without Se in order to understand if *F. graminearum* growth can be recovered.

2.5 *Effect of Se on alterations of protein components of* F. graminearum *strains*

One colony of *F. graminearum* from three replicates of each treatment was harvested, and each kind of the hyphae was moved into a 1.5-mL centrifuge tube. Then, 300 μL of purified water and 900 μL of absolute ethanol were added to mix. The mixture was centrifuged at 12,000 rpm for 2 min, and the supernatant was discarded. Then, 50 μL of formic acid aqueous solution (70%, v/v) and 50ul of acetonitrile were added to mix. Afterwards, the mixture was centrifuged at 12,000 rpm for 2 min, and the supernatant containing the protein extract of the hypha was transferred into a 1.5-mL centrifuge tube. To acquire the MS spectra, 1 μL of the protein extract was drop on MALDI plate. After the protein layer became dry at ambient condition, 1 μL of CHCA (MALDI matrix) was dropped on the dry protein layer and was left to dry. The MALDI plate was then inserted into the MALDI Biotyper instrument (Bruker, German) to acquire the full scan MS spectra of the hyphae protein extract.

3 RESULTS AND DISCUSSION

3.1 *Selenium can inhibit growth of* F. graminearum

Compared with control, different Se treatments significantly delayed the growth of the fungus, and the

Table 1. Inhibition of mycelial growth by different concentrations and chemical forms of Se *in vitro* for 96 h.

Treatment	Concentration (mg/L)	Colony diameter (mm)	Inhibition rate (%)
Ck	0	81.0	—
Na_2SeO_3	150	35.3	−56.5
	300	27.3	−66.3
Na_2SeO_4	150	13.0	−84.0
	300	11.0	−86.5
SeMet	150	9.7	−88.1
	300	6.0	−92.6
SeCys2	150	33.4	−58.8
	300	32.2	−60.3

diameter of the colonies became smaller. The diameter of the colony decreased with an increase of Se concentration of the same Se form. Under the same concentration, SeMet had the best inhibitory effect, followed by Na_2SeO_4, while inhibitory effects of $SeCys_2$ and Na_2SeO_3 were weaker than the former two (Table 1). The fungal blocks completely stopped growth on the medium containing the highest level of Se and then were transferred to a fresh medium free of Se; growth of the fungal blocks resumed after 24 h incubation. This observation suggests that Se inhibited the growth of *F. graminearum* hyphae, although the pathogen was not killed.

3.2 *Selenium inhibits mycotoxin production by* F. graminearum

After 96 h of culture, the DON content of the control was up to 600 ppb, whereas it was less than 65 ppb or not detected across all the Se treatments. This effect is probably due to low biomass of *F. graminearum* caused by high concentration of Se.

3.3 *Selenium changed the protein components of* F. graminearum

Compared with control, the patterns of protein components of *F. graminearum* under Se treatments changed significantly; however, the patterns were similar across the treatments with different Se forms Na_2SeO_3 and SeCys2 of 150 mg/L (Fig. 1).

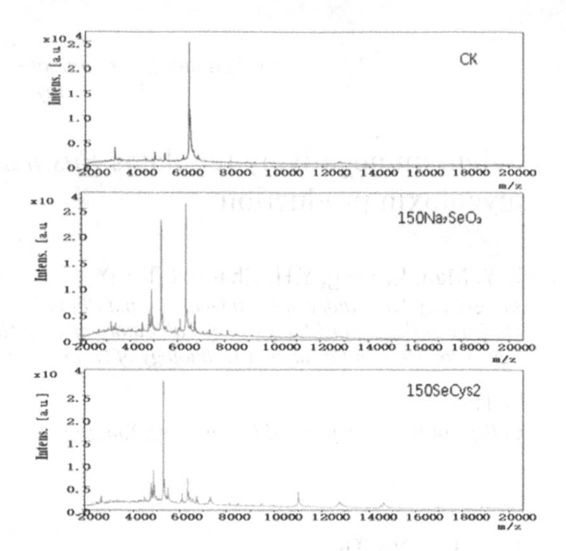

Figure 1. Changes in protein composition of strains treated with 150 mg/L Na_2SeO_3 or SeCys2.

4 CONCLUSIONS

Reducing DON accumulation produced by *F. graminearum* is critically important to ensure the safe production of wheat. At present, the prevention and control of wheat Fusarium head blight and DON accumulation have not been very effective. Selenium can inhibit the growth of *F. graminearum* and significantly reduce DON accumulation, but its effects varied with Se chemotypes. Currently, we are trying to understand the mechanism by which Se negatively regulates the growth and production of DON, which would be helpful for preventing and controlling the disease and reducing DON contamination, and for producing Se-fortified functional foods.

REFERENCES

Paranidharan, V., Abu-Nada, Y., Hamzehzarghani, H. & Kushalappa, A.C. 2008. Resistance-related metabolites in wheat against, *Fusarium graminearum*, and the virulence factor deoxynivalenol (DON). *Bot* 86: 1168–1179.

Sun, G.X., Liu, X., Williams, P.N. & Zhu, Y.G. 2010. Distribution and translocation of selenium from soil to grain and its speciation in paddy rice (*Oryza sativa* L.). *Environ Sci Tech* 44: 6706–6711.

Selenium Research for Environment and Human Health: Perspectives, Technologies and Advancements – Bañuelos, Lin, Liang & Yin (eds)
© 2020 Taylor and Francis Group, London, ISBN 978-1-138-39014-0

Feeding weaned beef calves Se-biofortified alfalfa hay as a preconditioning strategy

J.A. Hall
Oregon State University, Corvallis, Oregon, USA

A. Isaiah
Texas A&M University, College Station, Texas, USA

1 INTRODUCTION

Selenium (Se) is an essential trace mineral important for immune function and overall health of cattle. Optimal immune function is critical for calves undergoing the stresses of weaning, relocation to feedlots, and commingling with animals of different origins. The majority of Oregon-grown calves enter the feedlot. Even with good vaccination programs, producers often encounter significant health issues in the feedlot, including mortality. Reducing these losses by enhancing immune function would have a significant economic impact for cattle producers.

In plants, Se is primarily incorporated into methionine as selenomethionine, and when forage is consumed by livestock, Se from selenomethionine is incorporated into selenoproteins, whose functions range from antioxidant, anti-inflammatory, and detoxification to thyroid hormone activation. Nitrogenous fertilizers, widely hailed as one of the most important advances in agricultural technology, increase biomass but dilute essential minerals like Se, emphasizing the need for Se amendments. Application of Se directly to pastures and hayfields increases forage Se concentration in a dose-dependent manner (Filley et al. 2007) and improves blood Se concentrations, animal performance, and immunity (Hall et al. 2013a, Hall et al. 2013b). The objectives of this study were to show that feeding Se-biofortified hay increases whole-blood (WB) Se concentrations, enriches the nasal microbial diversity, improves calf performance, and aids in disease prevention in the feedlot.

2 MATERIALS AND METHODS

Study design: This was a prospective feeding trial of 9-week duration (October 11, 2017 through December 11, 2017) involving 30 weaned beef calves (steers), primarily of Angus breeding. The study design consisted of 2 treatment groups, with three pens of five animals per treatment. The weaned beef calves at baseline ranged in age from 6 to 9 months and originated from the Oregon State University beef ranch, Corvallis, OR, USA. Body weights at baseline ranged from 240 to 334 kg (286 ± 9.3 kg, mean ± SEM). Using a randomized complete block design, calves were blocked at the time of weaning by body weight, and fed a mixture of alfalfa and grass hay twice daily. The amount of hay fed was adjusted weekly to ensure that calves had all the hay they wanted for consumption yet with minimal wastage. The ration was formulated for growing beef calves in the 250 to 350 kg BW range to achieve a target average daily gain of 0.5 kg/day. The goal was to feed hay at a rate of 85% alfalfa and 15% grass hay. Calves were transitioned to their respective hay rations over a 3-week period. Also in Week 3, a medicated grain-based concentrate was fed (0.45 kg/calf/day). This amount was increased to 0.68 kg/calf/day in Weeks 4 through 9. The grain pellets contained a coccidiostat (monensin sodium, 35 g/ton). Pellets were fed once daily and consumed before hay was fed. After the 9-week preconditioning period, calves were shipped to a commercial feedlot. Calves were re-examined 23 days after transfer to the feedlot, and blood and fecal samples were collected. Finally, at the end of the 25-week feedlot period, calves were sent to a commercial meat packing plant, and hot carcass weights, carcass quality grade (no-roll, standard, select, choice, prime), and yield grade (1 to 4) were recorded at the time of slaughter.

Selenium-biofortified alfalfa hay: Third cutting alfalfa hay was enriched with Se by mixing inorganic sodium selenate with water and spraying it onto the foliage (approximately 10 cm height) of an alfalfa field at application rates of 0 or 89.9 g Se/ha. Plant samples were prepared for Se analysis as previously described (Davis et al. 2012), and Se concentrations were determined using an inductively coupled argon plasma emission spectrometry method by a commercial laboratory (Utah Veterinary Diagnostic Laboratory, Logan, UT).

Blood collection for Se analysis: Blood samples were collected from the jugular vein of calves, at baseline, after 3, 6, and 9 weeks of alfalfa hay consumption, and 3 weeks after transfer to the feedlot.

Nasal microbiota sample collection: Nasal swabs were collected from all calves 30 days before the feeding trial began, after 9 weeks of alfalfa hay consumption, and again 3 weeks after transfer to the feedlot. Sterile, individually wrapped, polyester tipped applicators were inserted approximately 10 cm into the nares, twirled to collect a mucosal swab, and then placed into individual sterile containers. Microbial DNA was extracted from nasal swab samples using MoBio Power soil DNA isolation kit (MoBio Laboratories, USA) as previously described (Hall et al. 2017). The V4 region of the 16S rRNA gene was amplified and sequenced at the MR DNA Laboratory (Shallowater, TX, USA). The sequencing data were processed using the QIIME 2 v 2018.6 (https://qiime2.org/) platform (Boylen et al. 2018).

3 RESULTS AND DISCUSSIONS

Agronomic biofortification: Fertilizing the alfalfa hay field with sodium selenate at 89.9 g/ha increased the Se content of third-cutting alfalfa from 0.06 (non-fertilized control) to 3.47 mg Se/kg dry matter. Calculated Se intake from alfalfa hay was 0.46 and 26.82 mg Se/calf/day, respectively, for calves consuming hay with Se concentrations of 0.06 and 3.47 mg Se/kg dry matter. The concentration of Se in the grass hay was 0.12 mg/kg. Calculated Se intake from grass hay was approximately 0.11 mg Se/calf/day. The measured Se concentration of the grain concentrate was 0.77 mg/kg dry matter. Calculated Se intake from grain concentrate was 0.52 mg Se/calf/day in Weeks 7 through 9 of the preconditioning feeding period. Thus, total dietary Se intake during Weeks 7 to 9 was 1.09 and 27.45 mg Se/calf/day, respectively, for calves in control and high-Se treatment groups.

Whole-blood Se concentrations: Feeding Se-fertilized alfalfa hay was effective at increasing WB-Se concentrations in weaned beef calves ($P_{Treatment}$, P_{Week}, and $P_{Interaction}$ all ≤ 0.004). The WB-Se concentrations continued to increase throughout the 9-week preconditioning period (Week 3: +50%; Week 6: +192%; Week 9: +292%). During the initial feedlot period (Week 12: +272%), WB-Se concentrations remained higher ($P_{Treatment} < 0.001$).

Body weights and carcass data: Feeding Se-fertilized alfalfa hay was effective at increasing BW in weaned beef calves ($P_{Treatment}$, $= 0.03$; P_{Week}, <0.001 and $P_{Interaction} = 0.42$) and tended to be effective at increasing hot carcass weight at Week 34 ($P_{Treatment} = 0.07$). At slaughter, no significant differences were observed for carcass quality grade levels, as both groups had all but 4 animals with choice grade. However, yield grade [1 (most desirable trim); 2, 3 (industry average), and 4] was improved in animals that were fed Se-enriched alfalfa hay during the preconditioning period ($P_{Treatment} = 0.008$), as more calves in the group fed Se-enriched alfalfa hay had 1 or 2 (86%) compared with calves in the control group (36%).

Nasal microbiota: Species richness (observed OTU) were significantly increased in calves fed Se-biofortified hay ($p < 0.001$) when compared with control calves. The alpha diversity measures, Chao1, and Shannon Index were also significantly increased in calves fed Se-biofortified hay ($p = 0.01$, $p = 0.03$, respectively). Principal component analysis plots showed that baseline samples formed a separate cluster when compared to samples collected after the 9-week feeding trial, and samples collected 3 weeks after entering the feedlot. ANOSIM tests of UniFrac distances found a significant difference in the overall bacterial composition of the nasal microbiota between control calves and calves that were fed Se-biofortified hay ($R_{unweighted} = 0.8836$, $R_{weighted} = 0.4543$ respectively, $p = 0.001$). Pairwise ANOSIM tests on weighted and unweighted UniFrac distances showed a significant difference in nasal samples collected at baseline and after consumption of Se-enriched hay and after entering feedlot ($p = 0.001$).

4 CONCLUSIONS

Feeding Se-enriched alfalfa hay during the weaning transition period improved growth and final carcass weights of beef calves. In addition, the nasopharyngeal microbial diversity was increased. This, in turn, may ameliorate a negative risk factor for bovine respiratory disease complex.

REFERENCES

Bolyen, E., Rideout, J.R., Dillon, M.R. et al. 2018. QIIME 2: Reproducible, interactive, scalable, and extensible microbiome data science. *PeerJ Prepr* 6: e27295v2.

Davis, T.Z., Stegelmeier, B.L., Panter, K.E. et al. 2012. Toxicokinetics and pathology of plant-associated acute selenium toxicosis in steers. *J Vet Diagn Invest* 24(2): 319–27.

Filley, S.J., Peters, A. & Bouska, C. 2007. Effect of selenium fertilizer on forage selenium content. *J Anim Sci* 85: 35.

Hall, J.A., Bobe, G., Hunter, J.K. et al. 2013a. Effect of feeding selenium-fertilized alfalfa hay on performance of weaned beef calves. *PLoS One* 8(3): e58188.

Hall, J.A., Bobe, G., Vorachek, W.R. et al. 2013b. Effects of feeding selenium-enriched alfalfa hay on immunity and health of weaned beef calves. *Biol Trace Elem Res* 156: 96–110.

Hall, J.A., Isaiah, A., Estill, C.T. et al. 2017. Weaned beef calves fed selenium-biofortified alfalfa hay have an enriched nasal microbiota compared with healthy controls. *PLoS One* 12(6): e0179215.

Selenium Research for Environment and Human Health: Perspectives, Technologies and Advancements – Bañuelos, Lin, Liang & Yin (eds)
© 2020 Taylor and Francis Group, London, ISBN 978-1-138-39014-0

Fed with selenium-enriched wheat: Impact on selenium distribution in rats

X.Q. Lu, Z.S. He & X.B. Yin*
School of Earth and Space Sciences, University of Science and Technology of China, Hefei, China

Y. Liu, L.X. Yuan & F. Li
Anhui Setek Co., Ltd, Chuzhou, Anhui, China

L.Q. Qin & X. Jiang
School of Public Healthy, Soochow University, Suzhou, China

L.X. Gao
Xiwang Agriculture Technology Station, Quanjiao, Anhui, China

1 INTRODUCTION

Selenium (Se) is an essential trace element for both animals and humans (Terry et al. 2000). Selenium deficiency has been linked to two kinds of diseases, which are endemic cardiomyopathy called Keshan disease and Kashin-Beck disease, a type of osteoarthritis. Therefore, Se supplements are necessary for those who are living in the areas with low Se. Selenium in Se-enriched foods is more bioavailable than inorganic Se and are gradually adopted for use as Se supplements in the last ten years (Ge et al. 1983). The distribution of Se in different tissues and its physiological effects on animals with various forms of supplemental Se have been investigated in many studies (Whanger & Butler 1988, Mahan et al. 1999). However, there are few reports related to Se-enriched wheat and its effects on animals. Therefore, the aim of this study is to evaluate the effect of Se-enriched wheat on Se distribution of rats.

2 MATERIALS AND METHODS

2.1 *Study design*

Selenium-enriched wheat was provided by Suzhou Setek Co., Ltd. The Se concentration of the wheat was 22 mg/kg, and more than 90% of the total was selenomethionine (SeMet). The Se concentration of basal feed (C0) for rats was 0.44 mg/kg. Appropriate amounts of Se-enriched wheat were added into basal feed and mixed to total Se of 0.8 mg/kg (C1), 1.2 mg/kg (C2), and 1.6 mg/kg (C3), respectively. 40 male Wistar rats (170–220 g) were randomly assigned in to four groups fed with C0, C1, C2, and C3. The back hair of the rats was collected for Se analysis every thirty days. At the end of three months, rats were sacrificed by exsanguination under ether anesthesia. Organs and femoral muscle were removed into polyethylene sample pack and kept at −20°C until detection. All animal experiments were approved by the Soochow University Animal Welfare Committee.

2.2 *Assessment of Se concentration in wheat, hair, muscle, and organs*

The procedure for determination of total Se was described in the previous study (Finley 2007).

3 RESULTS AND DISCUSSION

3.1 *Hair Se concentration*

The hair Se was at the same level of 0.5 mg/kg in the four groups at pre-treatment. Hair Se of Group C0 slightly decreased to 0.45 mg/kg at Month 1, maintained the same value at Month 2 and then decreased to 0.4 mg/kg at Month 3. Compared with C0, hair Se of Groups C1, C2, and C3 increased at Month 1 and Month 2, and it slightly decreased at Month 3 (Fig. 1). Hair is a pathway for animals to excrete Se, and hair Se is an important marker to measure the Se level in animal bodies. Hair Se was linearly increased with the Se supplementation. The concentrations of trace elements in animal's hair remain isolated from animal metabolic activities and indicate the concentrations of the elements at a particular time period (Ortac et al. 2006). The increase of hair selenium of rats in Group C1, C2, and C3 from the start of experiment to Month 2, and the slight decrease of hair Se at Month 3 illustrated that Se level in rats peaked in Month 2 with the supplementation of Se-enriched wheat.

3.2 *Muscle and organ Se*

Similar to the result of the hair, the Se concentrations in muscle and all the organs, except testis, have a fine linear relationship with the feed Se concentrations (Fig. 2). This result indicates that Se in Se-enriched wheat was easily absorbed by rats and distributed to muscle and different organs. However, testis Se

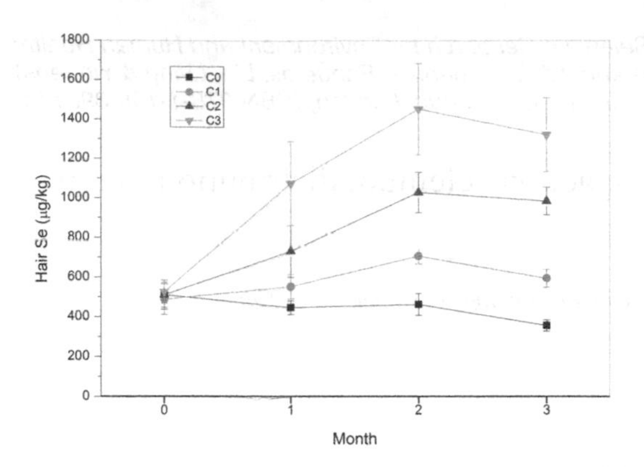

Figure 1. Hair Se of rats supplied with feed of different Se concentration mixed with Se-enriched wheat.

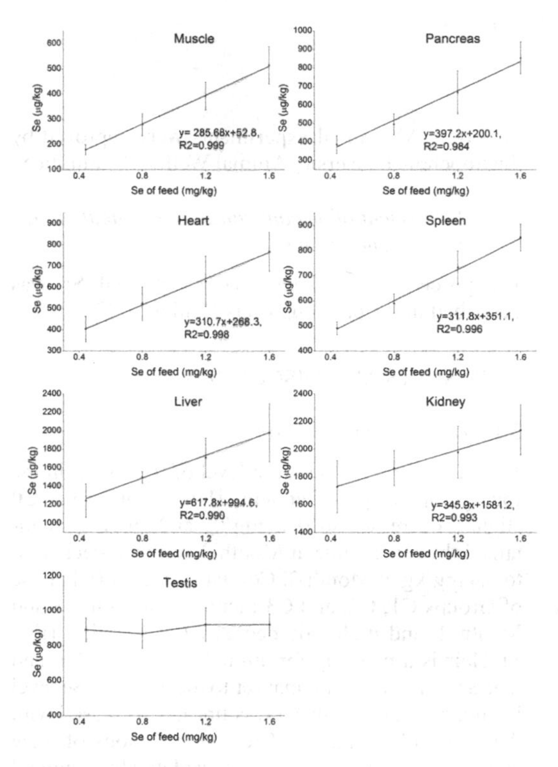

Figure 2. Muscle and organ Se of rats supplied with feed of different Se concentration mixing with Se-enriched wheat.

remained at the same level of 0.9 mg/kg, and liver and kidney had accumulated more Se than muscle and other organs in all groups of rats.

The Se distribution data in muscle and different organs of rats after supplement with Se-enriched wheat showed that the liver and kidneys accumulated the highest amount of Se, followed by pancreas, spleen, and heart. This rat study demonstrated that Se in muscle and tissues linearly increased with the supplement of Se-enriched wheat. Selenium in organs was the main Se pool for animals. Earlier studies showed that organ Se concentrations decreased slower in rats fed with SeMet (Se-enriched yeast) than selenite (Behne et al. 2009). Accordingly, we could consider that Se-enriched wheat, with SeMet as the main Se component, was an effective substance to maintain the Se level in animal body. Another interesting finding from this study was that the testis Se concentration remained constant after rats were supplied with feed of different Se concentration. This result may occur due to the abundant Se in the basal feed, which has already met the Se needs of rat testis. Testis is an important organ of the animal reproductive system. Excessive Se may be harmful for their next generation. Hence, the blood-testis barrier would prevent excessive Se from entering the testis.

4 CONCLUSIONS

Selenium-enriched wheat significantly improved muscle and most organs Se of rats. Hair Se increased in the first two months, and then slightly decreased. There was fine linear relationship between organ Se concentrations of rats and feed Se concentrations. Liver and kidneys accumulated more Se than other organs.

REFERENCES

Behne, D., Alber, D. & Kyriakopoulos, A. 2009. Effects of long-term selenium yeast supplementation on selenium status studied in the rat. *J Trace Elem Med Biol* 23(4): 258–264.

Finley, J.W. 2007. Increased intakes of selenium-enriched foods may benefit human health. *J Sci Food Agric* 87(9): 1620–1629.

Ge, K.Y., Xue, A., Bai, J. & Wang, S.Q. 1983. Keshan disease-an endemic cardiomyopathy in china. *Virchows Arch A Pathol Anat Histol* 401(1): 1–15.

Mahan, D.C., Cline, T.R. & Richert, B. 1999. Effects of dietary levels of selenium-enriched yeast and sodium selenite as selenium sources fed to growing-finishing pigs on performance, tissue selenium, serum glutathione peroxidase activity, carcass characteristics, and loin quality. *J Animal Sci* 77(8): 2172–2179.

Ortac, E., Ozkaya, O., Saraymen, R. et al. 2006. Low hair selenium and plasma glutathione peroxidase in children with chronic renal failure. *Pediatr Nephrol* 21(11): 1739–1745.

Terry, N., Zayed, A.M., de Souza, M.P. & Tarun, A.S. 2000. Selenium in higher plants. *Annu Rev Plant Biol* 51: 401–432.

Whanger, P.D. & Butler, J.A. 1988. Effects of various dietary levels of selenium as selenite or selenomethionine on tissue selenium levels and glutathione-peroxidase activity in rats. *J Nutr* 118(7): 846–852.

Selenium Research for Environment and Human Health: Perspectives, Technologies and Advancements – Bañuelos, Lin, Liang & Yin (eds)
© 2020 Taylor and Francis Group, London, ISBN 978-1-138-39014-0

Effect of melatonin and selenium on control of postharvest gray mold of tomato fruit

H. Zang, K. Jiang, J. Ma & M. Li
Key Laboratory of Agri-Food Safety of Anhui Province, School of Plant Protection, Anhui Agriculture University, Hefei, Anhui, China

X.B. Yin & L.X. Yuan
Advanced Lab for Selenium and Human Health, Suzhou Institute for Advanced Study, University of Science and Technology of China, Suzhou, Jiangsu, China

1 INTRODUCTION

Melatonin (MT) is a low-molecular-weight organic compound available in almost all living organisms from bacteria to mammals. It has many important functions in human, animals, and plants, which include regulating circadian rhythms in animals to controlling senescence in plants (Tan et al. 2012). Presently, available information is lacking on the presence of melatonin in different plant species, as well as on its biosynthesis and mechanisms of action, and its effects on commercial crop production (Reiter et al. 2015). Melatonin may improve storage life and quality of fruits and vegetables and have a role in vascular tissue reconnection during grafting. In addition, it may modify root architecture and influence nutrient uptake by roots. Another potentially important aspect related to melatonin is the production of melatonin-rich food crops (cereals, fruits, and vegetables) through the approach of biofortification. This strategy may improve nutrition and increase plant resistance against biotic and abiotic stresses and lead to improved crop with nutraceutical value (Yin et al. 2013, Li et al. 2016, Nawaz et al. 2016).

Selenium (Se) is an essential trace element for human and animal health (Zhu et al. 2009). Our previous work has demonstrated that Se can be used to control postharvest diseases in fruits and vegetables caused by the fungus *Botrytis cinerea* and *Penicillium expansum* (Wu et al. 2015). However, the relationship between Se and MT in promoting disease resistance in postharvest fruits and vegetables is a question for research.

Tomato fruit are susceptible to postharvest diseases caused by various pathogenic fungi (Liu et al. 2019). Gray mold, caused by *B. cinerea*, is considered the most important pathogen responsible for postharvest decay of fresh fruit and vegetables, having a wide range of hosts (Liu et al. 2019). The objective of this study was to evaluate the synergistic effects of combining Se with MT against gray mold caused by *B. cinerea* in tomato fruit. The possible mechanisms by which MT enhanced selenium-induced disease resistance to *Botrytis cinerea* in tomato fruit will also be discussed.

2 MATERIALS AND METHODS

Tomato fruit were harvested at commercial maturity. Fruit without physical injuries were surface sterilized with 2% (v/v) sodium hypochlorite for 2 min, washed with tap water, and air-dried prior to use. The *B. cinerea* was obtained from Key Laboratory of Plant Resources in North China, Institute of Botany, Chinese Academy of Sciences. *B. cinerea* was maintained on potato dextrose agar (PDA). Spores were obtained from two-week PDA cultures incubated at 23°C. The spores were scraped from the surface of the cultures and suspended in 5 mL of sterile distilled water containing 0.05% (v/v) Tween 80. Spore suspensions were filtered through four layers of sterile cheesecloth to remove any adhering mycelia. Spore concentration was determined with a hemacytometer and adjusted to 5×10^4 spores/mL with sterile distilled water.

Fruit were wounded with a sterile nail at the equator and 20 μL of the following treatment suspensions were added to each wound: Sodium selenite (0.1%, w/v) and melatonin (0.1–0.4 mmol/L). Fruit treated with sterile distilled water served as the control. After the wounds were air-dried, fruit were challenge-inoculated with 15 μL of a conidial suspension of *P. expansum* at 5×10^4 spores/mL. In another experiment, fruit were challenge-inoculated with a *B. cinerea* conidial suspension (15 μL) at 5×10^4 spores/mL before the treatment suspensions were added as described above. Treated fruit were stored at 25°C with high humidity (RH about 95%) and blue mold rot was measured daily after treatment. There were three replicates of each treatment with 20 fruits per replicate, and the experiment was repeated twice.

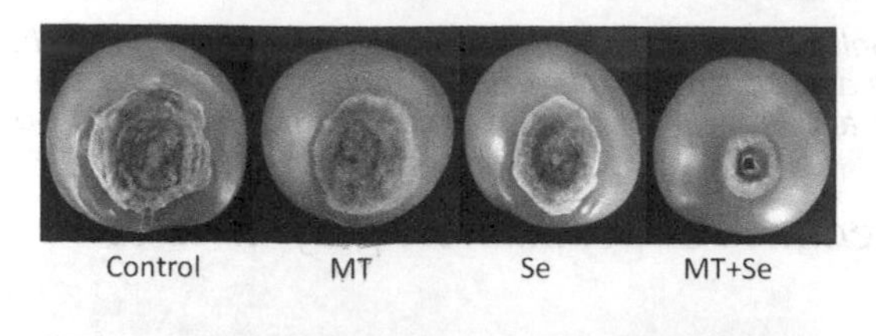

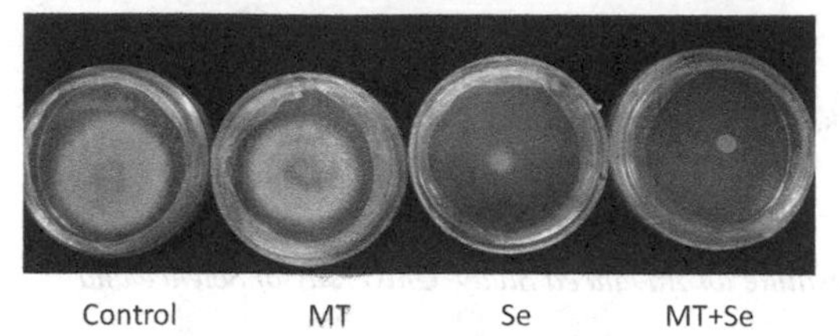

Figure 1. Inhibition effect of MT, alone or in combination with Se on the efficacy against gray mold of tomato fruit caused by *B. cinereal.*

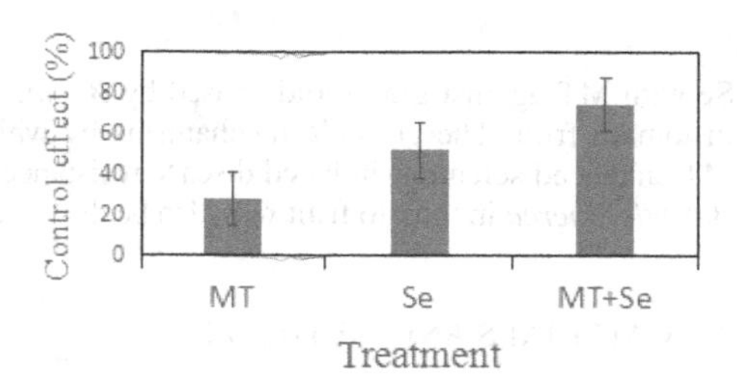

Figure 2. Control effect of gray mold of tomato fruit were treated with MT, Se, and MT+Se, followed by challenge inoculation with *B. cinerea*.

3 RESULTS AND DISCUSSION

A single application of Se was effective for controlling gray mold caused by *B. cinerea*, but better control was achieved when Se was used in combination with MT. The timing of Se and MT application affected the efficacy of disease control. Application of only Se alone before pathogen inoculation resulted in significantly higher incidence of *B. cinerea* disease and lesion diameters compared to the Se treatment after pathogen inoculation (Fig. 1).

In comparison, MT alone was ineffective for controlling *B. cinerea* when applied before pathogen inoculation and it showed little effect when applied after pathogen inoculation. Fruit decay was completely suppressed by a treatment of Se in combination with MT before *B. cinerea* inoculation (Fig. 2) after 7 days of storage at 25°C. Interestingly, although MT showed little effect on fruit decay when applied after pathogen inoculation, MT combined with Se significantly reduced the disease incidence and lesion diameter of blue mold rot.

4 CONCLUSIONS

In conclusion, we have shown that MT improved the efficacy of Se for the control of gray mold on tomato fruit. These results suggest that MT may be promising as an environmentally friendly additive to enhance the performance of Se against postharvest decay of fruit.

REFERENCES

Li, M.Q., Kamrul Hasan, M., Li, C.X. et al. 2016. Melatonin mediates selenium-induced tolerance to cadmium stress in tomato plants. *J Pineal Res* 61(3): 291–302.

Liu, C.X., Chen, L.L., Zhao, R.R. & Li, R. 2019. Melatonin induces disease resistance to *Botrytis cinerea* in tomato fruit by activating jasmonic acid signaling pathway. *J Agric Food Chem* 67(22): 6116–6124.

Nawaz M.A., Huang, Y., Bie, Z.L. et al. 2016. Melatonin: current status and future perspectives in plant science. *Front Plant Sci* 6:1230.

Reiter, R.J., Tan, D.X., Zhou, Z., Cruz, M.H.C., Fuentes-Broto, L. & Galano, A. 2015. Phytomelatonin: assisting plants to survive and thrive. *Molecules* 20(4): 7396–7437.

Tan, D.X., Hardeland, R., Manchester, L.C. et al. 2012. Functional roles of melatonin in plants, and perspectives in nutritional and agricultural science. *J Exp Bot* 63(2): 577–597.

Wu, Z.L., Yi, X.B., Bañuelos, G.S. et al. 2015. Effect of selenium on control of postharvest gray mold of tomato fruit and the possible mechanisms involved. *Front Microbio* 6: 1441.

Yin, L.H., Wang, P., Li, M.J. et al. 2013. Exogenous melatonin improves Malus resistance to Marssonina apple blotch. *J Pineal Res* 54(4): 426–434.

Zhu, Y.G., Pilon-Smits, E.A., Zhao, F.J., Williams, P.N. & Meharg, A.A. 2009. Selenium in higher plants: understanding mechanisms for biofortification and phytoremediation. *Trends in Plant Sci* 14(8): 436–442.

Selenium Research for Environment and Human Health: Perspectives, Technologies and Advancements – Bañuelos, Lin, Liang & Yin (eds)
 ISBN 978-1-138-39014-0

The effect of selenate stress on antioxidant enzyme activity of *Cardamine iolifolia* O.E. Schulz

J.J. Ming, Y.K. Yang, H.Q. Yin, Y. Kang, Y.F. Zhu, F.F. Chen & J.Q. Xiang
Enshi Tujia & Miao Autonomous Prefecture Academy of Agricultural Sciences, Enshi, Hubei, China
Hubei Selenium Industry Technology Research Institute, Enshi, Hubei, China

1 INTRODUCTION

Cardamine violifolia O.E. Schulz is a cruciferous plant, and it is mainly found in Enshi, Yichang, Hubei Province and Hupingshan, Hunan Province, China. *Cardamine violifolia* has the best ability for selenium (Se) accumulation and can accumulate Se up to 8000, 2000, and 2300 mg/kg in roots, stems, and leaves, respectively (Yuan et al. 2013). The plant has high medicinal value and can also be used as raw material for processing into Se food products, which has a great potential for future development.

Selenium is one of the essential trace elements for humans and animals. It participates in the synthesis of glutathione peroxidase (GSH-Px) and has the function of scavenging free radicals and participating in anti-aging activities (Brooks et al. 2001). An increase in Se content in plant tissues can regulate the balance of enzymatic system in plants (Han et al. 2013, Shahid et al. 2018, Zhang et al. 2012), reduce the accumulation of plasma membrane peroxidation from stress, and increase the plant's resistance to stress (Zhang et al. 2014). However, excessive Se can also cause toxic stress to plants. In the present study, the response changes of enzymatic detoxification mechanisms under selenate stress were investigated in *Cardamine violifolia* to provide additional information for cultivating this plant species.

2 MATERIALS AND METHODS

2.1 *Lettuce culture*

A greenhouse experiment with *Cardamine violifolia* was performed in Enshi, China. A substrate (peat soil/perlite = 4:1) was placed into rowing pots, and seven treatments with different Se concentrations added as sodium selenate (Na_2SeO_4), including CK (0 mg/kg), A1 (12.5 mg/kg), A2 (25 mg/kg), A3 (50 mg/kg), A4 (100 mg/kg), A5 (200 mg/kg), and A6 (400 mg/kg). Each treatment had three replicates. Inorganic fertilizer was applied as base fertilizer, and seedlings were transplanted to the pots when they grew four to five true leaves. After three months of growth, leaf samples were collected. For each treatment, five leaves were collected and washed with deionized water, transferred into liquid nitrogen, and then stored in a −80°C freezer.

Activity levels of SOD, POD, CAT, and GSH-Px in leaves were determined according to Zhang et al. (2012). About 0.5 g of fresh leaf sample were placed in a precooled mortar, extracted with 5 ml of precooled phosphate buffer (pH 7.0, fresh sample/extract = 1:10) along with some small quartz sand, and then centrifuged at 10,000 g for 15 min at 4°C. The final supernatant obtained was the crude extract. The four enzyme reactions were adjusted according to the experimental methods, and the sample absorbance measurements of the four enzymes were carried out at 560, 470, 240, and 412 nm, respectively. The activity of each enzyme was then calculated.

3 RESULTS AND DISCUSSION

3.1 *Growth of* Cardamine violifolia

The growth of *Cardamine violifolia* is shown in Figure 1. Plants in the 12.5, 25, and 50 mg/kg treatments were healthy, and their leaves were fully expanded compared with CK. Plants in the 200- and 400-mg Se/kg treatments were dwarfed, and their leaves were slightly curled, which was similar to the symptoms of heavy metal toxicity (Wahid et al. 2008). Therefore, we hypothesized that the Se tolerance threshold of *Cardamine violifolia* was about 400 mg/kg in this study.

3.2 *Antioxidant enzyme activity of* Cardamine violifolia

In general, plants maintain their stabile intracellular environment under external environmental stress, mainly by relying on their enzyme system to play a role and reduce the damage caused by environmental stress. The enzyme promoting system of antioxidant enzymes in plants consisted of SOD, POD, CAT, and GSH-Px. In addition, Se is an important component of GSH-Px synthesis. The activity of GSH-Px directly affects the content of GSH in downstream products and thereby regulates the intracellular environment

Figure 1. Effects of Se stress on the growth of *Cardamine iolifolia*. CK: 0 mg/kg, A1: 12.5 mg/kg, A2: 25 mg/kg, A3: 50 mg/kg, A4: 100 mg/kg, A5: 200 mg/kg, and A6: 400 mg/kg.

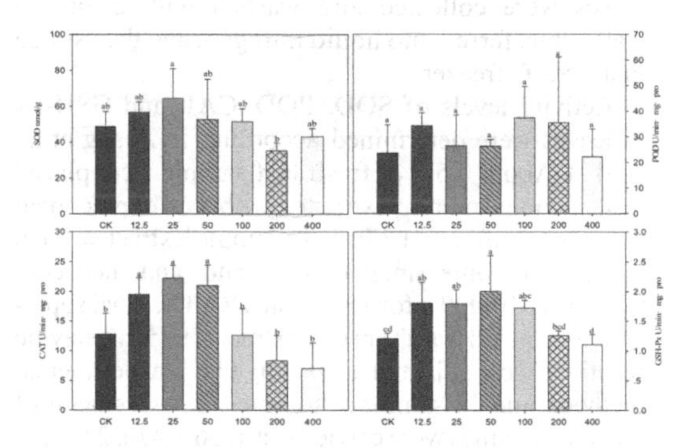

Figure 2. Effects of selenate stress on SOD, POD, CAT, and GSH-Px activities of antioxidant enzymes in the leaf of *Cardamine violifolia.*

(Rayman 2000). In this study, the activity of SOD, POD, CAT, and GSH-Px in the leaves of *Cardamine violifolia* increased first and then decreased with the increasing concentration of sodium selenate (Fig. 2) while there was no significant difference in the activity of POD among different treatments. Compared with CK, the 25- and 50-mg/kg treatments increased the activity of CAT and GSH-Px in leaves by 74.4% and 66.67%, respectively. It was possible that CAT and GSH-Px played a key role in the enzymatic system, and the enzyme activity was stronger when Se was lower than 50 mg/kg, whereas, under excessive Se stress, enzyme activity was inhibited (Yao et al. 2018).

4 CONCLUSIONS

In this greenhouse study, the Se tolerance threshold of *Cardamine violifolia* was about 400 mg/kg growing in the Se-rich substrate. CAT and GSH-Px played a key role under the condition of sodium selenate stress in the physiological regulation enzyme system of plant leaves, which provides new information on cultivating *Cardamine violifolia* in Se-rich soils.

ACKNOWLEDGEMENT

This work was supported by the NSFC (31560579). We thank the anonymous reviewers for their comments and suggestions.

REFERENCES

Brooks, J.D., Metter E.J., Chan, D.W. et al. 2001. Plasma selenium level before diagnosis and the risk of prostate cancer development. *J Urol* 166(6): 2034-2038.

Han, D., Li, X., Xiong, S. et al. 2013. Selenium uptake, speciation and stressed response of *Nicotiana tabacum* L. *Environ Exp Bot* 95: 6–14.

Rayman, M.P. 2000. The importance of selenium to human health. *Lancet* 356(9225): 233–241.

Shahid, M., Niazi, N.K., Khalid, S. et al. 2018. A critical review of selenium biogeochemical behavior in soil-plant system with an inference to human health. *Environ Pollut* 234: 915–934.

Wahid, A. & Ghani, A. 2008. Varietal differences in mung bean (*Vigna radiata*) for growth, yield, toxicity symptoms and cadmium accumulation. *Ann Appl Biol* 152(1): 59–69.

Yao, H., Ouyang, L., Liao, M.A. et al. 2018. Effects of different concentrations of selenium on antioxidant enzyme activity of *Perilla frutescens*. *Adv Eng Res* 162: 93–96.

Yuan, L., Zhu, Y., Lin, Z. et al. 2013. A novel selenocystine-accumulating plant in selenium-mine drainage area in Enshi, China. *PLOS One* 8(6): e65615.

Zhang, M., Hu, C., Zhao, X. et al. 2012. Molybdenum improves antioxidant and osmotic-adjustment ability against salt stress in Chinese cabbage (*Brassica campestris* L. ssp. Pekinensis). *Plant Soil* 355(1-2): 375–383.

Zhang, M., Tang S., Huang, X. et al. 2014. Selenium uptake, dynamic changes in selenium content and its influence on photosynthesis and chlorophyll fluorescence in rice (*Oryza sativa* L.). *Environ Exp Bot* 107: 39–45.

Selenium Research for Environment and Human Health: Perspectives, Technologies and Advancements – Bañuelos, Lin, Liang & Yin (eds)
© 2020 Taylor and Francis Group, London, ISBN 978-1-138-39014-0

Alleviating effects of exogenous nitric oxide donor sodium nitroprusside on selenium toxicity in rice

Z.H. Dai & S.X. Tu*
College of Resources and Environment, Huazhong Agricultural University, Wuhan, China
Hubei Research Center for Soil Remediation Engineering, Wuhan, China

1 INTRODUCTION

Selenium (Se) is a trace element in the environment and is an indispensable micronutrient for humans. Humans obtain most of their Se through diet (Rayman & Margaret 2008). Despite the nutritional benefits, high Se concentrations in soil can be toxic and poses threats to wildlife, livestock, and local populations (Mehdawi & Pilon-Smits 2012, Schiavon & Pilon-Smits 2017). Presently, the development of Se-enriched agricultural products is the most popular and effective way to increase Se intake. Production of Se-enriched agricultural products mainly depends on growing crops in Se-enriched soils.

Selenium deficiencies widely occur in many parts of the world. There are, however, some soils and mineral deposits that are naturally rich in Se, and anthropogenic activities such as mining, agriculture, and oil production can also lead to higher concentrations of Se in the environment.

Nitric oxide (NO) is a key signaling molecule in plants and is helpful to regulate plant growth and development (Rizwan et al. 2018). Many studies have found that NO plays a role in physiological processes of plants, such as seed dormancy and germination, root development, photosynthesis, stomatal closure, regulation of pollen tube growth, flowering, fruit ripening, etc. (Parvaiz et al. 2016, Wu et al. 2018). Kolbert et al. (2018) found that the nitro-oxidative processes were correlated with Se tolerance in *Astragalus* species, and Xiao et al. (2017) found that high levels of sodium nitroprusside (exogenous NO donor) reduced the Se content in rice shoots and roots, but further research is needed to explain the mechanism associated with the ability of NO to alleviate high Se stress in rice. We tested the hypothesis that Se toxicity in rice may be alleviated by NO, because nitric oxide is a key signaling molecule in plants and plays an important role in alleviating many other adverse and non-adverse stresses. Therefore, this study is aimed to investigate the role of exogenous NO in affecting Se tolerance in rice.

2 MATERIALS AND METHODS

Rice (*Oryza sativa L.*, cv. Yangliangyou 6) seeds were surface sterilized in 10% H_2O_2 for 10 min, rinsed thoroughly with deionized water, and soaked in deionized water for one night. Thereafter, the rice seeds were placed in plastic trays for germination at 30°C in the growth chamber cabinet. At the three-leaf stage, the rice seedlings were transplanted into the greenhouse at 27°C and subjected to lighting for 14 h per day. The rice seedlings were cultured with 1/2 international rice nutrient solution (Yoshida *et al.* 1971), which was changed every 3 d. After 7 d, uniform-sized rice seedlings were selected and transplanted into plastic pots. Selenium as Na_2SeO_3, NO as sodium nitroprusside (SNP), and cPTIO (NO scavenger) as 2-(4-carboxy-2-phenyl)-4,4,5,5-tetramethyl-imidazoline-1-oxyl-3-oxide were respectively added to full-strength international rice nutrient solution.

There were four treatments: (1) control (CK); (2) 25 μM Na_2SeO_3; (3) 100 μM SNP + 25 μM Na_2SeO_3; and (4) 100 μM SNP + 25 μM Na_2SeO_3 + 200 μM cPTIO. Rice seedlings were grown for 14 d. During this time, the nutrient solutions were replaced every 3 days and pH value was adjusted to 6.0 with NaOH or HCl. All chemicals were purchased from Sigma-Aldrich unless stated otherwise.

3 RESULTS AND DISCUSSION

In the high Se treatment, Se content in roots and shoots was 349.16 mg/kg and 97.38 mg/kg, respectively. The addition of SNP decreased Se content in roots by 10% compared to high Se treatment, while the addition of cPTIO eliminated the effects of SNP. There was no difference between high Se treatment and Se+SNP treatment on Se content in shoots.

Selenium exists in various forms such as Se(IV), Se(VI), SeCys, and SeMet in plants. Compared with high Se treatment, the SNP + Se treatment mainly decreased the inorganic Se content in rice roots, in which the Se(IV) decreased by 42% and Se(VI) decreased by 100%. Selenium only exists as organic forms in rice shoots, and organic Se increased by 21% compared with high Se treatment.

High Se treatment significantly decreased the activity of SOD and CAT, which was 73% and 62% lower than CK, respectively. Compared with the high Se treatment, the SNP + Se treatment positively increased the activity of SOD, POD, and CAT by 292%, 36%, and

207%, respectively, and the SNP + Se + cPTIO treatment increased the activity of SOD, POD, and CAT by 215%, 49%, and 145%, respectively.

High Se treatment inhibited the photosynthesis of rice seedlings. Net photosynthetic rate (P_n), intercellular CO_2 concentration (C_i), transpiration rate (T_r), and stomatal conductance (g_s) were decreased by 60%, 37%, 75%, and 81% compared with CK, respectively. The SNP + Se treatment effectively alleviated the inhibition effects of high Se treatment, and the photosynthesis of rice seedlings was significantly improved. The P_n, C_i, T_r, and g_s were increased by 138%, 71%, 284%, and 443%, respectively, compared with high Se treatment, but no differences were observed compared to CK. The SNP + Se + cPTIO treatment eliminated the effects of SNP on the photosynthesis of rice seedlings.

High Se treatment significantly inhibited the growth of roots and shoots of rice seedlings. Under high Se treatment, root length and shoot height were each 10% shorter than CK. The fresh weight and dry weight of shoots were decreased by 32% and 29%, respectively, compared with CK, while the fresh weight and dry weight of roots exhibited no difference from CK. The SNP+Se treatment alleviated the inhibition effects of high Se treatment on the growth of rice seedlings. Compared to SNP+Se treatment with high Se treatment, the root length and shoot height increased by 21% and 7% and increased the fresh and dry weights of shoots by 24% and 27%, respectively. Roots fresh weight increased by 13% with no effect on root dry weight.

Under SNP + Se + cPTIO treatment, root length and shoot height decreased by 4% and 16%, and fresh weights and dry weights of shoots decreased by 42% and 36%. Fresh and dry weights of roots decreased by 20% and 38% when compared with CK.

4 CONCLUSIONS

NO produced by SNP alleviated the toxicity of high Se treatment to rice seedlings and increased the tolerance of rice seedlings to Se. This effect was mainly reflected in promoting the growth of rice seedlings, enhancing the photosynthesis and antioxidant capacity of rice seedlings, and reducing the absorption and accumulation of Se in rice seedlings. Therefore, NO can be used to regulate Se content in rice. The present results provide theoretical guidance for the production of high-quality and safe Se-enriched rice in high-Se and Se-contaminated soils.

REFERENCES

Kolbert, Z., Molnár, Á., Szollosi, R., Feigl, G., Erdei, L. & Ördög, A. 2018. Nitro-oxidative stress correlates with Se tolerance of Astragalus species. *Plant Cell Physiol* 59 (9): 1827–1843.

Mehdawi, A.F.E. & Pilon-Smits, E.A.H. 2012. Ecological aspects of plant selenium hyperaccumulation. *Plant Biol* 14 (1): 1–10.

Parvaiz, A., Abdel, L.A.A., Abeer, H., Abd Allah, E.F., Salih, G. & Tran, L.S.P. 2016. Nitric oxide mitigates salt stress by regulating levels of osmolytes and antioxidant enzymes in chickpea. *Front Plant Sci* 7: 1–11.

Rayman, M.P. 2008. Food-chain selenium and human health: emphasis on intake. *Brit J Nutr* 100 (2): 254–268.

Rizwan, M., Mostofa, M.G., Ahmad, M.Z. et al. 2018. Nitric oxide induces rice tolerance to excessive nickel by regulating nickel uptake, reactive oxygen species detoxification and defense-related gene expression. *Chemosphere* 191: 23–35.

Schiavon, M. & Pilon-Smits, E.A.H. 2017. Selenium biofortification and phytoremediation phytotechnologies: a review. *J Environ Qual* 46 (1): 10–19.

Wu, S., Hu, C., Tan, Q. et al. 2018. Nitric oxide acts downstream of abscisic acid in molybdenum-induced oxidative tolerance in wheat. *Plant Cell Rep* 37 (4): 599–610.

Xiao, Q., Li, X.L., Gao, G.F. et al. 2017. Nitric oxide enhances selenium concentration by promoting selenite uptake by rice roots. *J Plant Nutr Soil Sci* 180 (6): 788–799.

Yoshida, S., Fornd, D. & Cock, J. 1971. *Laboratory Manual for Physiological Studies of Rice*: 61–66. Laguna, Philippines: The International Rice Research Institute.

Selenium Research for Environment and Human Health: Perspectives, Technologies and Advancements – Bañuelos, Lin, Liang & Yin (eds)
© 2020 Taylor and Francis Group, London, ISBN 978-1-138-39014-0

Maternal dietary selenium supply and offspring developmental outcomes

J.S. Caton
Department of Animal Sciences, North Dakota State University, North Dakota, USA

1 INTRODUCTION

Selenium (Se) is an essential trace element needed for normal growth and development (Sunde 1997, McDowell 2003, NASEM 2016). Over the past few decades, perceptions about Se have dramatically changed, ranging from Se being a nutrient that is potentially dangerous and should be controlled within narrow ranges to one that has many potential health benefits. Dietary biofortification is adding additional or elevated amounts of a dietary nutrient with the expectation of specific measurable positive outcomes. Regarding Se, dietary biofortification occurs either naturally with Se enriched ingredients or from direct dietary additions. Developmental programming is the concept that an insult or change during a critical window of fetal development can result in both short- and long-term consequences for the offspring. Recent data in developmental programming indicate that maternal nutrition plays a significant role in postnatal outcomes with lifelong consequences. Our laboratories have investigated Se biofortification of maternal diets during gestation with a focus on both maternal and offspring outcomes. Gestation is a critical time for livestock with the delivery of healthy offspring and a quick recovery of the mother being essential to successful livestock production enterprises. Our overarching hypothesis has been that additional Se in maternal diets would alter molecular, cellular, tissue, and whole animal events, and potentially mitigate compromised offspring outcomes resulting from compromised nutrient supply during gestation. This paper examines the impacts of maternal dietary Se supply (biofortification) during gestation on the offspring.

2 GROWTH AND DEVELOPMENT

2.1 *Maternal*

In general, providing additional Se into gestating livestock diets that are already meeting known requirements has little impact on body weight or average daily gain (Reed et al. 2007, Neville et al. 2008, Swanson et al. 2008). These findings are supported by other work with cattle and sheep fed growing and finishing diets. Other data have indicated that biofortification of ewe diets during gestation can increase average daily gain and feed use efficiency (Meyer et al. 2010); however, these authors also reported no changes in final ewe body weights at parturition.

Research reported preciously (Caton 2009, Caton et al. 2011, Taylor et al., 2014) at this conference series and elsewhere (Taylor et al. 2009) have demonstrated that maternal, fetal, and near adult offspring present with elevated Se serum and muscle concentrations in response to biofortification of maternal ruminant livestock diets during gestation. These data are taken to indicate that body protein can work as a functional nutrient reserve for Se. This reserve can be useful during times of low Se supply and provide for a de novo Se source when released through protein turnover.

Biofortification of diets during gestation in ruminant livestock can result in increased milk production and colostrum quality (Swanson et al. 2008, Meyer et al. 2011). These responses may be partially regulated by increases in mammary gland vascularity in response to biofortification of gestating ewe diets as reported by Vonnahme et al. (2011). Increased milk production and colostrum quality should result in improved offspring performance and health.

2.2 *Offspring*

Research from our laboratory investigating moderate changes in maternal nutrition during the first 50 d of gestation in beef heifers has shown changes in fetal liver gene expression. Specifically, a total of 80 expressed genes were detected and grouped into glutathione metabolism (n = 42), glutathione peroxidase (n = 11), and selenoprotein metabolism (n = 27). Of the glutathione metabolism genes, two (GSTA2 and GSR) were upregulated in controls, and two (ETHE1 and GGT5) were upregulated in fetal livers from moderately nutrient-restricted dams. For glutathione peroxidase, one gene (LTC4S) was upregulated in the restricted treatment compared with controls. In selenoprotein metabolism, one gene (TRNAU1AP) was upregulated in the restricted treatment compared with controls (Crouse et al. 2017). These data suggest that by d 50 of gestation, many genes associated with selenoproteins, glutathione metabolism, and glutathione peroxidase are being transcribed in the bovine fetal liver; of these, six genes were responsive to moderate changes in maternal nutrition and Se supply.

Growth and development of offspring can be impacted by biofortifying maternal diets with Se in some instances. Lamb body weight during late gestation or at birth has been increased via biofortification of maternal diets with Se during mid through late gestation in situations where maternal energy intake was limited (Reed et al. 2007, Meyer et al. 2010). Additionally, Meyer et al. (2010) reported that biofortification of maternal diets with Se can alter gain, feed use efficiency and diet digestibility in lambs nearly 180 days old. These lambs also had increased visceral adiposity and altered insulin sensitivity as measured by glucose tolerance.

3 CONCLUSIONS

Existing data are interpreted to imply that biofortification of maternal diets with Se during gestation can alter gene expression during early embryonic development, impact growth and development, alter metabolism, and have lasting impacts on the offspring. Additional work focused on determining mechanisms of action and practical management implications is needed.

REFERENCES

Caton, J.S. 2011. Dietary biofortification of gestating animal diets with natural sources of selenium: whole animal and specific tissue responses. In G. Banuelos, Z.Q. Lin, & X. Yin (eds.), *Selenium Deficiency, Toxicity, and Biofortification for Human Health*: 63–64. Hefei, China: University of Science and Technology of China Press.

Caton, J.S. 2009. Significance of elevated selenium in muscle tissue: Functional food and nutrient reserve? In G.S. Banuelos, Z.-Q. Lin, & X.B. Yin (eds.), *Selenium Deficiency, Toxicity, and Biofortification for Human Health*: 57–58. Hefei, China: University of Science and Technology of China Press.

Hammer, C.J., Thorson, J.F., Meyer, A.M. et al. 2011. Effects of maternal selenium supply and plane of nutrition during gestation on passive transfer of immunity and health in neonatal lambs. *J Animal Sci* 89: 3690–3698.

McDowell, L.R. 2003. *Minerals in Animal and Human Nutrition* (2nd ed). Amsterdam, The Netherlands: Elsevier Press.

Meyer, A.M., Reed, J.J., Neville, T.L. et al. 2010. Effects of nutritional plane and selenium supply during gestation on ewe and neonatal offspring performance, body composition, and serum selenium. *J Animal Sci* 88: 1786–1800.

Meyer, A.M., Reed, J.J., Neville, T.L. et al. 2011. Nutritional plane and selenium supply during gestation impact yield and nutrient composition of colostrum and milk in primiparous ewes. *J Animal Sci* 89: 1627–1639.

NASEM. 2016. Nutrient Requirements of Beef Cattle. Washington, DC: The National Academies Press.

Neville, T.L., Caton, J.S., Hammer, C.J. et al. 2010. Ovine offspring growth and diet digestibility are influenced by maternal Se supplementation and nutritional intake level during pregnancy despite a common postnatal diet. *J Animal Sci* 88: 3645–3656.

Neville, T.L., Ward, M.A., Reed, J.J. et al. 2008. Effects of level and source of dietary selenium on maternal and fetal body weight, visceral organ mass, cellularity estimates, and jejunal vascularity in pregnant ewe lambs. *J Animal Sci* 86: 890–901.

Reed, J.J., Ward, M.A., Vonnahme, K.A. et al. 2007. Effects of selenium supply and dietary restriction on maternal and fetal body weight, visceral organ mass, cellularity estimates, and jejeunal vascularity in pregnant ewe lambs. *J Animal Sci* 85: 2721–2733.

Sunde, R.A. 1997. Selenium. In B.L. O'Dell & R.A. Sunde (eds), *Handbook of Nutritionally Essential Mineral Elements*. New York, NY: Marcel Dekker.

Swanson, T.J., Hammer, C.J., Luther, J.S. et al. 2008. Effects of gestational plane of nutrition and selenium supplementation on mammary development and colostrum quality in pregnant ewe lambs. *J Animal Sci* 86: 2415–2423.

Taylor, J.B, Caton, J.S. & Larsen, R. 2014. Selenium biofortification in North America: Using naturally selenium-rich feeds for livestock. In G.S. Banuelos, Z.Q. Lin & X.B. Yin (eds), *Selenium in the Environment and Human Health:* 155–156. London: CRC Press, Taylor & Francis Group.

Taylor, J.B., Reynolds, L.P., Redmer D.A. & Caton, J.S. 2009. Maternal and fetal tissue selenium loads in nulliparous ewes fed supranutritional and excessive selenium during mid- to late pregnancy. *J Animal Sci* 87: 1828–1834.

Vonnahme, K.A., Wienhold, C.W., Borowicz, P.P. et al. 2011. Supranutritional selenium increases mammary gland vascularity in postpartum ewe lambs. *J Dairy Sci* 94: 2850–2858.

Interactions with heavy metals

Selenium Research for Environment and Human Health: Perspectives, Technologies and Advancements – Bañuelos, Lin, Liang & Yin (eds)
© 2020 Taylor and Francis Group, London, ISBN 978-1-138-39014-0

Effect of selenium on cadmium uptake, translocation and accumulation in rice (*Oryza sativa* L.)

H.F. Li, Y.N. Wan, Q. Wang & Y. Yu
College of Resources and Environmental Sciences, China Agricultural University, Beijing, China

1 INTRODUCTION

Cadmium (Cd) is one of the most harmful and widespread heavy metals in agricultural soils. The metal enters from anthropogenic processes, transfers to plants, accumulates in edible parts, and leads to chronic toxic effects in human beings (Templeton & Liu 2010). Paddy rice (*Oryza sativa* L.), the staple food for Chinese people, is a major source of Cd intake. Therefore, effective measures are needed to reduce rice Cd levels. Selenium (Se) is an essential trace element for humans and animals. Others have recently reported that Se amendments mitigate heavy metal accumulation and toxicity in various plants (Feng et al. 2013). In the present study, hydroponic culture and pot experiments were used to investigate the effects of Se on Cd uptake, translocation and accumulation in rice.

2 MATERIAL AND METHODS

Hydroponic culture experiments: Rice seedlings (*Oryza sativa* L., Zhunliangyou 608) were planted in plastic pots filled with 1/2 Kimura nutrient solution (pH 5.5) for 35 days.

(1) Cd uptake kinetics was carried out to investigate the interaction between Cd and Se on root surfaces: seedlings were totally soaked in a series of uptake solutions (1 L) containing 0, 1, 5, 10, 20, 40, 80, and 100 μM Cd ($3CdSO_4 \cdot 8H_2O$), with or without 5 or 10 μM Se (Na_2SeO_3 or Na_2SeO_4) for 1 h.

(2) This experiment was conducted to study the interaction between Cd and Se in plants by setting up the time course: seedlings were transferred to pots (2.5 L), in which Se (5 μM, Na_2SeO_3 or Na_2SeO_4) or Cd (5 μM, $3CdSO_4 \cdot 8H_2O$) was added to form two treatments: 1) Cd; and 2) Cd+Se (selenite or selenate). The plants were sampled at 1, 3, 6, 20, 30, 48, and 72 (120 for selenate) h after treatment.

(3) There was a subcellular distribution of Cd in rice root, and translocation occurred via xylem 6 days after applying the treatments (same as (2)). Rice root cell wall (F1), organelle (F2) and soluble cytosol (F3) were isolated by differential centrifugation (Feng et al. 2011). Rice shoots were cut 2 cm above the root using a pre-sterilized blade, and xylem sap was collected from the cut surfaces and placed into a 1 mL vessel using a pipette for 2 h.

Pot experiments: Surface soil was collected from paddy rice fields in Jiangxi Province (total Cd: 0.45 mg/kg, pH 5.25, gray fluvo-aquic paddy soil) and Hunan Province (total Cd: 5.12 mg/kg, pH 7.26, yellow paddy soil), China. Na_2SeO_3 was blended into each soil type at rates of 0, 0.5, and 1.0 mg/kg. The soils were either flooded or aerobic. Two pre-germinated rice seedlings (Zhunliangyou 608) were planted into each pot and harvested at grain maturity.

The plant samples were HNO_3-digested using microwave digestion system, and the total Cd content was determined by ICP-MS (Agilent 7700ce, Santa Clara, USA). All results were expressed as an average of three replications. Treatment effects were determined by analysis of variance using SAS.

3 RESULTS AND DISCUSSION

3.1 *Effect of Se on Cd uptake kinetics and Cd uptake at different times by rice seedlings*

The parameters obtained from the different treatments showed that the presence of selenite slightly improved the influx of Cd into rice roots. The additions of 5 and 10 μM selenite increased the V_{max} of Cd uptake by 5.9% and 13.8%, while K_m decreased by 15.8% and 16.1%, respectively. However, selenate had no significant effect on the influx of Cd.

When the exposure time was ≤ 6 h, selenite slightly improved Cd uptake. Compared with Cd-alone treatment, selenite addition increased the Cd uptake by 17.5%, 8.4%, and 0.15% after 1, 3, and 6 h, respectively. However, selenite showed a significant reductive effect on Cd uptake after 20 h, which was enhanced by 18.6–47.6% with time (20–72 h exposure). A similar tendency was seen with selenate addition: At exposure times of ≤ 20 h, Cd uptake slightly increased. When treated for over 30 h, the presence of selenate solution also inhibited Cd uptake. After longer exposure (120 h), selenate significantly suppressed Cd uptake by 29.7%. We also found that selenite was more effective in decreasing Cd uptake by plants than selenate.

3.2 *Effect of Se on subcellular distribution of Cd in rice roots and translocation to xylem*

Compared with the Cd treatment alone, the application of selenite significantly decreased the Cd levels in the cell wall, organelle, and cytosol. However, the selenate addition did not influence the Cd content in cytosol or organelle significantly but only decreased it in the cell wall. The addition of selenite had a significant effect on the distribution of Cd; namely, the percentage of Cd in cytosol increased by 23.4% and decreased in the organelle by 28.7%. Regarding the addition of selenate, the Cd distribution ratio of cytosol increased by 13.1%. A significant decrease (55.9%) was also observed in the Cd content in xylem sap after adding selenite.

3.3 *Effect of Se on Cd accumulation in rice grain*

The results showed that in the acidic soil (pH 5.25), the Cd content in the grain of rice grown in aerobic conditions was 12.7–13.5 times higher than that grown in flooded conditions. In the slightly alkaline soil (pH 7.26), however, flooding had no significant effect on grain Cd content. Selenite additions to the slightly alkaline soil decreased the rice grain Cd contents by 45.2% and 67.7–77.4% under aerobic and flooding conditions, respectively, compared to the control (CK) treatment. However, no significant differences in grain Cd content were found among the selenite treatments in the acidic soil.

4 CONCLUSIONS

The addition of Se slightly increased the uptake of Cd by rice seedlings at the initial Cd exposure, but decreased Cd uptake and translocation when the time of exposure was increased. The presence of selenite decreased Cd content more effectively than selenate. The selenite addition increased the proportion of Cd distributed to soluble cytosol, while decreased both the Cd distribution in the organelle in rice root and the Cd in the xylem sap. Our results showed that the effect of Se on Cd accumulation in rice can be partly ascribed to the increase in the compartmentalization of Cd in the vacuole and the decrease in the root-to-shoot Cd translocation. Furthermore, the effects of selenite on Cd accumulation in rice grain varied to some extent with soil condition (physicochemical properties, Cd content, and water management).

REFERENCES

Feng, R.W., Wei, C.Y. & Tu, S.X. 2013. The roles of selenium in protecting plants against abiotic stresses. *Environmental and Experimental Botany* 87: 58–68.

Feng, R.W., Wei, C.Y., Tu, S.X., Tang, S.R. & Wu, F.C. 2011. Simultaneous hyperaccumulation of arsenic and antimony in Cretan brake fern: evidence of plant uptake and subcellular distributions. *Microchemical Journal* 97: 38–43.

Templeton, D.M. & Liu, Y. 2010. Multiple roles of cadmium in cell death and survival. *Chemico-Biological Interaction* 188: 267–275.

Selenium Research for Environment and Human Health: Perspectives, Technologies and Advancements – Bañuelos, Lin, Liang & Yin (eds)
© 2020 Taylor and Francis Group, London, ISBN 978-1-138-39014-0

The interaction between selenium and cadmium in lettuce

J.P. Song
University of Science and Technology of China, Hefei, China
Jiangsu Bio-Engineering Research Center for Selenium, Suzhou, China

Z.Z. Zhang, Z.D. Long, Z.M. Wang & L.X. Yuan
Jiangsu Bio-Engineering Research Center for Selenium, Suzhou, Jiangsu, China

Q.Q. Chen
College of Life Science, Resources and Environment, Yichun University, Jiangxi, China

X.B. Yin
University of Science and Technology of China, Hefei, Anhui, China
Jiangsu Bio-Engineering Research Center for Selenium, Suzhou, China
National-local Joint Engineering Laboratory of Se-enriched Food Development, Ankang, China

1 INTRODUCTION

Cadmium (Cd) is a harmful agricultural pollution element that produces reactive oxygen species (ROS) and causes oxidative damage to plants (Bari et al. 2019) that negatively impacts plant growth and photosynthesis (Rizwan et al. 2017). For humans, Cd intake can increase the risk of kidney failure and cancer (Chaney et al. 2015, Jarup et al. 2002).

Selenium (Se) is an essential element for humans and animals (Long et al. 2018). In soil, Se is absorbed by plants and then transferred via food chain to humans. The adequate supplementation of Se may reduce risk of some types of cancer, protect the liver and reduce oxidation activity (Brown et al. 2001). In recent years, others found that Se and Cd have a significant antagonistic relationship in the pancreas of chicken (Jin et al. 2018). In addition, Se may significantly reduce Cd toxicity in crops (Wan et al. 2018).

To examine this relationship in plants, we conducted a hydroponic experiment in lettuce (*Lactuca sativa* L. var. *ramose* Hort) and subjected it to different concentrations of Se and Cd. The objective of this study was to investigate the relationship between Se and Cd and measure the effects on the accumulation of Se and Cd.

2 MATERIALS AND METHODS

2.1 *Lettuce culture*

A hydroponic experiment was performed in a greenhouse in Jiangsu Bio-Engineering Research Center for Selenium. A 10% Hoagland nutrient solution was added to growing pots, and selenite was added at three concentrations: 0.5, 1.0, and 1.5 μM. Cadmium chloride was added at two concentrations: 0.5 and 1.0 μM. Twelve replicates were established for each treatment. Lettuce seedlings of the same age were washed with deionized water and then planted into hydroponic pots. The illumination time was 12 h/d with 300 $\mu Mol/m^2/s$ of artificial light. The temperature of day and night was 25°C and 18°C, respectively. Plants were harvested after one month of growth under these growing conditions.

2.2 *Selenium and Cd concentrations analysis*

The harvested root and shoot samples were dried at 65°C, ground, duplicated, and digested overnight with 8 mL nitric acid and 2 mL perchloric acid. The digestion solution was heated on the electric heating plate until white fumes were observed. After cooling, one of the duplicated samples was brought up to 25 mL volume with deionized water, then total Cd concentration was determined by Atomic Absorption Spectroscopy (AAS). The other duplicated sample was washed with deionized water and reheated until 2 ml solution was left. Then we added 5 mL 12 mol/L HCl to sample solution to reduce selenate to selenite and brought to 25 mL volume for Se analysis by Hydride Generation Atomic Fluorescence Spectrometry (HG-AFS).

3 RESULTS AND DISCUSSION

With 1.0 μM Cd, higher concentrations of Se were observed in roots, but no difference was observed with 0.5 μM Se compared with 0.5 μM Cd (Fig. 1a). With 0.5 μM Cd and 1.5 μM Se, there was a significant reduction of Se accumulation in shoot compared with lower Se concentrations ($p < 0.05$). With 1.0 μM Cd treatment and increase of Se concentration, there was an increase of accumulated Se in shoot (Fig. 1b).

Compared with 0.5 μM Se, 1.5 μM Se increased the Cd concentration in root by 13.2 and 4.5 times with 0.5 and 1 μM Cd treatments, respectively (Fig. 1c).

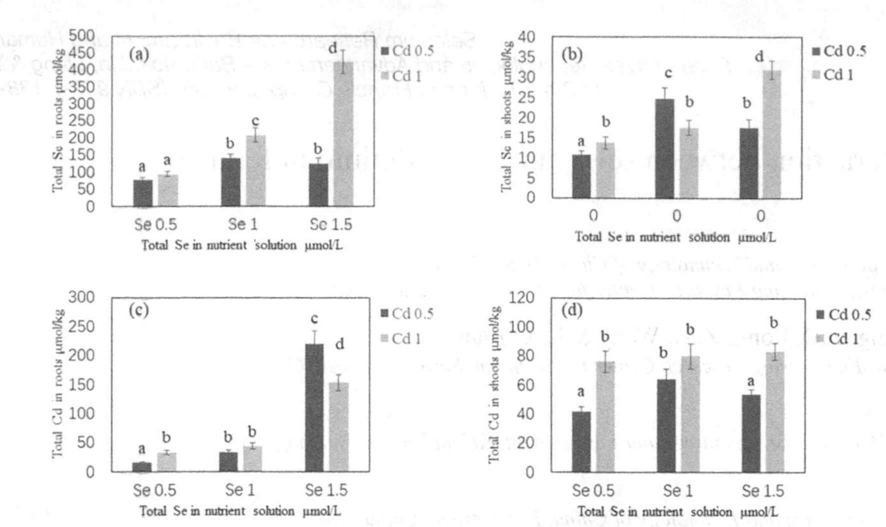

Figure 1. Treated lettuce with different additions (μM) of Se and Cd and measured concentration of Se and Cd. Selenium concentration in roots (a); Se concentration in shoots (b); Cd concentration in roots (c); Cd concentration in shoots (d).

With 1.0 μM Cd treatment, there were no significant differences on Cd accumulation in shoots (Fig. 1d). Overall, Se increased the accumulation of Cd in the roots of lettuce. Under low (0.5 μM) Cd concentration, Se had an antagonistic effect on Cd in the shoots of lettuce. With high Cd concentration, the interactions between Se and Cd were not significant.

The optimal Se level may alleviate Cd toxicity in crops (Hu et al. 2014), whereas the promotion of Se on Cd toxicity was also found in other studies (Fargasova et al. 2006). Others have observed that Se inhibited Cd accumulation in cabbage stems at low levels of Cd, and Se facilitated Cd accumulation at high levels of Cd (Yu et al. 2018). These observations are consistent with our present results. We hypothesized that with low Cd exposure, lettuce preferentially stored Cd. With addition of Se, there was an antagonistic effect on Cd at the interface of root and shoot. Furthermore, some studies indicated this effect might occur depending on a specific ratio of Se and Cd (Feng et al. 2013).

4 CONCLUSIONS

Selenium improved Cd accumulation in roots but inhibited Cd accumulation in shoots of lettuce at low Cd concentration. To provide a theoretical basis for production of Cd-decreased crops, further studies are required to identify speculation and explain the mechanism of this study phenomenon.

REFERENCES

Bari, M.A., Akther, M.S., Abu, R.M. et al. 2019. Cadmium tolerance is associated with the root-driven coordination of cadmium sequestration, iron regulation, and ROS scavenging in rice. *Plant Physiol Biochem* 136: 22–33.

Brown, K.M. & Arthur, J.R. 2001. Selenium, selenoproteins and human health: a review. *Public Health Nutr* 4: 593–9.

Chaney, R. L. 2015. How does contamination of rice soils with Cd and Zn cause high incidence of human Cd disease in subsistence rice farmers. *Curr Pollution Rep* 1: 13–22.

Fargasova, A., Pastierova, J. & Svetkova, K. 2006. Effect of Se-metal pair combinations (Cd, Zn, Cu, Pb) on photosynthetic pigments production and metal accumulation in *Sinapis alba* L. seedlings. *Plant Soil Environ* 52: 8–15.

Feng, R.W., Wei, C.Y., Tu, S.X. et al. 2013. A dual role of Se on Cd toxicity: Evidences from the uptake of Cd and some essential elements and the growth responses in paddy rice. *Biol Trace Elem Res* 151: 113–21.

Hu, Y., Norton, G.J., Duan, G.L. et al. 2014. Effect of selenium fertilization on the accumulation of cadmium and lead in rice plants. *Plant Soil* 384: 131–40.

Jarup, L. 2002. Cadmium overload and toxicity. *Nephrol Dial Transplant* 17: 35–39.

Jin, X. Jia, T., Liu, R. et al. 2018. The antagonistic effect of selenium on cadmium-induced apoptosis via PPAR-gamma/PI3K/Akt pathway in chicken pancreas. *J Hazard Mater* 357: 355–62.

Long, Z.D. Yuan, L.X., Hou, Y.Z. et al. 2018. Spatial variations in soil selenium and residential dietary selenium intake in a selenium-rich county, Shitai, Anhui, China. *J Trace Elem Med Biol* 50: 111–16.

Rizwan, M.S., Ali, M., Adrees, M. et al. 2017. A critical review on effects, tolerance mechanisms and management of cadmium in vegetables. *Chemosphere* 182: 90–105.

Wan, Y., Camara, A.Y., Yu, Y. et al. 2018. Cadmium dynamics in soil pore water and uptake by rice: Influences of soil-applied selenite with different water managements. *Environ Pollut* 240: 523–33.

Yu, Y., Yuan, S., Zhuang, J. et al. 2018. Effect of selenium on the uptake kinetics and accumulation of and oxidative stress induced by cadmium in Brassica chinensis. *Ecotoxicol Environ Saf* 162: 571–80.

Selenium Research for Environment and Human Health: Perspectives, Technologies and Advancements – Bañuelos, Lin, Liang & Yin (eds)
© 2020 Taylor and Francis Group, London, ISBN 978-1-138-39014-0

Detoxification of mercury in soil by selenate and related mechanisms

T.A.T. Tran & Q.T. Dinh
College of Natural Resources and Environment, Northwest A&F University, Yangling, Shaanxi, China
Faculty of Natural Science, Thu Dau Mot University, Thu Dau Mot city, Binh Duong, Vietnam

F. Zhou & D.L. Liang
College of Natural Resources and Environment, Northwest A&F University, Yangling, Shaanxi, China

1 INTRODUCTION

Mercury (Hg) is an extremely toxic pollutant that threatens ecosystem balance and human health by persisting and accumulating in the environment and the food chain. High concentrations of Hg in plants result in the excessive accumulation of reactive oxygen species, which trigger oxidative stress, inhibit growth, and even cause death. Selenium (Se) is a trace element essential to human and animals. The application of Se in soil can inhibit Hg uptake by plants through the formation of Hg–Se insoluble complexes in the rhizosphere and/or roots. We propose that Se also can mitigate the toxicity of Hg by regulating the physiological mechanism that mediates the opposing effects of the plant antioxidant system and resulting peroxidation products. Earlier studies have investigated Se accumulation in plants by treating plant growth media or soil with selenate (Se[VI]). To the best of our knowledge, no study has determined the effect of exogenous Se(VI) on the physiological mechanisms underlying the changes in plant growth under Hg stress. Therefore, the main objective of this study was to investigate the effects of Se(VI) application on the detoxification of Hg via the physiological and biochemical responses of pak choi (*Brassica chinensis*).

2 MATERIALS AND METHODS

2.1 *Experiment design*

Eum-Orthic Anthrosol, a typical agricultural soil with an old cultivation history in northern China, was collected at depth of 0 cm to 20 cm at the Northwest A&F University farm in Shaanxi Province, China. The basic physicochemical properties of the soil were as follows: pH 7.75, cation exchange capacity (CEC) 23.34 cmol/kg, 39.5% clay, calcium carbonate 55.0 g/kg, organic matter 16.33 g/kg, total nitrogen 1.11 g/kg, total Se 0.207 mg/kg and total Hg 0.05 mg/kg. Exogenous sodium selenate (Na_2SeO_4) and mercuric chloride ($HgCl_2$) were purchased from a reagent factory in Tianjin, China.

The dosages of metal exposure to soil samples in this study were established as 0, 0.5, 1, and 2.5 mg/kg soil for Se, and 0, 1, 2, and 3 mg/kg soil for Hg. Ten pak choi seeds were sowed in each pot, and the seedlings were thinned to five in each pot 10 days after germination. The pots were grown in the greenhouse and watered periodically to keep soil moisture at 70% of the field capacity. Plants were harvested after 38 days, cleaned, dried at 65°C for 3 days, ground, and analyzed later.

2.2 *Determination methods*

The dried samples were subsequently measured for dry weights. The activity of SOD was measured for its ability to inhibit the photochemical reduction of nitro blue tetrazolium (NBT) and the activity of CAT was determined by UV absorption method (Zhang et al. 2012). The MDA contents were determined by the thiobarbituric acid (TBA) reaction (Zhang et al. 2012).

3 RESULTS AND DISCUSSION

3.1 *Effects of applied Se(VI) on the antioxidant system under Hg stress*

When soils were treated with Se(VI) 0.5 mg/kg, the SOD and CAT activities significantly increased by 43.6–45.3% and 41.6–42.5% ($p < 0.01$), respectively, in comparison with Hg-only treatment. No significant differences were found on SOD and CAT activities ($p > 0.05$) when Se (VI) was applied at 1.0 mg/ kg. However, the SOD and CAT activities of plant significantly decreased by 57.3–60.3% and 56.9–61.0% ($p < 0.01$), respectively, when soils were treated with the high concentration of Se(VI) (2.5 mg/kg). This result may be attributed to antioxidative effects in pak choi under Hg stress with respective Se application (0.5 mg/kg).

3.2 *Effects of applied Se(VI) on peroxidant under Hg stress*

The MDA contents in pak choi significantly decreased by 39.6–40.5% when Se(VI) was applied at 0.5 mg/kg in comparison with respective Hg treatment ($p < 0.01$). However, a high significant

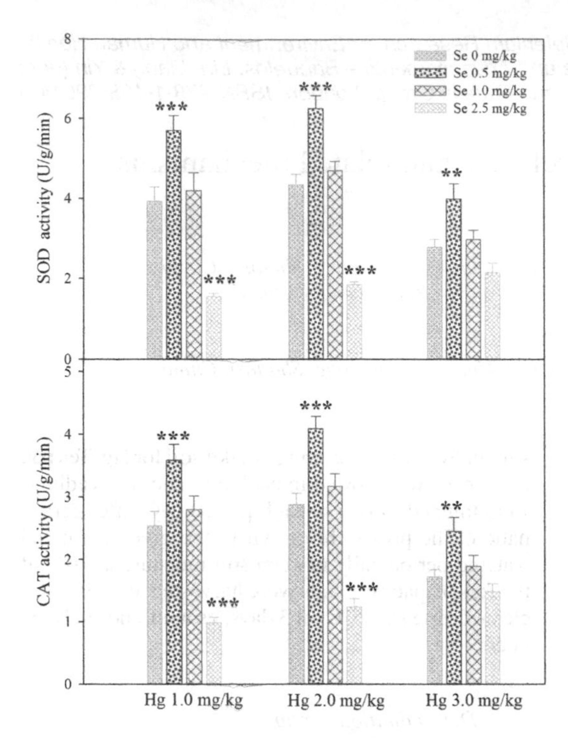

Figure 1. The effect of applied Se(VI) on the antioxidant enzymes activities of pak choi under Hg stress. ** and *** indicate significant difference from the respective Hg treatment.

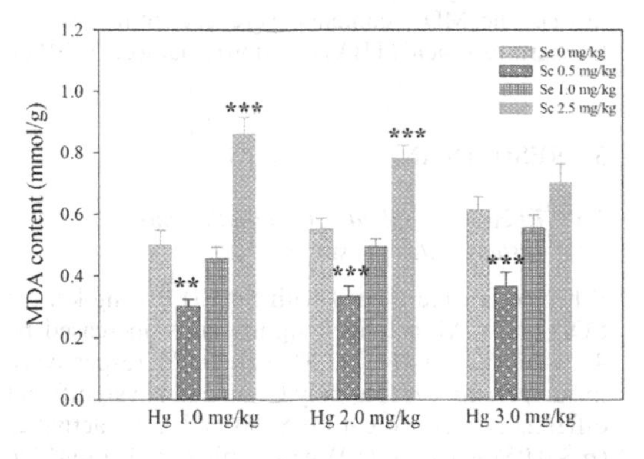

Figure 2. The effect of applied Se(VI) on peroxidant of pak choi under Hg stress. ** and *** indicate significant difference from a respective Hg treatment.

increase (41.4–71.7%) was observed when Se(VI) treatments were applied at 2.5 mg/kg for 1.0 and 2.0 mg Hg/kg ($p < 0.001$). No significant decrease ($p > 0.05$) in MDA contents was observed when Se(VI) was applied at 1.0 mg/kg. As a result, an application of low-concentration Se (0.5 mg/kg) can decrease lipid peroxidation levels against oxidative stress under Hg stress.

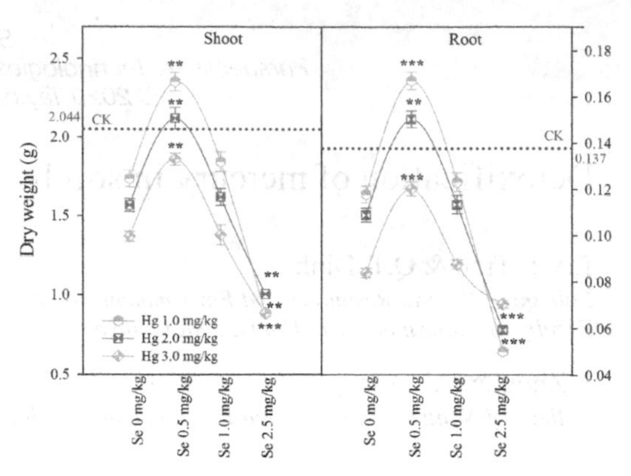

Figure 3. The effect of applied Se(VI) on pak choi dry weight: shoot and root under Hg stress. *, ** or *** indicate significant difference from the respective Hg treatment.

3.3 *Effect of applied Se(VI) on pak choi growth under Hg stress*

The shoot and root dry weight of pak choi significantly increased by 35.2–35.7% and 38.2–43.3%, respectively, when Se(VI) was applied at 0.5 mg/kg in comparison with a respective Hg treatment ($p < 0.01$). In contrast, no significant difference ($p > 0.05$) was found when Se(VI) was applied at 1.0 mg/kg. Notably, the shoot and root dry weight of pak choi significant decreased by 35.9–49.1% and 45.4–57.4%, respectively, when Se(VI) was applied at 2.5 mg/kg for 1.0 and 2.0 mg Hg/kg treatment ($p < 0.01$). These results suggested that pak choi developed protective responses to regulate oxidative stress by modifying the activities of antioxidant enzymes and prevented the production of peroxidant when Se was applied at low concentration (0.5 mg/kg). Thus, growth of pak choi was stimulated.

4 CONCLUSIONS

The selenate application at low concentration (0.5 mg/kg) has an important role in promoting the detoxification of Hg under oxidative stress via significantly improving the growth of pak choi via the elevation of antioxidant enzyme activities (SOD and CAT), as well as via the suppression of the content of MDA.

REFERENCES

Tran, T.A.T., Zhou, F. & Yang, W.X. 2018. Detoxification of mercury in soil by selenite and related mechanisms. *Ecotoxicol Environ Safe* 159: 77–84.

Zhang, M., Hu, C.X., Zhao, X.H. et al. (2012). Molybdenum improves antioxidant and osmotic-adjustment ability against salt stress in Chinese cabbage (*Brassica campestris* L. *ssp. Pekinensis*). *Plant Soil* 355: 375–383.

Selenium Research for Environment and Human Health: Perspectives, Technologies and Advancements – Bañuelos, Lin, Liang & Yin (eds)

Effects of selenium-zinc interaction on zinc bioavailability of pak choi

M.Y. Xue, M. Wang, F. Zhou, Q.T. Dinh & D.L. Liang
College of Natural Resources and Environment, Northwest A&F University, Yangling, Shaanxi, China

1 INTRODUCTION

Zinc (Zn) is an essential trace element for humans (Krężel & Maret 2016), and about a third of the world's population is estimated to be at risk of Zn deficiency (Li et al. 2015). The usefulness of a Zn fertilizer will depend on its bioavailability in soil (Cakmak & Kutman 2018). However, 90% of inorganic Zn fertilizers ($ZnSO_4 \cdot 7H_2O$) applied to calcareous soils quickly become unavailable to plants due to the physicochemical properties of soil, including high pH and high levels of carbonate (Lu et al. 2012).

Selenium (Se) is an essential element for human and animal health. There were many studies reporting on the existence of selenium-zinc interactions. Fargasova et al. (2006) reported that the combination of Se and Zn inhibited the accumulation of Zn in mustard. However, Souza et al. (2014) and Longchamp et al. (2016) confirmed that the application of Se can promote the absorption and transport of Zn in hydroponic crops. Therefore, the effects of selenium-zinc interaction on growth and Zn absorption of plants have not yet been clarified.

The specific objectives of this study were as follows: (1) clarify the effect of the interaction of selenium-zinc on the growth of pak choi and (2) explore the effect of selenium-zinc interaction on the uptake and translocation of Zn in pak choi.

2 MATERIALS AND METHODS

The soil used in the experiment was collected from the surface layer (0–20 cm depth) of a farm located at the Northwest A&F University in Shaanxi Province, China. The soil type is Eum-Orthic Anthrosol that represents a typical agricultural soil with an old cultivation history in Northern China. Soil samples were initially spiked with selenite (0, 1.0 and 2.5 mg/kg soil) and with $ZnSO_4 \cdot 7H_2O$ (0, 20, and 50 mg/kg) and then incubated for 30 d to homogenize under natural environment before conducting the pot experiment. The pot experiment was conducted in a greenhouse Northwest A&F University using pak choi. Soil moisture was maintained at 70% of the field capacity by quantitative watering once every 2 to 3 days. The plants were harvested 40 d after planting, dried, ground, and prepared for Zn analyses. The Zn concentration in plant tissue was determined by an atomic absorption spectrophotometer (AA320CRT; Shanghai Analytical Instrument Overall Factory, Shanghai, China) after acid digestion with 4:1 (v/v) HNO_3-$HClO_4$ at 170°C.

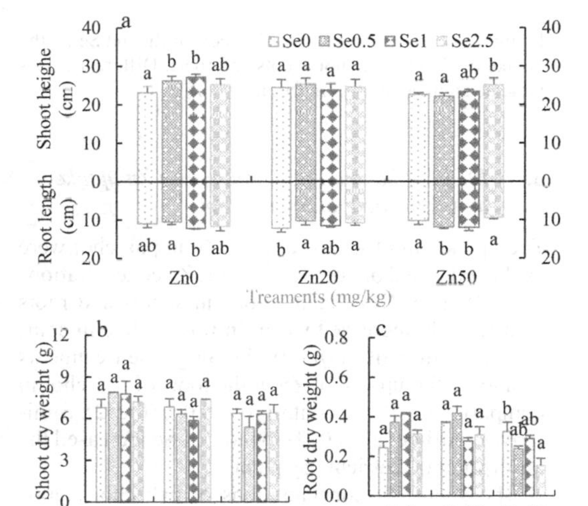

Figure 1. The effect of single and combined Zn and Se on pak choi length (a), and pak choi dry weight: shoot (b) and root (c). Different letters indicate significant differences ($p < 0.05$).

3 RESULTS AND DISCUSSION

3.1 *Effect of Se and Zn interaction on the growth of pak choi*

The shoot height and biomass of pak choi were promoted by a single Se application at low concentration (≤1 mg/kg), while they were inhibited at high concentration of Se (≥2.5 mg/kg) in comparison with the control. However, single Zn application had no significant effect on shoot height, root length, shoot dry weight, or root dry weight of pak choi (Fig. 1).

The shoot height and root dry weight of pak choi also significantly decreased either when Se was applied at 2.5 mg/kg for Zn at 50 mg/kg or Se at 1.0 and 2.5 mg/kg for Zn at 20 mg/kg in comparison with corresponding single Zn 20 mg/kg treatment (Fig. 1).

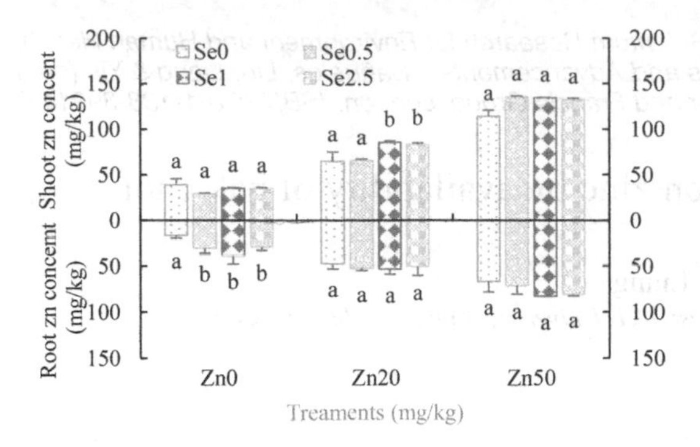

Figure 2. Single and combined effects of Zn and Se on the uptake of Zn by pak choi shoots and roots. Different letters indicate significant differences ($p < 0.05$).

3.2 *Effect of Se and Zn interaction on Zn uptake and translocation in pak choi*

The uptake and translocation of Zn in pak choi were evaluated based on roots and shoots Zn concentrations (Fig. 2). The Zn concentrations in shoots and roots significantly increased when Zn was applied in comparison with control ($p < 0.05$). Single Se treatments promoted the uptake of Zn in the roots of pak choi in comparison with the control ($p < 0.05$) (Fig. 2). Similarly, Boldrin et al. (2013) found that Se increased the absorption of nutrients by plants.

The shoot Zn concentrations of pak choi significantly increased when Se was applied at 1.0, 2.5 mg/kg, and Zn 20 mg/kg in comparison to a single Zn 20 mg/kg treatment (Fig. 2). Hence, the application of Se can increase the Zn content in the edible parts of pak choi by promoting the translocation of Zn (Longchamp et al. 2016).

4 CONCLUSIONS

Applying high concentrations of Se (≥1 mg/kg) combined with Zn inhibited the growth of pak choi. This effect included a significant decrease in root length and root biomass of pak choi, but the shoot height and shoot biomass of pak choi were not significantly decreased. The application of Se promoted the uptake of Zn by pak choi. This promotion may, however, only significantly occur when Se is at appropriate levels (1.0–2.5 mg/kg) along with the Zn concentration of 20 mg/kg.

REFERENCES

Boldrin, P.F., Faquin, V., Ramos, S.J., Boldrin, K.V.F., Ávila, F.W. & Guilherme, L.R.G. 2013. Soil and foliar application of selenium in rice biofortification. *J Food Compos Anal* 31(2): 238–244.

Cakmak, I. & Kutman, U. 2018. Agronomic biofortification of cereals with zinc: a review. *Eur J Soil Sci* 69(1): 172–180.

Fargasova, A., Pastierova, J. & Svetkova, K. 2006. Effect of Se-metal pair combinations (Cd, Zn, Cu, Pb) on photosynthetic pigments production and metal accumulation in *Sinapis alba* L. seedlings. *Plant Soil Environ* 52(1): 8–15.

Krężel, A. & Maret, W. 2016. The biological inorganic chemistry of zinc ions. *Arch Biochem Biophys* 611: 3–19.

Li, M., Wang, S., Tian, X. et al. 2015. Zn distribution and bioavailability in whole grain and grain fractions of winter wheat as affected by applications of soil N and foliar Zn combined with N or P. *J Cereal Sci* 61: 26–32.

Longchamp, M., Angeli, N. & Castrec-Rouelle, M. 2016. Effects on the accumulation of calcium, magnesium, iron, manganese, copper and zinc of adding the two inorganic forms of selenium to solution cultures of Zea mays. *Plant Physiol Biochem* 98: 128–137.

Lu, X., Cui, J., Tian, X., Ogunniyi, J.E., Gale, W.J. & Zhao, A. 2012. Effects of zinc fertilization on zinc dynamics in potentially zinc-deficient calcareous soil. *Agron J* 104(4): 963–969.

Souza, G.A., Hart, J.J., Carvalho, J.G. et al. 2014. Genotypic variation of zinc and selenium concentration in grains of Brazilian wheat lines. *Plant Sci* 224(13): 27–35.

Selenium Research for Environment and Human Health: Perspectives, Technologies and Advancements – Bañuelos, Lin, Liang & Yin (eds)
© 2020 Taylor and Francis Group, London, ISBN 978-1-138-39014-0

Detoxification of exogenous selenate in cadmium-contaminated soil with pak choi

M.X. Qi, M.K. Wang, Y. Liu, N.N. Liu, Q.T. Dinh & D.L. Liang*
College of Natural Resources and Environment, Northwest A&F University, Yangling, Shaanxi, China

1 INTRODUCTION

Cadmium (Cd) has adverse health effects on human beings and plants (Ismael et al. 2019). Cadmium can inhibit growth and biomass, interfere with the home-ostasis of zinc, iron, and calcium, and decrease photosynthesis and carbon assimilation (Shanying et al. 2017). Moreover, Cd from contaminated soil enters the food chain by plant uptake, thereby affecting human health. Therefore, it is very important to reduce the accumulation of Cd in edible parts of crops grown in Cd-contaminated soil.

Selenium (Se) is not essential for plant growth but it can stimulate plant growth, improve the quality of agricultural products at low dose, and also alleviate a variety of environmental pressures, including heavy metal accumulation. At higher concentrations, Se can be used as oxidant and cause damage to plants (Saidi et al. 2014). In contrast, studies have demonstrated that the addition of Se is also a safe and effective way to reduce the accumulation of Cd in plants (Gao et al. 2018, Zhao et al. 2019, Rahman et al. 2019). Bian et al. (2018) showed that Se application in Cd-contaminated soil can increase biomass production and reduce the absorption of Cd in peanuts. Selenate is commonly used in biofortification strategies because of its availability for absorption by plants. Therefore, the aim of this study was to explore the possible detoxification ability of selenate in native Cd-contaminated soil.

2 MATERIALS AND METHODS

The Cd-contaminated soil samples were collected from the Xi'an, Shaanxi, China. A pot experiment was conducted in a greenhouse located at Northwest A&F University in Shaanxi Province, China. For the treatments, 5 kg of air-dried soil was placed in pots. For each kilogram of soil, 0.30, 0.15, and 0.15 g of N, P, and K fertilizers, respectively, were added as base fertilizers. The soil samples were spiked with selenate (0, 0.5, 1, and 2.5 mg Se/kg soil). Each treatment was prepared in four replicates (three planted with pak choi and one without). Twenty-five pak choi seeds were sown in each pot, and seven days after germination, the seedlings were thinned to 5 per pot. Meanwhile, the soil moisture was maintained at 70% of the maximum water-holding capacity with deionized water. Plants were harvested after 35 d.

For Se determination, the plants were successively rinsed with tap water and deionized water for three times until no more soil particles remained on roots. Afterward, roots and shoots of pak choi were separated, oven-dried at 50°C for 3 days until samples reach constant weight. The plant samples were weighed, ground, and analyzed by HG-AFS after digestion for Se analysis. The Cd concentration in plant tissue was determined by ICP-MS after digestion with 4:1 (v/v) HNO_3-$HClO_4$ at 160°C.

3 RESULTS

The effects of exogenous selenate on pak choi shoot length and root height on cadmium-contaminated soil are shown in Figure 1. Selenate has no significant effect on root length and shoot height of pak choi ($p > 0.05$), except at the application of selenate concentration of 1 mg/kg. Moreover, shoot height of pak choi was 22.3% higher than that of the control treatment.

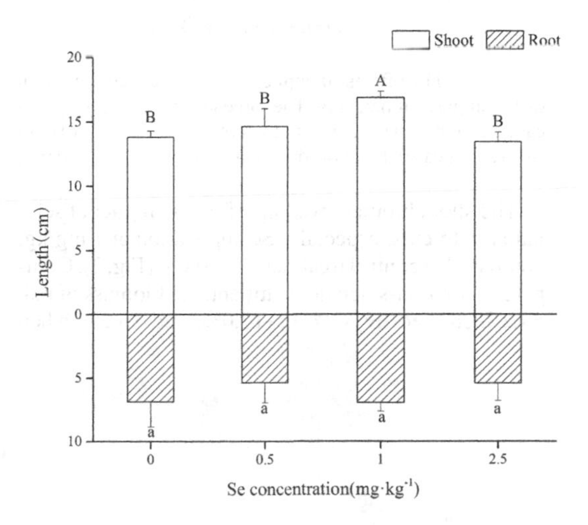

Figure 1. The effect of applied Se on pak choi shoot length and root height on Cd-contaminated soil. The letters (A, a) represent the result of significance analysis in pak choi (shoot height and root length) under the different Se treatments ($p < 0.05$).

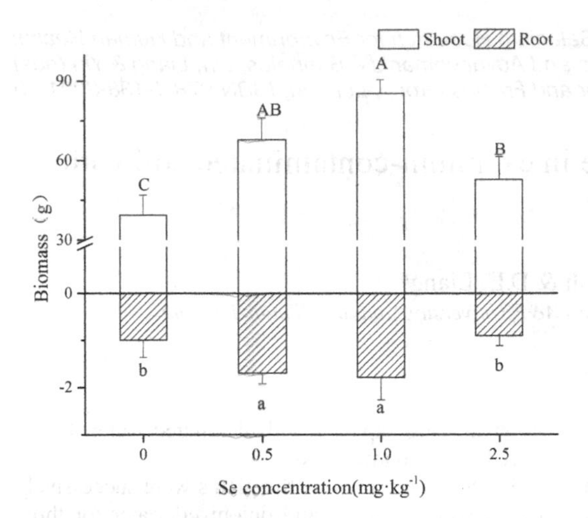

Figure 2. The effect of applied Se on pak choi biomass on Cd-contaminated soil. The letters (A, a) represent the result of significance analysis in pak choi (shoot biomass, root biomass) under the different Se treatments ($p < 0.05$).

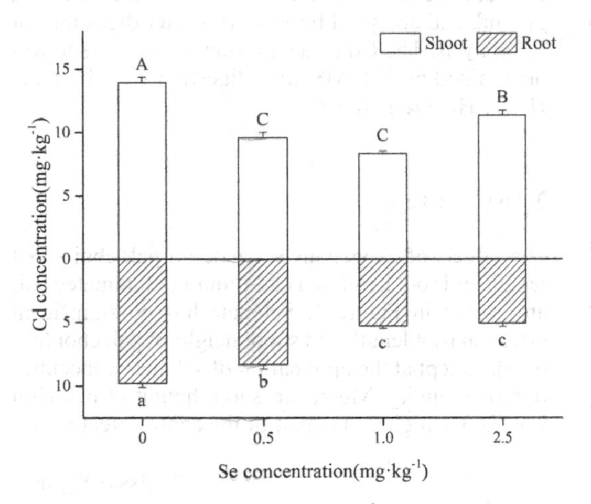

Figure 3. The effects of applied Se on Cd concentration in different parts of pak choi. The same letters mean no significant difference in Cd concentration in shoot (lower case) or in root (upper case) under the different Se treatments ($p < 0.05$).

The shoot biomass was significantly higher at selenate treatments, especially Se application at 1 mg/kg, compared to control treatment ($p < 0.05$) (Fig. 2). Compared with non-selenate treatment, the biomass of pak choi increased by 87.7%, 117.0%, and 55.2% when selenate was applied at the concentration of 0.5, 1, and 2.5 mg/kg, respectively. The effect of selenate application on root biomass of pak choi was similar to that of shoot.

Applying exogenous selenate reduced Cd accumulation in plant tissues (Fig. 3). The concentration of Cd accumulated in pak choi significantly decreased when selenate was applied, especially at 1 mg/kg. This result is consistent with the effect of selenate on shoot height, root length, and biomass of pak choi. This study shows selenate reduced Cd concentrations in plant tissue, which is beneficial for the growth of pak choi in Cd-contaminated soils.

4 CONCLUSIONS

Low concentration of selenate (0.5~1 mg/kg) can promote plant growth and reduce Cd concentration in plants. However, the mechanism involved with the selenate and Cd interaction in the soil and Cd transport within the plant needs further research.

REFERENCES

Bian, W., Tian, X., Ding, X., Wang, K. & Zhang, L. 2017. Effects of selenium application on cadmium and selenium absorption and photosynthetic efficiency in peanut. *Environ Chem* 36(11): 2349–2356.

Gao, M., Zhou, J., Liu, H. et al. 2018. Foliar spraying with silicon and selenium reduces cadmium uptake and mitigates cadmium toxicity in rice. *Sci Total Environ* 631: 1100–1108.

He, S., Yang, X., He, Z. & Baligar, V.C. 2017. Morphological and physiological responses of plants to cadmium toxicity: a review. *Pedosphere* 27(3): 421–438.

Ismael, M.A., Elyamine, A.M., Moussa, M.G., Cai, M., Zhao, X. & Hu, C. 2019. Cadmium in plants: uptake, toxicity, and its interactions with selenium fertilizers. *Metallomics* 11(2): 255–277.

Rahman, M.M., Hossain, K.F.B., Banik, S. et al. 2019. Selenium and zinc protections against metal-(loids)-induced toxicity and disease manifestations: a review. *Ecotoxicol Environ Saf* 168:146–163.

Saidi, I., Chtourou, Y. & Djebali, W. 2014. Selenium alleviates cadmium toxicity by preventing oxidative stress in sunflower (*Helianthus annuus*) seedlings. *J Plant Physiol* 171(5): 85–91.

Zhao, Y., Hu, C., Wu, Z. et al. 2019. Selenium reduces cadmium accumulation in seed by increasing cadmium retention in root of oilseed rape (*Brassica napus* L.). *Environ Exp Bot* 158: 161–170.

Selenium Research for Environment and Human Health: Perspectives, Technologies and Advancements – Bañuelos, Lin, Liang & Yin (eds)
© 2020 Taylor and Francis Group, London, ISBN 978-1-138-39014-0

Effects of selenate on the uptake and transportation of cadmium in radish (*Raphanus sativus*)

Y. Liu, M.X. Qi, M.K. Wang, N.N. Liu & D.L. Liang*
College of Natural Resources and Environment, Northwest A&F University, Yangling, Shaanxi, China

1 INTRODUCTION

Cadmium (Cd) is one of the most harmful and widespread heavy metals in agricultural soils. Cadmium in soil is easily taken up by crop roots and rapidly transported to the edible parts. Long-term Cd intake can cause chronic toxicity diseases in livestock and humans (Templeton & Liu 2010). Thus, it is imperative to find effective and low-cost strategies to reduce Cd toxicity for the health of humans. Selenium (Se), as a necessary beneficial element for humans and animals, is involved in the synthesis of various proteins and enzymes in mammals and plays an important role in anti-oxidation, anti-mutation, anti-cancer and other aspects (Brown & Arthur 2001). Moreover, many studies have demonstrated that Se may promote antioxidant capacity and enhance plant tolerance to abiotic stresses, such as heavy metal exposure and UV-induced oxidation. Ding et al. (2014) reported that Se significantly reduced the Cd content in rice grown hydroponically. Using the antagonistic effects of Se on Cd in an effective way, Se may reduce the amount of Cd entering the human body through the food chain. Therefore, the present research used a pot experiment to determine if the application of exogenous Se can alleviate cadmium-induced plant growth inhibition, Cd uptake, and Cd distribution and translocation.

2 MATERIALS AND METHODS

A pot experiment was conducted in greenhouse Northwest A&F University using radish (*Raphanus sativus* L.). Soil was collected from the Cd contaminated farm site in Yangling, Shaanxi province. The basic physicochemical properties of the soil were as follows: pH 8.61, CEC 13.88 cmol/kg, organic matter 15.83 g/kg, total Se 0.46 mg/kg and total Cd 5.36 mg/kg. Selenate (as Na_2SeO_4) exposure levels were set at 0.5, 1.0, and 2.5 mg/kg soil. In addition, one treatment without Se was prepared as the control. Pots were maintained under greenhouse conditions and were watered to maintain soil moisture at 70% field capacity. Plants were harvested after 35 days. Radish samples were collected from each pot, cleaned, and dried to constant weight at 90°C for 30 min and 55°C for 72 h. The dry samples were ground into fine powder. Plant samples were acid-digested with 4:1 (v/v) HNO_3 - $HClO_4$ at 160°C. The Cd concentration of the digestion solution was measured by ICP-MS. BCF, TF, and DF factors were used to characterize the distribution of Cd in radish:

$$BCF = C_{plant}/C_{soil} \tag{1}$$

$$TF = C_{shoot}/C_{root} \tag{2}$$

$$DF = M_{edible}/M_{total} \tag{3}$$

3 RESULTS AND DISCUSSION

3.1 *Effect of applied Se on Cd uptake in radish under Cd stress*

Effect of selenate application on Cd concentration in radish tissues is shown in Figure 1. The concentration of Cd in shoots and roots is lower than that without Se treatment. At the application level of 1 mg/kg selenate, the concentration of Cd in radish was lowest in both shoot and root, where the greatest inhibitory effect on Cd absorption took place in radish. Shanker et al. (1996) reported that selenate can be reduced to Se^{2-}, thus forming an insoluble Cd-Se complex with low availability with Cd and resulting in reduced the absorption of Cd by plants.

3.2 *Effect of applied Se on Cd uptake and translocation in radish under Cd stress*

The BCF value decreased with an increase of Se treatment (Table 1), indicating that the absorption of Cd by plant was inhibited by Se application. The TF values of the plant were all greater than 1, indicating that the crop had strong ability to transport Cd to the aboveground part. Compared with the control, the application of Se increased the TF of crops, indicating that Se promoted Cd transport from root to the shoot. With the increase of Se level, the proportion of Cd in edible parts of radish decreased. The results showed that the application of Se promoted the distribution of Cd in shoot.

Some studies have shown that Se can significantly reduce the accumulation of Cd in rice shoot but has

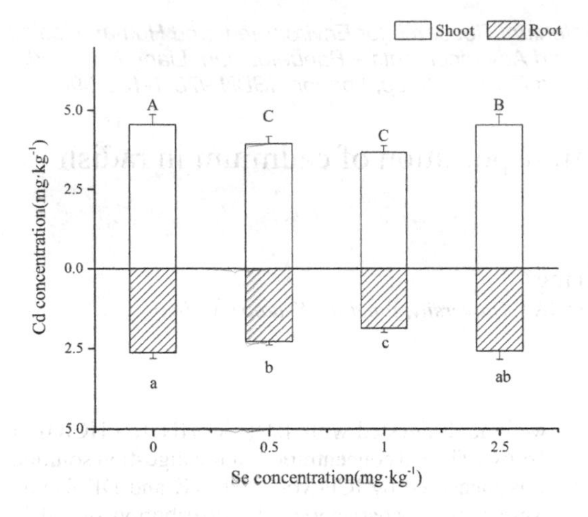

Figure 1. The effect of applied Se on Cd uptake in radish under Cd stress.

Table 1. Effect of Se on the BCF, TF, and DF of Cd in radish.

Se Concentration (mg/kg)	BCF	TF	DF (%)
CK	0.49	1.73	7.25
0.5	0.42	1.73	5.51
1	0.37	1.95	4.79
2.5	0.50	1.77	3.02

no significant effect on the accumulation of Cd in the root (Chen et al. 2014). However, in this study, most of the Cd absorbed by plants was accumulated in shoots, and the distribution ratio of Cd in radish shoots increased with the increase of Se level, indicating that Se application promoted the distribution of Cd in shoot.

4 CONCLUSIONS

The application of Se at the appropriate concentration (1 mg/kg) into the soil can reduce the accumulation of Cd in plants, and the application of exogenous Se can regulate the transport of Cd in crops. However, the mechanism of Se inhibiting Cd in plants still needs to be further explored.

REFERENCES

Brown, K.M. & Arthur, J.R. 2001. Selenium, selenoproteins and human health: A review. *Public Health Nutr* 4(2b): 593-599.

Chen, M., Cao, L., Song, X., Wang, X., Qian, Q. & Liu, W. 2014. Effect of iron plaque and selenium on cadmium uptake and translocation in rice seedlings *(Oryza sativa)* grown in solution culture. *Int J Agr Biol* 16(6):1159-1164

Ding, Y., Feng, R., Wang, R., Guo, J. & Zheng, X. 2014. A dual effect of Se on Cd toxicity: Evidence from plant growth, root morphology and responses of the antioxidative systems of paddy rice. *Plant Soil* 375(1-2): 289-301.

Shanker, K., Mishra, S., Srivastava, S. et al. 1996. Effect of selenite and selenate on plant uptake of cadmium by maize *(Zea mays)*. *Bull Environ Contam Toxicol* 56(3): 419-424.

Templeton, D. M. & Liu, Y. 2010. Multiple roles of cadmium in cell death and survival. *Chem Biol Interact* 188(2): 267-275.

Zhang, H., Feng, X., Zhu, J. et al. 2012. Selenium in soil inhibits mercury uptake and translocation in rice (*Oryza sativa* L.). *Environ Sci Tech* 46(18): 10040-10046.

Selenium Research for Environment and Human Health: Perspectives, Technologies and Advancements – Bañuelos, Lin, Liang & Yin (eds)
© 2020 Taylor and Francis Group, London, ISBN 978-1-138-39014-0

Responses of roots of rice exposed to selenium and cadmium

G. Liao, Y. Wei & M. Gu*
Cultivation Base of Guangxi Key Laboratory for Agro-Environment and Agro-Products Safety, College of Agriculture, Guangxi University, Nanning, China

1 INTRODUCTION

Recently, the use of selenium (Se) to alleviate the toxicity of cadmium (Cd) in living organisms has attracted increased attention. Studies have shown that Se can reduce the uptake of Cd by rice (Saidi et al. 2014, Liao et al. 2016). However, most studies have only focused on the antioxidant function of Se, Se-regulated photosynthetic system and Se repairing damaged cells, while research on how the Se inhibits Cd uptake and transfer in rice roots has not yet been systematically conducted.

In this study, hydroponic techniques were used to culture rice at seedling stage to investigate the root morphology and cell wall components in rice as affected by the interaction of Se and Cd.

2 MATERIALS AND METHODS

Rice (*Oryza sativa*) seedlings of Y-Liangyou No. 2 of uniform size were selected and transplanted into a bottle containing 1 L Hoagland's nutrient solution. There were two seedlings per bottle, and the pH of the solution was adjusted to 5.8-6.0. After acclimation for two weeks, seedlings were subsequently exposed to different concentrations of Se (Na_2SeO_3) and Cd ($CdCl_2$) for the following treatments: CK, 0.2 mg/kg Se (Se0.2), 1 mg/kg Se (Se1), 2 mg/kg Cd (Cd2), 5 mg/kg Cd (Cd5), 0.2 mg/kg Se and 2 mg/kg Cd (Se0.2Cd2), 0.2 mg/kg Se and 5 mg/kg Cd (Se0.2Cd5), 1 mg/kg Se and 2 mg/kg Cd (Se1Cd2), and 1 mg/kg Se and 5 mg/kg Cd (Se1Cd5). The solution was vigorously aerated and replaced twice weekly. Two weeks later, plants were harvested, and root scanning was performed. Roots were cleaned with deionized water, and 1.0 g of fresh samples was stored in liquid nitrogen for determination of Cd content in root sub-cells. The remaining roots were placed in an oven and dried at 70°C to constant weight to determine the Cd concentrations. Plant sample digestion and data analysis was conducted following the methods described by Liao et al. (2016).

3 RESULTS AND DISCUSSION

3.1 *Root morphology responses to Se and Cd*

Selenium reduced the proportion of fine roots and increased the proportion of medium roots in rice. This effect was more pronounced with an increase in Se concentration (Fig. 1). Selenium changed the root morphology of rice, suggesting that these changes can affects the uptake of Cd by rice (Feng et al. 2016).

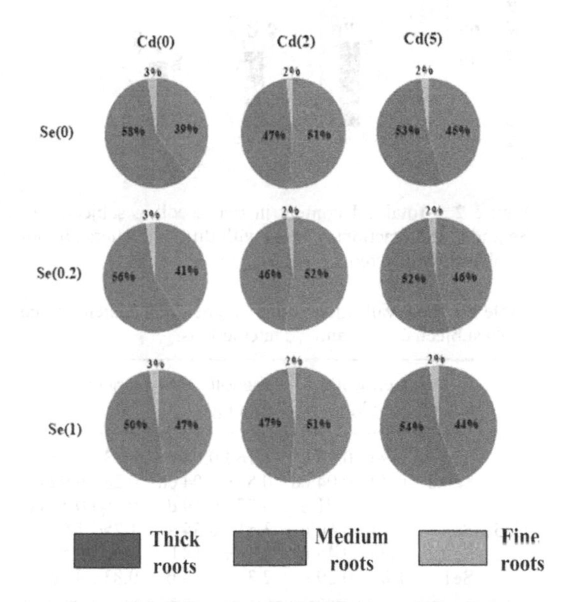

Figure 1. The percentage of fine roots, medium roots and thick roots of rice subjected to Se and Cd interactions.

3.2 *Cadmium content in rice seedlings*

The content of Cd in roots increased significantly with an increase in the concentration. A certain concentration of Se addition significantly reduced the content of Cd in roots, except for Se0.2. The variation of Cd content in the aboveground plant is similar to that of the root system (Fig. 2).

3.3 *Cadmium content in cell wall, vacuolar and organelle of roots*

Cadmium in the root system is mainly present in the cell wall and vacuoles, and the remaining organelles contain low concentrations (Table 1). Selenium reduced the Cd content of the subcellular components of rice roots under the same Cd concentration treatments (Table 1). These results show that Se changes the subcellular distribution of root Cd by

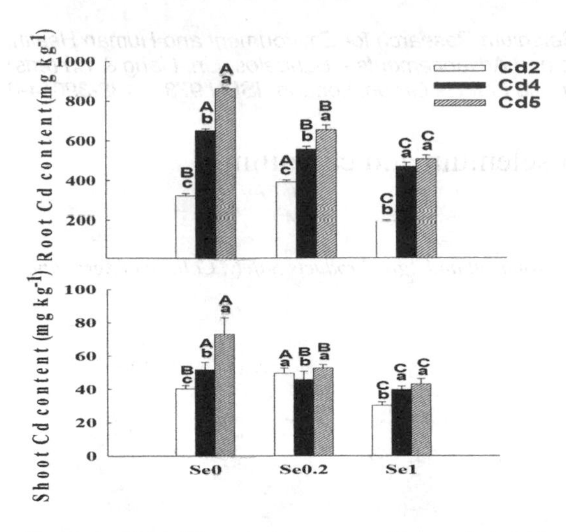

Figure 2. Total Cd content in rice seedling subjected to Se and Cd interactions. Means with different letters are not significantly different (p > 0.05).

Table 1. Cell wall, vacuole and organelle Cd content in rice roots subjected to Se and Cd interactions.

Treatments		Cell wall (mg/kg)	Vacuole (mg/kg)	Organelle (mg/kg)
Cd2	Se0	0.96±0.03 c*	1.16±0.04 c	0.33±0.01 c
	Se0.2	0.79±0.04 cd	0.85±0.04 cd	0.28±0.92 c
	Se1	0.61±0.01 e	0.77±0.19 d	0.16±0.57 cd
Cd5	Se0	2.74±0.66 a	3.51±0.34 a	1.25±1.00 a
	Se0.2	1.96±0.45 b	2.40±0.18 b	0.96±0.57 ab
	Se1	1.07±0.29 c	2.36±0.13 b	0.83±0.38 b

*The presented data are mean values ± SE. Lowercases letters in the same column indicate significant differences among different treatments, respectively ($p < 0.05$).

increasing the Cd content of the cell wall component and decreasing the Cd content in the cell vacuole and organelle components. In addition, Se may increase the GSH content in rice, promote the binding of Cd to sulfhydryl groups, and thereby reduce the toxicity of Cd (Schützendubel et al. 2001).

4 CONCLUSIONS

These results support the hypothesis that Se changes the root morphology to inhibit Cd uptake and simultaneously changes the subcellular distribution of root Cd content to inhibit the transfer of Cd to aboveground parts of rice.

ACKNOWLEDGEMENT

This research work was financially supported by the Innovation Driven Special Foundation of Guangxi (No. AA17202038), Natural Science Foundation of Guangxi (No. 2016GXNSFAA380308).

REFERENCES

Feng, R.W. Liao, G.J. Guo, J.K. et al. 2016. Responses of root growth and antioxidative systems of paddy rice exposed to antimony and selenium. *Environ Exp Bot* 122: 29–38.

Liao, G.J. Wu, Q.H. Feng, R.W. et al. 2016. Efficiency evaluation for remediating paddy soil contaminated with cadmium and arsenic using water management, variety screening and foliage dressing technologies. *J Environ Manag* 170: 116–122.

Saidi, I. Chtourou, Y. & Djebali, W. 2014. Selenium alleviates cadmium toxicity by preventing oxidative stress in sunflower (*Helianthus annuus*) seedlings. *J Plant Physiol* 171: 85–91.

Schützendubel, A. Schwanz, P. Teichmann, T. et al. 2001. Cadmium-induced changes in antioxidative systems, hydrogen peroxide content, and differentiation in Scots pine roots. *J Plant Physiol* 127(3): 887–898.

Selenoproteins

Selenium Research for Environment and Human Health: Perspectives, Technologies and Advancements – Bañuelos, Lin, Liang & Yin (eds)

The role of selenium and selenoproteins in immunity

P.R. Hoffmann
John A. Burns School of Medicine, University of Hawaii, Honolulu, Hawaii, USA

1 INTRODUCTION

1.1 *Selenium and selenoprotein biology*

The immune system relies on adequate dietary selenium (Se) intake and this nutrient exerts its biological effects mostly through its incorporation into selenoproteins (Huang et al. 2012). The selenoproteome contains 25 members in humans that exhibit a wide variety of functions. While many members of the selenoprotein family function as enzymes involved in redox reactions, some are likely not enzymes themselves, and functions are gradually becoming better understood for these non-enzymatic members (Reeves & Hoffmann 2009). Understanding how levels of dietary Se intake modulate immunity through different selenoproteins has been a challenge, but progress has been made in this field of study.

2 MATERIALS AND METHODS

2.1 *Cell line studies, rodent models, human studies*

Cell lines established from different immune cell types have been extensively used. While Se concentration can vary between different batches of complete media in which these cell lines are grown (mainly due to sources of fetal bovine sera), the optimal growth and functioning of these immune cell lines has found to be approximately 80-100 nM. The minimum Se intake and supernutritional levels in rodents has been established to be 0.1 ppm and approximately 0.8 ppm, respectively (RA 2012). For many of our studies, experiments are designed to include moderately low (0.08 μg/g), adequate (0.25 μg/g), and supernutritional (1.0 μg/g) levels of Se intake. Defined diets are provided by Research Diets, Inc. (New Jersey, USA). Transgenic mouse models for studying selenoproteins in our laboratory have focused on SELENOK, but many others have been established by other investigators. For studies involving human immune cells, peripheral blood mononuclear cells were purified from buffy coats of blood from normal healthy donors. Human macrophages were cultured from monocytes and Miltenyi columns used for purifying T cells.

3 RESULTS AND DISCUSSION

3.1 *Selenium levels can affect immunity and inflammation*

Selenium deficiency can give rise to immune-incompetence that leads to increased susceptibility to infections and possibly to cancers. There is some evidence that Se can modulate the pathology that accompanies chronic inflammatory diseases in the gut and liver as well as in inflammation-associated cancers (Avery & Hoffmann 2018, Huang et al. 2012). Selenium deficiency and suppressed selenoprotein expression have been implicated in higher levels of inflammatory cytokines in a variety of tissues including the gastrointestinal tract and others tissues. A mouse model of allergic asthma showed that Se deficiency reduced airway inflammation while adequate Se intake produced higher levels of inflammation that were then decreased when supra-nutritional levels of Se were used (Hoffmann et al. 2007).

Selenium supplementation, for the most part, is immuno-stimulatory, depending on the baseline Se status. This is measured by a wide range of parameters including T cell proliferation, NK cell activity, innate immune cell functions, and many others. The activation of human blood leukocytes has been shown to increase in response to Se-enriched food (Bentley-Hewitt et al. 2014). Vaccine responses against pathogens such as poliovirus have been shown to improve with Se supplementation (Broome et al. 2004), even though results were mixed when analyzing the influenza vaccine in older adults. Similarly, integrated-omics analyses of pathways affected by the Se status in rectal biopsies from 22 healthy adults showed reduced inflammatory and immune responses and cytoskeleton remodeling in the suboptimal Se status group (Meplan et al. 2016). Similarly, Se supplementation was shown to modulate the inflammatory response in respiratory distress syndrome patients by restoring the antioxidant capacity of the lungs, which moderated the inflammatory responses through IL-1β and IL-6 levels and meaningfully improved the respiratory mechanics (Mahmoodpoor et al. 2019). Adaptive immunity is affected by Se intake including the activation and functions of T and B cells. There are several reports of the skewing of T cell immunity toward Th1 phenotypes. For example, our laboratory used a mouse model of viral antigen vaccination to test effects of low (0.087 μg/g), medium (0.25 μg/g), and high (1.0 μg/g) Se diets and found that Th1 immunity was enhanced along with the T cell receptor signal strength (Hoffmann et al. 2010). In a separate study, oral administration of synthetic Se nanoparticles induced a robust Th1 cytokine pattern after a hepatitis B surface antigen vaccination in a mouse model (Mahdavi et al. 2017). Less information is available regarding the effects of Se on cytotoxic $CD8^{+}$T cells even though cytotoxic T cells from aged mice (24 months old) showed enhanced mitogen-induced proliferation when treated with Se supplementation (Roy et al. 1995). The deletion of all selenoproteins using the tRNA deficient model developed in the Hatfield laboratory has shown that the lack of selenoproteins in macrophages and T cells impairs a variety of functions (Carlson et al. 2010). Also, models targeting individual selenoproteins have been revealing including the MSRB1 and SELENOK knockout mice (Lee et al. 2013, Verma et al. 2011).

4 CONCLUSIONS

Under conditions of Se deficiency, innate and adaptive immune responses are impaired. The benefits of supranutritional Se to boost immunity against pathogens, vaccinations, or cancers have been explored and have not provided entirely clear results. Manipulation of individual selenoproteins may offer a more precise approach for enhancing the immune system or mitigating chronic inflammation.

REFERENCES

Avery, J.C. & Hoffmann, P.R. 2018. Selenium, selenoproteins, and immunity. *Nutrients* 10(9): 1203.

Bentley-Hewitt, K.L., Chen, R.K., Lill, R.E. et al. 2014. Consumption of selenium-enriched broccoli increases cytokine production in human peripheral blood mononuclear cells stimulated ex vivo, a preliminary human intervention study. *Mol Nutr Food Res* 58(12): 2350–2357.

Broome, C.S., McArdle, F., Kyle, J.A. et al. 2004. An increase in selenium intake improves immune function and poliovirus handling in adults with marginal selenium status. *Am J Clin Nutr* 80(1): 154–162.

Carlson, B.A., Yoo, M.H., Shrimali, R.K. et al. 2010. Role of selenium-containing proteins in T-cell and macrophage function. *Proc Nutr Soc* 69(3): 300–310.

Hoffmann, F.W., Hashimoto, A.C., Shafer, L.A., Dow, S., Berry, M.J., & Hoffmann, P.R. 2010. Dietary selenium modulates activation and differentiation of CD4+ T cells in mice through a mechanism involving cellular free thiols. *J Nutr* 140(6): 1155–1161.

Hoffmann, P.R., Jourdan-Le Saux, C., Hoffmann, F.W. et al. 2007. A role for dietary selenium and selenoproteins in allergic airway inflammation. *J Immunol* 179(5): 3258–3267.

Huang, Z., Rose, A.H., & Hoffmann, P.R. 2012. The role of selenium in inflammation and immunity: from molecular mechanisms to therapeutic opportunities. *Antioxid Redox Signal* 16(7): 705–743.

Lee, B.C., Peterfi, Z., Hoffmann, F.W. et al. 2013. MsrB1 and MICALs regulate actin assembly and macrophage function via reversible stereoselective methionine oxidation. *Mol Cell* 51(3): 397–404.

Mahdavi, M., Mavandadnejad, F., Yazdi, M.H. et al. 2017. Oral administration of synthetic selenium nanoparticles induced robust Th1 cytokine pattern after HBs antigen vaccination in mouse model. *J Infect Public Health* 10(1): 102–109.

Mahmoodpoor, A., Hamishehkar, H., Shadvar, K. et al. 2019. The effect of intravenous selenium on oxidative stress in critically Ill patients with acute respiratory distress syndrome. *Immunol Invest* 48(2): 147–159.

Meplan, C., Johnson, I.T., Polley, A.C. et al. 2016. Transcriptomics and proteomics show that selenium affects inflammation, cytoskeleton, and cancer pathways in human rectal biopsies. *FASEB J* 30(8): 2812–2825.

RA, S. 2012. Selenoproteins: hierarchy, requirements, and biomarkers. In D.L. Hatfield, M.J. Berry & V.N. Gladyshev (eds), *Selenium: Its Molecular Biology and Role in Human Health* (3 ed.): 137–152. New York: Springer.

Reeves, M.A. & Hoffmann, P.R. 2009. The human selenoproteome: recent insights into functions and regulation. *Cell Mol Life Sci* 66(15): 2457–2478.

Roy, M., Kiremidjian-Schumacher, L., Wishe, H.I., Cohen, M.W. & Stotzky, G. 1995. Supplementation with selenium restores age-related decline in immune cell function. *Proc Soc Exp Biol Med* 209(4): 369–375.

Verma, S., Hoffmann, F.W., Kumar, M. et al. 2011. Selenoprotein K knockout mice exhibit deficient calcium flux in immune cells and impaired immune responses. *J Immunol* 186(4): 2127–2137.

Selenium Research for Environment and Human Health: Perspectives, Technologies and Advancements – Bañuelos, Lin, Liang & Yin (eds)
© 2020 Taylor and Francis Group, London, ISBN 978-1-138-39014-0

New developments in recombinant selenoprotein production

Q. Cheng & E.S.J. Arnér
Division of Biochemistry, Department of Medical Biochemistry and Biophysics, Karolinska Institutet, Stockholm, Sweden

1 INTRODUCTION

1.1 *Selenoproteins*

Selenoproteins are found in all kingdoms of life, albeit not all organisms, and confer unique features to such proteins due to the chemical nature of selenocysteine (Sec), and the defining entity of selenoproteins (Arnér 2010, Metanis & Hilvert 2014, Reich & Hondal 2016).

1.2 *Challenges in selenoprotein production*

Because Sec is cotranslationally inserted at an in-frame UGA stop codon by dedicated selenoprotein translation machinery interacting with a secondary structure of the mRNA (called a Sec Insertion Sequence [SECIS] element), heterologous selenoproteins cannot be expressed in recombinant form in *E. coli* using traditional methods typically compatible with non-selenoproteins (Arnér 2002, Johansson et al. 2005, Metanis & Hilvert 2014).

2 RECOMBINANT SELENOPROTEIN PRODUCTION IN *E. COLI*

2.1 *Natural translation of endogenous selenoproteins in E. coli*

The endogenous selenoprotein synthesis machinery of *E. coli* was characterized by August Böck and coworkers, who also showed that it is fully compatible with an overexpression system for endogenous selenoproteins, i.e. bacterial formate dehydrogenases (Zinoni et al. 1990). This research originally set the stage for our attempts to produce recombinant mammalian selenoproteins in *E. coli.*

2.2 *Recombinant mammalian selenoprotein production in E. coli between 1999 and 2017*

In 1999 the first recombinant production in *E. coli* of a mammalian selenoprotein was reported. This was thioredoxin reductase 1 (TrxR1) of rat, which was produced using the engineering of an *E. coli*-compatible SECIS element enabled by the close to C-terminal location of the Sec residue in the enzyme, resulting in about 15% Sec content (Arnér et al. 1999). Different measures subsequently increased Sec contents, and many different forms of proteins with penultimate Sec residues were produced in the coming years, as summarized elsewhere (Cheng & Arner 2018). It was not until 2017, however, that we developed a novel recombinant approach.

3 NEW DEVELOPMENTS IN RECOMBINANT SELENOPROTEIN PRODUCTION

3.1 *Improved yield and specificity in production of thioredoxin reductases*

We used a new strain of *E. coli* that has its 321 chromosomal TGA codons exchanged for TAG and its release factor RF1 genetically deleted. Together with the redefinition of the tRNA for Sec with the corresponding anticodon, we produced recombinant TrxR1 with nearly full Sec content when combined with a variant of the previously designed and tailored SECIS element (Cheng & Arner 2017).

3.2 *Production of glutathione peroxidases and other selenoproteins with internal Sec residues*

Using the production conditions mentioned in 3.1, we also found that the SECIS element could be completely removed, thereby enabling production of selenoproteins containing internal (not close the C-terminus) Sec residues. This consequence was exemplified with production of human glutathione peroxidase, GPx, which contained approximately 15% Sec, mixed with Lys- or Gln-mediated suppression of the UAG codon (Cheng & Arner 2017). Although full Sec content was not achieved, this further development of the methodology was a major improvement and potentially opened the field for production of many novel recombinant selenoproteins having internal Sec residues.

4 OUTLOOK AND REMAINING CHALLENGES

Even if the production of GPx as mentioned in section 3.2 was a major long-awaited methodological breakthrough in recombinant selenoprotein production, the following challenges still remain:

- How can the Sec contents be further improved in recombinant GPx's and other selenoproteins with internal Sec residues?
- Are other yet unknown but nonetheless distinct translational events restricting the use of *E. coli* for production of recombinant selenoproteins?
- Can the Sec-containing variants of recombinant selenoproteins be purified from non-Sec-containing species?

Continued efforts will address these topics, aiming to establish recombinant protein production in *E. coli* as a general method also in the field of selenoprotein studies.

ACKNOWLEDGEMENT

The authors acknowledge funding from Karolinska Institutet, The Swedish Research Council, The Swedish Cancer Society, and The Knut and Alice Wallenberg Foundations.

REFERENCES

Arnér, E.S.J. 2002. Recombinant expression of mammalian selenocysteine-containing thioredoxin reductase and other selenoproteins in *Escherichia coli*. *Methods Enzymol* 347: 226–235.

Arnér, E.S.J. 2010. Selenoproteins-What unique properties can arise with selenocysteine in place of cysteine? *Exp Cell Res* 316(8): 1296–1303.

Arnér, E.S.J., Sarioglu, H., Lottspeich, F., Holmgren, A. & Böck, A. 1999. High-level expression in *Escherichia coli* of selenocysteine-containing rat thioredoxin reductase utilizing gene fusions with engineered bacterial-type SECIS elements and co-expression with the selA, selB and selC genes. *J Mol Biol* 292: 1003–1016.

Cheng, Q. & Arner, E.S. 2017. Selenocysteine insertion at a predefined UAG codon in a release factor 1 (RF1) depleted Escherichia coli host strain bypasses species barriers in recombinant selenoprotein translation. *J Biol Chem* 292: 5476–5487.

Cheng, Q. & Arner, E.S.J. 2018. Overexpression of recombinant selenoproteins in *E. coli*. *Methods Mol Biol* 1661: 231–240.

Johansson, L., Gafvelin, G. & Arnér, E.S.J. 2005. Selenocysteine in proteins – properties and biotechnological use. *Biochim Biophys Acta* 1726(1): 1–13.

Metanis, N. & Hilvert, D. 2014. Natural and synthetic selenoproteins. *Curr Opin Chem Biol* 22: 27–34.

Reich, H.J. & Hondal, R.J. 2016. Why nature chose selenium. *ACS Chem Biol* 11: 821–841.

Zinoni, F., Heider, J. & Böck, A. 1990. Features of the formate dehydrogenase mRNA necessary for decoding of the UGA as selenocysteine. *Proc Natl Acad Sci USA* 87: 4660–4664.

Selenium Research for Environment and Human Health: Perspectives, Technologies and Advancements – Bañuelos, Lin, Liang & Yin (eds)
© 2020 Taylor and Francis Group, London, ISBN 978-1-138-39014-0

Selenium nucleic acids for nuclei acid structure and function studies

V.G. Vandavasi, B. Hu, Y.Q. Chen, Q.W. Zhao, H.H. Liu, J.H. Gan, A. Kovalevsky, L. Hu & Z. Huang*
SeNA Research Institute, Life Science College, Sichuan University, Chengdu, China
Department of Chemistry, Georgia State University, Atlanta, Georgia, USA

1 INTRODUCTION

1.1 *Selenium-functionalized nucleic acids (SeNA)*

Selenium (Se) atom-specific functionalization can offer nucleic acids with many unique and novel properties (such as facilitated crystallization and phase determination) without significant perturbation of 3D structures of nucleic acids and their protein complexes.

1.2 *Structures and functions of nucleic acids*

Nucleic acids possess not only the ability to store genetic information and participate in transcription and translation, but also the capacity to adopt well-defined 3D structures, which can be readily adjusted to meet various functional needs (such as catalysis and therapeutics, see Hu et al. 2019, Vandavasi et al. 2018, Chen et al. 2019, Zhao et al. 2018, Liu et al. 2017). Although the importance of numerous nucleic acids in catalysis, gene expression, protein binding, and therapeutics has been acknowledged by the entire scientific society, the current understanding of nucleic acid-protein functions and structures is still limited, especially high-resolution structures.

This presentation will focus on the most recent Se-atom functionalization of nucleic acids and their potential applications in 3D structure-and-function studies and anticancer therapeutics in molecular medicine.

2 MATERIALS AND METHODS

2.1 *Crystallization of Se-functionalized nucleic acids*

Syntheses of Se-modified nucleotides and the DNA oligonucleotides incorporating $SeCH_3$ substituent at the 2′ position of the sugar moiety have been described previously (Vandavasi et al. 2018). Solution of the purified d[GTGG(C^{Se})CAC] oligonucleotide was first heated to 90°C for 1 min, and then allowed to cool slowly to room temperature (20°C) to give d[GTGG(C^{Se})CAC]$_2$, as the sequence is self-complimentary.

2.2 *X-ray and neutron data collection*

Room- and cryo-temperature X-ray data were collected on an in-house Rigaku HomeFlux system equipped with the R-AXIS IV^{++} image plate detector, the Rigaku MicroMax-007 HF Cu rotating-anode generator, Osmic VariMax HR optics, and the Oxford CryoStream operating at 100K. Quasi-Laue neutron data to 2.0 Å resolution (room temperature) and to 1.9 Å resolution (100 K) were collected from the 0.4 and 0.2 mm^3 DNA crystals, respectively (Vandavasi et al. 2018). During the neutron diffraction experiment, the crystal was held stationary at different ϕ settings for each exposure. In total, 22 diffraction images were collected (with an exposure time of 2 h per image) from 3 different crystal orientations for the room-temperature experiment. A total of 25 diffraction images (6 h per image) were collected for a crystal at 100K.

2.3 *X-ray and neutron data processing structure refinement*

Both room- and cryo-temperature joint XN structures of d[GTGG(C^{Se})CAC]$_2$ were determined using *nCNS*. Initial rigid-body refinement was followed by several cycles of positional, atomic displacement parameter (B factor), and D occupancy refinement. Between each cycle the structures were examined, and water molecule orientations were built based on the omit F_O-F_C difference neutron scattering length density map with meaningful hydrogen bonding interactions in mind.

3 RESULTS AND DISCUSSION

3.1 *Structure determination*

The octameric self-complimentary oligonucleotide d[GTGG(C^{Se})CAC]$_2$ crystallizes in a tetragonal unit cell ($P4_32_12$) with one DNA oligomer in the asymmetric unit and the double helix generated through the crystallographic 2-fold axis. Its neutron structures obtained at 2.0 Å and 1.9 Å resolutions at room (RT) and cryo (LT) temperatures, respectively, were refined jointly with the 1.56 Å (RT) and 1.65 Å (LT) X-ray crystallographic data to produce joint X-ray/neutron (XN) structures.

3.2 *Protonation of the phosphate at the room temperature*

Unexpectedly, we detected protonation of the R_P-oxygen of Ade7 phosphate in the room temperature XN structure. A strong peak was observed in the difference F_O-F_C neutron scattering density length map located at a distance about 1 Å from the R_P-oxygen. The density peak was interpreted as D, because no extra electron density was seen near this oxygen. The D atom occupancy refined to 67%; thus, the R_P-oxygen atom is ⅔ protonated.

3.3 *Mg^{2+}-bound phosphate at the cryo-temperature*

When the crystal of the oligonucleotide was flash-frozen in liquid nitrogen and the XN structure determined, to our surprise, we have found that the Ade7 backbone phosphate was no longer protonated. Instead, Mg^{2+} ion was hydrated with five D_2O molecules bound to the R_P-oxygen, completing its octahedral coordination sphere. The metal and water were clearly visible in the neutron scattering length density and electron density maps, while there was no indication of a D atom presence near the backbone phosphate. Neutrons are the perfect probe to visualize H and D atoms in biological macromolecules. With the neutron scattering power of H (and its heavier isotope D) being as good as that of C, N, and O, positions of virtually all H and D atoms can be determined in a neutron structure, whereas X-rays often cannot provide this information even at ultra-high resolutions, especially for water molecules. In the room temperature XN structure obtained using a crystal grown at pH of 5.6, we unequivocally observed protonation of the R_P oxygen atom of Ade7 backbone phosphate.

3.4 *Metal and proton binding*

Metal ions are believed to have essential biological roles in nucleic acid folding and enzymatic reactions. Metal ion interactions with nucleic acids are important for counterbalancing the high concentration of charged phosphate groups in DNA and RNA, but usually only a handful of metal ions are sufficiently ordered to be observed in crystal structures. In the structure, two Mg^{2+} ions are observed as hexahydrated $Mg(D_2O)_6^{2+}$ complexes bound in the major groove. Magnesium ions interact through the outersphere contacts with the nucleobases of Gua3 and Gua4, which is typical for A-DNA oligonucleotide structures. Thus, Mg^{2+} ions and the protonated Ade7 backbone phosphates provide six positive charges to balance the fourteen negative charges present in the oligonucleotide double helix, with the rest of charge balancing presumably coming from the metal ions that are disordered in the structure. Surprisingly, in the low-temperature XN structure, the Ade7 backbone phosphate protonation is replaced with metal coordination.

A backbone phosphate oxygen in an A-DNA nucleotide crystal can be protonated. The protonation state is altered and replaced with metal coordination when the temperature is decreased to 100 K. For a more complete understanding of a nucleic acid structure and function, it may be necessary to obtain DNA and RNA structures at both room and low temperatures. Selenium modification of the ribose sugar 2′-position in d[GTGG(C^{Se})CAC]$_2$ allowed us to obtain high-resolution neutron diffraction data from DNA crystals that are an order of magnitude smaller than those required previously. This work makes future studies of DNA and RNA structure and function using neutron crystallography possible, including mechanistic studies of ribozymes, DNAzymes and riboswitches.

4 CONCLUSIONS

This novel Se-atom-specific functionalization will provide important tools to investigate nucleic acid structure/folding, recognition and catalysis, to study nucleic acids and their protein interactions, to improve biochemical and biophysical properties of nucleic acids, and to explore potential nucleic acid therapeutics and diagnostics.

REFERENCES

Chen, Y., Liu, H., Yang, C. et al. 2019. Structure of the error-prone DNA ligase of African swine fever virus identifies critical active site residues. *Nat Commu* 10: 387.

Hu, B., Wang, Y.T., Sun, S.C. et al. 2019. Synthesis of selenium-triphosphates (dNTPaSe) for more specific DNA polymerization. *Angew Chem Int Ed* 58: 7835–7839.

Liu, H., Yu, X., Chen, Y. et al. 2017. Crystal structure of an RNA-cleaving DNAzyme. *Nat Commu* 8: 2006.

Vandavasi, V.G., Blakeley, M.P., Keen, D.A., Hu, L.R., Huang, Zhen & Kovalevsky, A. 2018. Temperature-induced replacement of phosphate proton with metal ion captured in neutron structures of A-DNA. *Structure* 26: 1–6.

Zhao, Q., Yang, W., Qin, T. & Huang, Z. 2018. Moonlighting phosphatase activity of klenow DNA polymerase in the presence of RNA. *Biochem* 7: 5127–5135.

Selenium Research for Environment and Human Health: Perspectives, Technologies and Advancements – Bañuelos, Lin, Liang & Yin (eds)
© 2020 Taylor and Francis Group, London, ISBN 978-1-138-39014-0

Genomic analysis of selenoproteins and the expression levels in patients with Kashin-Beck disease

R.Q. Zhang
Key Laboratory of Trace Elements and Endemic Diseases of National Health Commission of the People's Republic of China, School of Public Health, Xi'an Jiaotong University Health Science Center, Xi'an, China
School of Public Health, Shaanxi University of Chinese Medicine, Xianyang, Shaanxi, China

D. Zhang, X.L. Yang, D.D. Zhang, Z.F. Li, B.R. Li, Q. Li, C. Wang, X.N. Yang & Y.M. Xiong*
Key Laboratory of Trace Elements and Endemic Diseases of National Health Commission of the People's Republic of China, School of Public Health, Xi'an Jiaotong University Health Science Center, Xi'an, China

1 INTRODUCTION

Selenium (Se), as an essential trace element in human and animal life, is the active center of many important selenoenzymes (Adadi et al. 2019). At present, there are 25 kinds of human selenoproteins and their key functions in the occurrence and development of complex diseases are still research hotspots (Papp et al. 2007). Our previous work suggested that selenoproteins, as important antioxidants, may have an impact on the occurrence and development of Kashin-Beck Disease (KBD). In the present study, the function of selenoproteins was systematically analyzed by bioinformatics techniques in a genome-wide perspective to identify the core selenoproteins, which can be used in re-recognizing the role of selenoproteins in complex diseases. The mRNA levels of core members of selenoproteins in KBD patients were test by RT-qPCR. Through the above analyses, we aim to identify the core selenoproteins, their main biological functions, and their key signaling pathways. Furthermore, the mRNA levels of core selenoproteins were detected in KBD patients to provide an experimental basis for exploring the pathogenic mechanism of KBD.

2 MATERIALS AND METHODS

2.1 *Bioinformatics analyses of selenoproteins*

This study used bioinformatics analysis technology to systematically analyze the biological functions of 25 selenoproteins (*GPX1, GPX2, GPX3, GPX4, GPX6, DIO1, DIO2, DIO3, TXNRD1, TXNRD2, TXNRD3, SELENOF, SELENOH, SELENOI, SELENOK, SELENOM, SELENON, SELENOO, SELP, SELENOS, SELENOT, SELR, SELENOV, SELENOW,* and *SEPHS2*). Gene Ontology Term (GO; https://www.geneontology.org), including molecular function, biological process and cellular component) and Kyoto Encyclopedia of Genes and Genomes (KEGG; https://www.genome.ad.jp/kegg) pathway enrichment analyses were performed towards 25 selenoproteins by Cytoscape 3.5.1. Fisher's exact test (two-side) or x^2 test were performed to classify the pathway category. The false discovery rate (FDR) was used for the P-value correction. The PPI network relationships of the 25 selenoproteins in the whole network was established based on the biological network from GeneMANIA by STRING10.0. The screened networks were visualized with Cytoscape 3.7.1.

2.2 *RT-qPCR analysis*

Total RNA was extracted from peripheral blood mononuclear cells (PBMCs) of 12 KBD patients and 12 control patients. Revert Aid RT Reverse Transcription Kit was used to convert the RNA into complementary DNA (cDNA). RT-qPCR was used to test the mRNA levels of the core selenoproteins both in KBD and controls according the PPI network. RT-qPCR was conducted with the CFX96 Real-Time PCR system. The PCR cycling conditions were as follows: Initial denaturation at 95° for 30s, followed by 40 cycles of denaturation at 95° for 5 s, and annealing at 57.0° for 30 s. All primers were synthesized by Beijing Huada Genetic Engineering Company. The relative mRNA level for each gene was calculated through the comparative cycle threshold (Ct) equation as follows: $2^{-\Delta\Delta Ct}$ ($\Delta\Delta Ct =$ mean $\Delta Ct_{KBD\ sample}$-mean $\Delta Ct_{control\ sample}$; $\Delta Ct = Ct_{target\ gene} - Ct_{\beta\text{-}ACTIN}$, in which Ct values of target genes were normalized to Ct values of β-actin.

The *T-Test* was performed to determine significance levels of expression differences for the selected genes between KBD patients and healthy controls.

3 RESULTS AND DISCUSSION

3.1 *GO and KEGG enrichment analysis*

The selenoproteins are mainly involved in the biological processes such as response to oxidative stress, thyroid hormone generation, cellular homeostasis, response to reactive oxygen species, hormone biosynthetic process, selenocysteine incorporation, etc. They are mainly accomplished by signaling pathways such as selenocompound metabolism, glutathione metabolism, thyroid hormone synthesis, etc.

3.2 *PPI network*

A PPI network was constructed based on the biological interactions of the 25 selenoproteins to further elucidate their associations at the protein level. Twenty-two selenoproteins were included in the PPI network. SELENOS was the core protein of the PPI network, followed by TXNRD1, TXNRD2, and TXNRD3.

3.3 *The mRNA levels of core selenoproteins in KBD patients*

According to the results of PPI network, four genes (*SELENOS*, *TXNRD1*, *TXNRD2*, and *TXNRD3*) were selected for RT-qPCR analysis. The mRNA level of *SELENOS* in the PBMCs of KBD patients were not significantly different from those of the controls, while the mRNA levels of *TXNRD1*, *TXNRD2*, *TXNRD3* in the PBMCs of KBD patients were significantly lower than those of the controls ($p < 0.05$).

3.4 *Discussion*

Selenium is an important component of several antioxidant enzymes and selenoproteins in humans. Others have found that selenoproteins play an important role in balancing the redox reaction, which exerts biological functions such as improving immunity and anti-apoptosis (Yang et al. 2016). There are currently 25 known selenoproteins in humans. In present study, enrichment analyses of biological functions and KEGG signaling pathways towards 25 selenoproteins were conducted, and the results suggested that selenoproteins were mainly involved in the biological processes, such as oxidative stress, cell homeostasis, response to reactive oxygen species, and lipid peroxidation. They are also involved in the signaling pathways, such as compound metabolism, glutathione metabolism, and thyroid hormone synthesis, which provide clues for further exploring the mechanism of selenoproteins in KBD development. The PPI of the selenoproteins showed that SELENOS was the core protein of the PPI network, followed by TXNRD1, TXNRD2, and TXNRD3, suggesting that SELENOS, TXNRD1, TXNRD2, and TXNRD3 may play more important roles in 25 selenoproteins.

RT-qPCR results showed that mRNA levels of *SELENOS, TXNRD1, TXNRD2, and TXNRD3* in PBMCs of KBD patients were decreased compared with controls. Selenoproteins are known to be involved as an antioxidant, anti-tumor, have anti-heavy metal properties, and also show an excellent regulation effect on immune response and hormone levels. Studies have shown that SELENOS may regulate the redox balance of cells by virtue of its reductase activity (Qazi et al. 2018). SELENOS can effectively eliminate the level of oxygen free radicals in cells, reduce the damage of cells, and protect the normal physiological functions of cells and organisms. Experiments showed that the cytoplasmic region of SELENOS has thioredoxin (TXNRD)-dependent reductase activity, and both of them play an alternate oxidation-reduction-oxidation-reduction process in the redox process. The thioredoxin (Trx) system is important to the survival ability of cells. Recently, increasing evidence has shown that mammalian thioredoxin reductase (TrxR) is a promising therapeutic target in terms of anti-oxidation, anti-tumor, and anti-apoptosis in some diseases (Liu et al. 2019). In KBD, TrxRs show an ability to protect the apoptosis of articular chondrocytes due to its biological functions.

4 CONCLUSIONS

SELENOS was the core protein of the PPI network from selenoproteins, followed by TXNRD1, TXNRD2, and TXNRD3. Selenoproteins involves multiple signaling pathways, such as glutathione metabolic pathway and selenoprotein compound metabolic pathway. The mRNA levels of *TXNRD1*, *TXNRD2*, and *TXNRD3* are decreased in KBD patients. Selenoproteins may have important value in the prevention and treatment of KBD by virtue of their anti-oxidative damage and anti-inflammatory properties. Knowledge of selenoproteins biological functions still need to be further explored.

ACKNOWLEDGEMENT

This study was funded by the National Natural Science Foundation of China (No. 81773372 & 81573104).

REFERENCES

Adadi, P., Barakova, N.V., Muravyov, K.Y. et al. 2019. Designing selenium functional foods and beverages: A review. *Food Res Int* 120: 708–725.

Papp, Lv., Lu, J., Holmgren, A. et al. 2007. From selenium to selenoproteins: synthesis, identity, and their role in human health. *Antioxid Redox Signal* 9(7): 775–806.

Qazi, I.H., Angel, C., Yang, H. et al. 2018. Selenium, Selenoproteins, and female reproduction: A review. *Molecules* 23(12): pii: E3053.

Sun, L.Y., Meng, F.G., Li, Q. et al. 2014. Effects of the consumption of rice from non-KBD areas and selenium supplementation on the prevention and treatment of paediatric Kaschin-Beck disease: An epidemiological intervention trial in the Qinghai Province. *Osteoarthritis Cartilage* 22(12): 2033–40.

Yang, L., Zhao, G.H., Yu, F.F. et al. 2016. Selenium and iodine levels in subjects with Kashin-Beck disease: A Meta-analysis. *Biol Trace Elem Res* 170(1): 43–54.

Selenium Research for Environment and Human Health: Perspectives, Technologies and Advancements – Bañuelos, Lin, Liang & Yin (eds)
© 2020 Taylor and Francis Group, London, ISBN 978-1-138-39014-0

Impact of high dietary selenium on the selenoprotein transcriptome, selenoproteome, and selenometabolites in multiple species

R.A. Sunde
Department of Nutritional Sciences, University of Wisconsin, Madison, Wisconsin, USA

1 INTRODUCTION

Selenium (Se) has long been known as a toxic element for animals, and yet we lack good biomarkers for assessing high Se status. Little is known about the mechanism(s) of Se toxicity at a molecular level. In contrast, in Se deficiency, levels of selenoproteins decrease dramatically as tissue Se decreases, and this is accompanied by dramatic decreases in transcript levels for a subset of selenoproteins in most species (Sunde et al. 2016). To explore the effect of high Se status and identify potential biomarkers for high Se status, weanling rodents and day-old turkeys were fed truly Se-deficient diets (<0.005 μg Se/g) supplemented with graded levels of inorganic Se (as selenite) up to 5 μg Se/g.

2 MATERIALS AND METHODS

2.1 *Diets and animals*

For rats, the basal Se-deficient diet was a torula-yeast diet containing 0.005 μg Se/g by analysis, supplemented with 100 mg/kg all-rac-α-tocopherol acetate and 0.4% L-methionine to prevent liver necrosis and ensure adequate growth. Rats were fed the basal Se-deficient diet supplemented with graded levels of Se, with 0, 0.08, 0.24, 0.8, 2, or 5 μg Se/g diet as Na_2SeO_3 (n = 4/group) for 28 days, as described previously in detail (Raines & Sunde 2011).

For turkeys, male 1-day old poults were fed the basal 20% torula yeast diet with 7% additional crystalline amino acids, containing 0.005 μg Se/g and 0.93% L-methionine, and supplemented with 150 mg/kg all-rac-α-tocopherol acetate to prevent gizzard myopathy and ensure adequate growth. This diet was supplemented with 0, 0.4, 2 or 5 μg Se/g diet as sodium selenite (n = 4/group), as described previously (Taylor & Sunde 2016, 2017). Care and treatment protocols were approved by the University of Wisconsin Institutional Animal Care and Use Committee.

2.2 *Enzyme activity and Se analysis*

Glutathione peroxidase-1 (Gpx1) activity in tissues and Gpx3 activity in plasma were measured by the coupled assay procedure using 120 μM H_2O_2. Gpx4 activity was measured using 78 μM phosphatidylcholine hydroperoxide, its specific substrate. Neutron activation analysis was kindly conducted by the University of Missouri Research Reactor to determine tissue and diet Se concentrations.

2.3 *RNA transcript analysis*

Total RNA was isolated with TRIzol Reagent (Invitrogen, Carlsbad, CA) following the manufacturer's protocol. The rat studies used GeneChip Rat Genome 230 2.0 arrays (Affymetrix), which contain over 31,000 probe sets that target transcripts representing over 28,700 rat genes. The turkey transcripts were analyzed by RNA-Seq, using Illumina HiSeq 2500 paired-end analysis at UW Madison, which generated 28-M reads per sample. These reads were aligned on Build 102 of the turkey genome assembly 5.0 at the University of Minnesota.

3 RESULTS AND DISCUSSION

3.1 *High Se in rats*

In rats, growth is decreased significantly by feeding 5 but not 2 μg Se/g. Diets with 5 μg Se/g increased liver Se to 4.4X levels in rats fed Se-adequate diet (0.24 μg Se/g), but activities of plasma Gpx3, liver Gpx1, and liver Gpx4 activities are little changed at 102%, 77%, and 115%, respectively, of levels in Se-adequate rats, illustrating that selenoenzyme activities are not good biomarkers of high Se status (Raines & Sunde 2011).

While rat liver transcripts for *Gpx1*, *Selenoh*, and *Selenow* fall dramatically in Se deficiency to <25% of levels in Se-adequate rats, there was no significant effect of 5 vs. 0.24 μg Se/g on levels of the 18 most abundant liver selenoprotein transcripts. This result shows that selenoprotein transcripts are also not good biomarkers of high Se status in rats (Raines & Sunde 2011).

Using microarrays, we found that 1193 general transcripts (4% of the transcriptome) are significantly altered by feeding 5 but not 2 μg Se/g to rats. There was, however, considerable overlap with sets of transcripts altered by calorie restriction or drug overload with known Nrf2 targets, suggesting that these changes are more general and downstream secondarily to molecular site(s) of Se toxicity in the rat.

3.2 *High Se in turkeys*

We have now explored the effect of Se deficient and high Se status in turkeys because Se deficiency diseases and Se requirements are distinctly different from those in rodents. Feeding up to 5 μg Se/g to turkey poults has no significant effect on growth or health. In poults fed 5 μg Se/g, liver Se increases to 5.6X Se-adequate (0.4 μg Se/g) levels whereas kidney Se only increases 2X. Selenium response curves for selenoenzyme activity demonstrate that the minimum turkey poult Se requirement should be raised to 0.4 μg Se/g, 4X that of rodents. Above this level, supplementation up to 5 μg Se/g only increases enzyme activities to ≤2X the levels in Se-adequate liver, as well as other tissues (Taylor et al. 2019). This observation illustrates that selenoenzymes are also not good biomarkers for high Se status in avians.

Similar to rats, no selenoprotein transcripts are significantly increased > 2X nor decreased to < 0.5X Se-adequate levels by feeding up to 5 μg Se/g. This result demonstrates that selenoprotein transcripts cannot serve as good biomarkers for high Se status in avians (Taylor et al. 2019).

More recently, we used RNA-Seq analysis to examine the effect of high Se status on the full turkey transcriptome. We found that that only 2 or 45 transcripts, none selenoprotein transcripts, were significantly altered by 2 or 5 (respectively) vs. 0.4 μg Se/g, affirming the lack of toxicity of 5 μg Se/g as selenite in turkey poults (Taylor, Mendoza, Reed & Sunde, unpubl.). Differences in tissue expression of selenoproteins may underlie the lack of Se toxicity in the turkey.

3.3 *Selenometabolites with high dietary Se*

If neither altered selenoprotein expression, selenotranscript expression, nor altered general transcript expression are associated with adaptation to high Se status, how do animals adapt (homeostatically?) to high intake of Se? In collaboration with the CNRS/UPPA, Institute for Analytical Sciences and Physical Chemistry for the Environment and Materials (IPREM), in Pau, we have begun to use HPLC followed by ICP tandem MS to identify and quantitate selenometabolites in turkey liver. These analyses show that there is no selenomethionine in turkeys fed our truly Se-deficient diets supplemented with selenite. In addition, selenocysteine levels in poults fed 2 or 5 μg Se/g are the same as in Se-adequate liver, confirming that the accumulating Se in turkey liver is not present as selenoproteins. Lastly, HPLC-ICP tandem MS has identified several <1000 Da novel water-soluble selenometabolites, that can account for the near 6-fold increase in liver Se with 5 vs 0.4 μg Se/g treatment in turkey poults (Bierla, Taylor, Szpunar, Sunde & Lobinski, unpubl.).

4 CONCLUSIONS

These studies indicate that modulation of selenoprotein expression is likely not a component of the homeostatic response to high Se in rodents and avians. As a result, neither selenoprotein activity nor transcript levels are good biomarkers of high Se status. New analyses using HPLC-ICP MS, however, has identified several selenometabolites in liver which may provide new insight into how avians and perhaps other animals adapt to high Se status.

REFERENCES

Raines, A.M. & Sunde, R.A. 2011. Selenium toxicity but not deficient or super-nutritional selenium status vastly alters the transcriptome in rodents. *BMC Genomics* 12: 26.

Sunde, R.A., Li, J/L. & Taylor, R.M. 2016. Insights for setting of nutrient requirements, gleaned by comparison of selenium status biomarkers in turkeys and chickens versus rats, mice, and lambs. *Adv Nutr* 7: 1129–1138.

Taylor, R.M., Bourget, V.G. & Sunde, R.A. 2019. High dietary inorganic selenium has minimal effects on turkeys and selenium status biomarkers. *Poult Sci* 98: 855–865.

Taylor, R.M. & Sunde, R.A. 2016. Selenoprotein transcript level and enzyme activity as biomarkers for selenium status and selenium requirements of turkeys (*Meleagris gallopavo*). *PLoS. ONE* 11: e0151665.

Taylor, R.M. & Sunde, R.A. 2017. Selenium requirements based on muscle and kidney selenoprotein enzyme activity and transcript level in the turkey poult (*Meleagris gallopavo*). *PLoS ONE* 12: e0189001.

Selenium Research for Environment and Human Health: Perspectives, Technologies and Advancements – Bañuelos, Lin, Liang & Yin (eds)
© 2020 Taylor and Francis Group, London, ISBN 978-1-138-39014-0

Selenium-mediated epigenetic regulation of selenoprotein expression in colorectal cancer

P. Tsuji
Towson University, Towson, Maryland, USA

1 INTRODUCTION

1.1 *Colorectal cancer and selenium*

Colorectal cancer is the second leading cause of cancer-related deaths in the United States with an expected 51,020 deaths in 2019 from colon cancer alone (American Cancer Society 2019). Evidence from epidemiological, early clinical, and preclinical studies suggested that dietary supplementation with the essential trace mineral selenium (Se) reduces the incidence of and mortality from colon cancer (Jacobs et al. 2004). However, a more recent human clinical trial did not conclude protection against colon or other cancers in a population with high baseline plasma Se levels (Lippman et al. 2009). This finding clearly demonstrates the need for further basic research on molecular mechanism behind the potential effects of Se in cancer (Hatfield & Gladyshev 2009). Primarily, Se appears to mediate its biological functions through selenoproteins, whereas the effect of Se intake on tissue expression of most selenoproteins has been established, if and how Se exerts epigenetic effects on or via selenoproteins, especially those that have been implicated in both prevention and promotion of cancers, remains to be elucidated.

Interestingly, bioinformatics analyses suggest that there may be an inverse correlation in gene expression of several selenoproteins and their DNA methylation status in colon cancer cell lines. This analysis suggests that at least some selenoprotein genes may undergo gene expression regulation via DNA methylation, likely mediated by DNA-methyltransferases (DNMT). The general objective of this project is to assess whether the potential cancer-regulatory effects of Se via selenoproteins may be, at least in part, regulated via epigenetic mechanisms.

2 MATERIALS AND METHODS

2.1 *Materials and reagents*

Dulbecco's Minimum Essential Medium (DMEM), Gibco fetal bovine serum, and TRIzol reagent were purchased from Invitrogen (Carlsbad, CA, USA), iScript cDNA synthesis Kit and SYBR green supermix from Bio-Rad Laboratories (Philadelphia, PA, USA), primers for real-time PCR from Integrated DNA Technologies (Coralville, IA, USA), and sodium selenite from Sigma-Aldrich (St. Louis, MO, USA). All other reagents used were commercially available.

2.2 *Culture of mammalian cells*

HCT116 and HT29 human colorectal cancer cells were cultured in growth medium (DMEM supplemented with 5% heat-inactivated fetal bovine serum [FBS]) in a humidified atmosphere with 5% CO_2 at 37°C. Cells were split at 80% confluency, and only low passage numbers were used for the assays. The colorectal cancer cells were incubated with sodium selenite (200 nM) for 24 or 48 hours. Cells were maintained at 5% FBS for at least one week before starting experiments to minimize the Se contribution by FBS.

2.3 *Real-time RT-PCR analysis*

Total RNA was extracted from HCT116 and HT29 cells using the TRIzol/chloroform method following manufacturer's instructions. cDNA was generated using iScript with 1.0 μg of total RNA. For real-time quantitative RT-PCR (qPCR), 1 μL of cDNA was used in 10 μL reactions by employing the CFX Real-Time PCR detection system (BioRad Laboratories, Hercules, CA, USA). Melting curves were analyzed to verify specificity of the amplifications. mRNA expression levels were normalized to the expression of *GAPDH* as the internal control.

2.4 *Statistical analyses*

Real-time RT-PCR data are presented as means ± SE and were analyzed using GraphPad Prism (v.4, La Jolla, CA, USA). Differences with $p < 0.05$ are considered significant. The CellMiner's NCI-60 Analysis Tool, which integrates datasets of 60 human cancer cell lines routinely used for comparative molecular analyses (Reinhold et al. 2012), was used to compare average transcript expression patterns of genes encoding for selenoproteins and DNA methyltransferases, as well as to compare gene-specific methylation levels.

3 RESULTS AND DISCUSSION

3.1 *Selenoprotein expression*

Differential selenoprotein transcript levels assessed via qPCR in HCT116 and HT29 cells at baseline correlated well with the relative levels in the NCI60 CellMiner database for these cell lines. As expected, additional Se in the form of sodium selenite added to the cell culture medium for 24 or 48 hours resulted in increased mRNA expression of stress-related selenoproteins, such as *GPX1* and *SELENOW*, in both HCT116 and HT29 cells, whereas mRNA expression levels of housekeeping selenoproteins, such as *TXNRD1*, remained unaffected. We are currently assessing the protein expression of select selenoproteins in both HCT116 and HT29 cancer cell lines exposed to low and high Se levels for 6–96 hours, followed up by catalytic activity assays whenever possible.

3.2 *DNA-methyltransferase expression*

mRNA expression of DNA-methyltransferases 1, 3A, and 3B were quantitated using qPCR and normalized to *GAPDH*. Selenium is known to generally decrease DNA methylation (Jablonska & Rezka 2017), presumably because of decreased *DNMT* expression or activity. Decreased *DNMT1* and *3A* expression was observed in HT29 cells after 24 h exposure to 200 nM Se. However, HCT116 cells appeared resistant to Se-mediated decrease of *DNMT1* and *3A* mRNA expression. DNMT protein expression and catalytic activity are currently being evaluated.

3.3 *DNA methylation*

CellMiner indicated that several selenoprotein genes, e.g. *GPX4* and *SELENOF*, are highly methylated across all 60 cancer cell lines. Other selenoprotein genes appear highly methylated only in some cancers but not others, such as TXNRD1 in CNS- and melanoma-derived cell lines, or appear to have highly variable DNA methylation levels across cell lines, such as SELENOP, or within cell lines from one tissue origin, such as GPX3. We are currently assessing the global DNA methylation in both HCT116 and HT29 cancer cell lines exposed to low and high Se levels, followed up by gene-specific methylation assessment of selenoproteins.

4 CONCLUSIONS

Little is known about the mechanism of action of the selenoproteins that have been implicated to function in cancer etiology and prevention. Our bioinformatics and preliminary *in vitro* analyses suggest that epigenetic regulation of at least some selenoproteins in colorectal cancer may be Se-mediated. Knowledge of epigenetic regulation via an essential trace mineral would allow for individualized and targeted nutritional intervention strategies in a common malignancy.

REFERENCES

American Cancer Society. 2019. *Cancer Facts & Figures 2019*. Atlanta: American Cancer Society.

Hatfield, D.L. & Gladyshev, V.N. 2009. The outcome of Selenium and Vitamin E Cancer Prevention Trial (SELECT) reveals the need for better understanding of selenium biology. *Mol Interv* 9: 18–21.

Jablonska, E. & Reszka, E. 2017. Selenium and epigenetics in cancer: Focus on DNA methylation. *Adv Cancer Res* 136: 193–234.

Jacobs, E.T., Jiang, R., Alberts, D.S., Greenberg, E.R., Gunter, E.W. et al. 2004. Selenium and colorectal adenoma: results of a pooled analysis. *J Natl Cancer Inst* 96: 1669–1675.

Lippman, S.M., Klein, E.A., Goodman, P.J. et al. 2009. Effect of selenium and vitamin E on risk of prostate cancer and other cancers: The Selenium and Vitamin E Cancer Prevention Trial (SELECT). *JAMA* 301(1): 39–51.

Reinhold, W.C., Sunshine, M., Liu, H. et al. 2012. CellMiner: A web-based suite of genomic and pharmacologic tools to explore transcript and drug patterns in the NCI-60 Cell Line Set. *Cancer Res* 72: 3499–3511.

Selenium Research for Environment and Human Health: Perspectives, Technologies and Advancements – Bañuelos, Lin, Liang & Yin (eds)
© 2020 Taylor and Francis Group, London, ISBN 978-1-138-39014-0

Cellular selenium as the molecular target of electrophiles and pathogens

N.V.C. Ralston
Earth System Science and Policy, University of North Dakota, Grand Forks, North Dakota, USA
Sage Green NRG, Grand Forks, North Dakota, USA

L.J. Raymond
Sage Green NRG, Grand Forks, North Dakota, USA

1 INTRODUCTION

1.1 *Selenium's biochemical role in physiology*

Selenium (Se) is required in selenocysteine (Sec, U), the 21st genetically encoded amino acid and the most potent physiological nucleophile. Humans possess 25 selenoprotein genes expressed in tissue-dependent distributions with functions that are particularly important in brain and neuroendocrine tissues. Selenoenzymes prevent and reverse oxidative damage to lipids and proteins, support immune functions, assist in regulating calcium and thyroid hormone metabolism, tubulin polymerization, protein folding, regulate Se transport in the body, and one is required for Sec synthesis (Ralston & Raymond 2018).

The "oxygen family" (Group 16 of the periodic table), includes oxygen, sulfur (S), and Se elements with similar chemicophysical properties and chemical reactions (Wessjohann et al. 2007). Known as chalcogens, they have 6 valence electrons, leaving them two short of a full outer shell and oxidation states of -2, but can also occur in $+2$, $+4$, and $+6$ states. Neuroendocrine tissues are selectively supplied with Se to support Sec synthesis, so dietary deprivation is usually without overt consequences but does accentuate toxicity of soft electrophiles (Ralston & Raymond 2018). Poor Se status diminishes host immunocompetence and increases retroviral and bacterial virulence (Steinbrenner et al. 2015). Cooperative effects of soft electrophiles in populations with low-Se-intakes and the contributing influences on disease prognosis is the subject of the current assessment.

2 ELECTROPHILES AND SELENIUM

Few environmental stressors impair selenoenzyme activities in the brain. However, as the strongest intracellular nucleophile, Sec is vulnerable to binding by electron poor soft electrophilic chalcophiles (E*), i.e. metals, metalloids, nonmetals, and certain organic molecules. Upon their uptake by the body, mass action effects initially result in their binding to cysteine (Cys) and similar ligands that donate an electron pair and form covalent Cys-E*. Since the Cys of various thiomolecules are selenoenzyme substrates or cofactors for selenoenzymes, Cys-E* is delivered into precise proximity with the active site Sec, resulting in formation of Sec-E*, irreversibly inhibiting the enzyme. This biochemical mechanism appears to be shared by E*, such as mercury (Hg), silver (Ag), cadmium (Cd), and possibly lead (Pb). They are likely to share the same toxicodynamics, but tissue toxicokinetic distinctions result in different tissues being affected (Ralston & Raymond 2018, Ralston 2018). Aurothioglucose and related pharmacologic agents use gold to irreversibly inhibit selenoenzymes in a similar manner (Gromer et al. 1998). Heavier metalloids/nonmetal E* such as arsenic (As), S, and Se itself display increasing chalcogen affinity with the rank order: selenide ≫ sulfide ≫ oxide (Sodhi 2000). Organic E* include aldehydes, alkanals, aromatics, and $\alpha\beta$-unsaturated carbonyls, which occur at low levels in foods and in the environment, but their exposures vary substantially (LoPachin & Gavin 2012, 2016). Although their thioaffinities are less than those of metal E*, the higher prevalence of organic E* in dietary components, pharmaceuticals, metabolites, and environmental co-exposures may contribute to Se attrition and affect health of certain populations.

3 PATHOGENS AND SELENIUM

Numerous unresolved issues regarding Se-dependent functions in the immune system require further study to define how Se status relates to immune functions. Thioredoxin reductase is critical for preventing DNA damage and cell cycle arrest in activated T-cells during viral and parasite infections because it donates reducing equivalents to ribonucleotide reductase in nucleotide biosynthesis (Muri et al. 2018).

The Se contents of gastric mucosa of patients infected with *H. pylori* were found to be ~10 times higher than in control tissues and correlated with inflammation severity (Ustündağ et al. 2001), potentially reflecting increased biosynthesis of selenoproteins to alleviate oxidative stress (Touat-Hamici et al. 2014). The biochemical mechanisms involved remain incompletely characterized, but redistribution of Se may explain the wide range of pathologies reported with inverse associations between blood Se and severity of inflammation in disease states.

Among patients infected with human immunodeficiency virus (HIV), disease progression (Hurwitz et al. 2007, Baum et al. 2013) and lethality (Kupka et al. 2004, Jiamton et al. 2003) have been reported as being inversely related to Se status, although effects are not uniformly observed (Kupka et al. 2008). Keshan disease, a congestive cardiomyopathy involving Se-poor individuals infected with a mutated strain of Coxsackie virus, has been successfully treated with dietary Se-augmentation (Zhou et al. 2018). Originally observed in China, cases may also arise in regions with notably low intake of Se, since even avirulent strains of the Coxsackie B virus are prone to spontaneous mutations in Se-deficient hosts (Beck et al. 1995).

4 CONCLUSIONS

Epidemiological and environmental studies of the effects of metallic, metalloid, and organic E* from diet, pharmaceuticals, and/or exposures to environmental contaminants may benefit from considering the concomitant contributions of all members of this class of agents in relation to the Se status of study populations. Additional research is required to establish possibilities of inherited, acquired, or degenerative neurological disorders of Se homeostasis that may also influence vulnerability to E* exposures, immunocompetence, and resistance to pathogens.

Exacerbating effects of accentuated E* intakes, particularly in Se-poor populations, may affect pathogenic disease progression and could increase their mutation rates, thus contributing to promulgation of novel strains in Se-poor regions of the world. Adjuvant therapy with augmented dietary Se may be helpful in restoring immune function, but if other nutrients required for host immunocompetence are not also augmented, the observed benefits would be reduced. Recognizing the importance of dietary Se status in relation to E* exposures may enable greater differentiation of the roles of Se and other nutrients in immune responses to pathogenic diseases.

REFERENCES

Baum, M.K., Campa, A., Lai, S. et al. 2013. Effect of micronutrient supplementation on disease progression in asymptomatic, antiretroviral-naive, HIV-infected adults in Botswana: a randomized clinical trial. *JAMA* 310: 2154–2163.

Beck, M.A., Shi, Q., Morris, V.C. & Levander, O.A. 1995. Rapid genomic evolution of a non-virulent Coxsackie virus B3 in selenium-deficient mice results in selection of identical virulent isolates. *Nat Med* 1: 433–436.

Gromer, S., Arscott, D., Williams, C.H. et al. 1998. Human placenta thioredoxin reductase: Isolation of the selenoenzyme, steady state kinetics, and inhibition by therapeutic gold compounds. *JBC* 273(32): 20096–20101.

LoPachin, R.M. & Gavin T. 2016. Reactions of electrophiles with nucleophilic thiolate sites: relevance to pathophysiological mechanisms and remediation. *Free Radic Res* 50(2): 195–205.

LoPachin, R.M., Gavin, T., DeCaprio, A.P. & Barber, D.S. 2012. Application of the hard and soft acids and bases (HSAB) theory to toxicant-target interactions. *Chem Res Toxicol* 25: 239–251.

Muri, J., Heer, S., Matsushita, M., Pohlmeier, L. et al. 2018. The thioredoxin-1 system is essential for fueling DNA synthesis during T-cell metabolic reprogramming and proliferation. *Nat Commu* 9: 1851.

Hurwitz, B.E., Klaus, J.R., Llabre, M.M. et al. 2007. Suppression of human immunodeficiency virus type 1 viral load with selenium supplementation: a randomized controlled trial. *Arch Intern Med* 167: 148–154.

Jiamton, S., Pepin J., Suttent, R., Filteau, S. et al. 2003. A randomized trial of the impact of multiple micronutrient supplementation on mortality among HIV-infected individuals living in Bangkok. *AIDS* 17: 2461–2469.

Kupka, R., Msamanga, G.I., Spiegelman, D. et al. 2004. Selenium status is associated with accelerated HIV disease progression among HIV-1-infected pregnant women in Tanzania. *J Nutr* 134: 2556–2560.

Kupka, R., Mugusi, F., Aboud, S. et al. 2008. Randomized, double-blind, placebo-controlled trial of selenium supplements among HIV-infected pregnant women in Tanzania: maternal and child outcomes. *Am J Clin Nutr* 87: 1802–1808.

Nicholls, A.C. & Thomas, M. 1977. Coxsackie virus infection in acute myocardial infarction. *Lancet* 1: 883–884.

Ralston, N.V.C. 2018. Effects of soft electrophiles on selenium physiology. *Free Radic Biol Med* 127: 134–144.

Ralston, N.V.C. & Raymond, L.J. 2018. Mercury's neurotoxicity is characterized by its disruption of selenium biochemistry. *Biochim Biophys Acta Gen Subj.* 1862: 2405–2416.

Sodhi, G.S. 2000. *Fundamental Concepts of Environmental Chemistry.* Oxford: Alpha Science.

Steinbrenner, H., Al-Quraishy, S., Dkhil, M.A. et al. 2015. Dietary Selenium in Adjuvant Therapy of Viral and Bacterial Infections. *Adv Nutr* 6(1): 73–82.

Touat-Hamici, Z., Legrain, Y., Bulteau, A.L. et al. 2014. Selective up-regulation of human selenoproteins in response to oxidative stress. *J Biol Chem* 289: 14750–14761.

Ustündağ, Y., Boyacioğlu, S., Haberal, A. et al. 2001. Plasma and gastric tissue selenium levels in patients with Helicobacter pylori infection. *J Clin Gastroenterol* 32(5): 405–8.

Wessjohann, L.A., Schneider, A., Abbas, M. & Brandt, W. 2007. Selenium in chemistry and biochemistry in comparison to sulfur. *Biol Chem* 388: 997–1006.

Zhou, H., Wang, T., Li, Q. & Li, D. 2018. Prevention of Keshan disease by selenium supplementation: A systematic review and meta-analysis. *Biol Trace Elem Res* 186(1): 98–105.

Selenium Research for Environment and Human Health: Perspectives, Technologies and Advancements – Bañuelos, Lin, Liang & Yin (eds)

Association of selenium status and selenoprotein genetic variations with cancer risk

D.J. Hughes
Cancer Biology and Therapeutics Group, Conway Institute, University College Dublin, Dublin, Ireland

V. Fedirko
Department of Epidemiology, Rollins School of Public Health, Emory University, Atlanta, GA, USA

L. Schomburg & S. Hybsier
Institute for Experimental Endocrinology, University Medical School, Berlin, Germany

C. Méplan
School of Biomedical Sciences, Newcastle University, Newcastle upon Tyne, UK

M. Jenab
Section of Nutrition and Metabolism, International Agency for Research on Cancer, Lyon, France

1 INTRODUCTION

In humans, adequate dietary selenium (Se) intake is essential for synthesizing 25 selenoproteins, of which several are important in countering oxidative and inflammatory processes linked to carcinogenesis (Labunskyy et al. 2014). Experimental and observational studies suggest that suboptimal Se intake, as found across Europe, for example, and genetic variations in several selenoprotein genes may contribute to cancer development, particularly at gastrointestinal anatomical sites (Méplan 2015).

We previously reported in the European Prospective Investigation into Cancer and Nutrition (EPIC) cohort that a higher Se status (as assessed by circulating levels of Se and its major transport protein, Selenoprotein P; SELENOP) was associated with lower risks of colorectal cancer (CRC) (Hughes et al. 2015) and hepatocellular cancer (Hughes et al. 2016). Additionally, we recently showed that several common single nucleotide polymorphisms (SNPs) in Se-related genes (selenoprotein and Se metabolic pathway genes) alone or in combination with Se status may affect CRC development (Fedirko et al. 2019). We are currently conducting or collaborating in prospective studies of Se status, selenoprotein gene variants and risk of breast, gastric, pancreatic, and prostate cancers (*new data to be presented*).

2 MATERIALS AND METHODS

2.1 *EPIC study participants*

The cases and controls in these nested case-control studies (see Table 1) are participants within the EPIC study. Cases were primary tumours for the relevant studied cancer site, while controls were selected by incidence density sampling from all cohort members alive and cancer-free at the time of matching to cases (1:1), including age and sex as described previously for the CRC study (Hughes et al. 2015).

2.2 *Selenium status assays*

Serum Se status was assessed by pre-diagnostic serum measures of Se (reflection X-ray fluorescence spectroscopy), SELENOP (colorimetric enzyme-linked immunoassay), as described by Hughes et al. (2015, 2016), and GPX3 (enzyme activity assay).

2.3 *Genotyping*

In the CRC study, we designed and simultaneously assessed *Illumina Goldengate* assays for 1264 candidate functional and common tagging SNPs in 164 Se pathway genes (comprising selenoprotein genes, selenoprotein biosynthesis/transport genes, and metabolic pathway genes). From these assays, 1040 variants in 154 genes were successfully genotyped in DNA samples from 1420 CRC cases and 1421 controls within EPIC (Fedirko et al. 2019).

2.4 *Statistical analysis*

Multivariable odds ratios and 95% confidence intervals were calculated using logistic regression to determine the association of Se status biomarkers with cancer risk and of SNPs in the Se pathway with CRC risk. In the latter genetic study, pathway and gene-based analyses were also performed using the PIGE package Adaptive Rank Truncated test (Yu et al. 2009).

Interactions with Se status were tested at the SNP, gene, and pathway levels (p-values < 0.05 were considered nominally statistically significant). Multiple testing corrections were performed by the Benjamini–Hochberg (BH) and adjustment for correlated tests (p_{ACT}) procedures due to the SNP data in biologically related pathways. Analyses were conducted using SAS

Table 1. Summary of design of Se and cancer risk nested case-control studies conducted within the EPIC study.

Cancer	Selenium status *(pre-diagnostic serum or plasma)*	Case/control pairs*	Selenium-related SNPs *(germlineDNA)*	Case/controls	Progress
Colorectal	Se SELENOP	966	1040 Se-pathway SNPs	1420/1421	Hughes et al. 2015, 2019
Hepato-biliary	Se SELENOP	261	–	–	Hughes et al. 2016
Breast	Se SELENOP GPX3	2237	GWAS	~1500 each	Ongoing
Gastric	Se SELENOP GPX3	780	70 (GWAS in subset)	780 each	Ongoing

*Matching: at least 1 control per case, by age (within 2.5 years), sex (where applicable), study centre and date + timing of blood collection (women by menopausal status). Se, selenium; SELENOP, Selenoprotein P; GPX3, Glutathione Peroxidase 3; SNP, Single nucleotide polymorphism; GWAS, Genome-wide association studies.

version 9.2 (SAS Institute, Cary, NC, USA) and R (R Foundation for Statistical Computing, Vienna, Austria; http://www.R-project.org/) statistical packages.

3 RESULTS AND DISCUSSION

3.1 *Selenium status and cancer risk*

Higher serum levels of SELENOP were associated with a significantly lower risk of developing CRC (more evident in females; Hughes et al. 2015). Overall, there was an 11% reduction in CRC risk per 0.806 mg/L serum SELENOP increase (OR = 0.89, 95% CI: 0.82–0.98). Higher serum levels of both Se and SELENOP were associated with a lower hepatocellular carcinoma risk (OR = 0.37, 95% CI: 0.21–0.63 for a 1.5-mg/L increase in SELENOP; Hughes et al. 2016). Selenium status was not associated with the development of gallbladder or biliary tract tumours.

Preliminary data show no association of circulating SELENOP levels with overall breast cancer risk (p_{trend} across quintile groups = 0.53). A collaborative study with the Danish Diet, Cancer and Health cohort also observed no major association for toenail Se, plasma SELENOP, selenoprotein gene SNPs, and risk of advanced prostate cancer (Outzen et al., pers. comm.).

3.2 *Selenium pathway genetic variation and colorectal cancer risk*

In the primary Se pathway 1 (Se and selenoprotein transport/biosynthesis genes) 40 SNPs in 20 genes were nominally associated with CRC risk ($p < 0.05$), while 5 of these variants from 5 genes passed the more rigorous p-value < 0.01 in at least one genetic model prior to multiple testing adjustment (Fedirko et al. 2019). However, among these, only the Thioredoxin Reductase 1 (*TXNRD1*) rs11111979 variant retained borderline statistical significance after adjustment for correlated tests ($p_{ACT} = 0.10$; p_{ACT} significance threshold was $p < 0.1$). SNPs in Wingless/Integrated (Wnt) and Transforming growth factor (TGF) beta-signalling genes (*FRZB*, *SMAD3*, *SMAD7*) from the wider tested pathways affected by Se intake were also associated with CRC risk after the more stringent BH adjustments.

Pathway analyses indicated that for genes in antioxidant/redox and apoptotic pathways the influence of SNPs on the disease risk is also dependent on interaction with Se status.

4 CONCLUSIONS

Overall, our studies suggest that gastrointestinal cancer risk may be modified by Se status, genotype, sex, and gene variation interactions within biological pathways. Detailed investigation of Se intake levels and metabolism is needed to more fully elucidate the relevance for cancer etio-pathogenesis, especially for populations with diverse Se status levels and/or individuals with potentially at-risk or cancer protective Se pathway genotypes.

ACKNOWLEDGEMENT

M. Jenab was involved in this study on behalf of EPIC. Funding was provided by the Health Research Board of Ireland project grants HRA-PHS-2013-397, HRA-PHR-2015-1142, and HRB-ILP-2017-021 (to DJH). The EPIC study was supported by various funders (as detailed in Hughes et al. 2015).

Conflict of interest statement: LS is founder of selenOmed GmbH, a company involved in improving Se diagnostics. The other authors declare no conflicts of interest.

REFERENCES

Labunskyy, V.M., Hatfield, D.L. & Gladyshev, V.N. 2014. Selenoproteins: molecular pathways and physiological roles. *Physiol Rev* 94: 739–777.

Méplan, C. 2015. Selenium and chronic diseases: A nutritional genomics perspective. *Nutrients* 7: 3621–3651.

Hughes, D.J., Fedirko, V., Jenab, M. et al. 2015. Selenium status is associated with colorectal cancer risk in the European prospective investigation of cancer and nutrition cohort. *Int J Cancer* 136(5): 1149-1161.

Hughes, D.J., Duarte-Salles, T., Hybsier, S. et al. 2016. Prediagnostic selenium status and hepatobiliary cancer risk in the European prospective investigation into cancer and nutrition cohort. *Am J Clin Nutr* 104: 1–9.

Fedirko, V., Jenab, M., Méplan, C. et al. 2019. Association of selenoprotein and selenium pathway genotypes with risk of colorectal cancer and interaction with selenium status. *Nutrients* 11(4): E935.

Selenium Research for Environment and Human Health: Perspectives, Technologies and Advancements – Bañuelos, Lin, Liang & Yin (eds)
© 2020 Taylor and Francis Group, London, ISBN 978-1-138-39014-0

Molecular biology and pathophysiology of inborn errors of selenoprotein biosynthesis

U. Schweizer
Rheinische Friedrich-Wilhelms-Universität Bonn, Germany

1 INTRODUCTION

Selenoproteins contain the rare amino acid selenocysteine (Sec). There are 25 genes encoding selenoproteins in humans and 24 in mice (Kryukov et al. 2003). Despite the small number of selenoproteins, they are important for many biological processes, and some selenoproteins are essential for life. Accordingly, mutations in selenoproteins, or in any of the genes required for selenoprotein biosynthesis, lead to pathology in humans carrying such mutations, as well as in mouse models designed to be deficient for these genes (Schweizer & Fradejas-Villar 2016).

The biosynthesis of selenoproteins, i.e. the co-translational incorporation of Sec, and the biosynthesis of Sec-tRNA^{Sec} are largely known. The UGA codon is translated as Sec codon if the mRNA contains a selenocysteine insertion sequence (SECIS). The SECIS interacts with SECIS-binding protein 2 (SECISBP2) and instructs the ribosome not to terminate at the UGA/Sec codon but translates Sec using Sec-tRNA^{Sec}, which in turn is accompanied by the specific elongation factor EFSEC. The tRNA^{Sec} is peculiar in many respects, and its modification is thought to play a role in translation. If, however, selenium (Se) supply is scarce, some selenoprotein mRNAs become unstable and appear to be degraded by a mechanism that remains incompletely understood.

Because of the fundamental role of codon-reassignment in selenoprotein biosynthesis, it is believed that understanding selenoprotein biosynthesis leads to a better understanding of translation and many more molecular biological processes involved in selenoprotein expression.

2 OVERVIEW OF HUMAN MUTATIONS IN SELENOPROTEIN GENES AND ASSOCIATED PATHOLOGY

2.1 *Selenoprotein genes*

The first selenoprotein gene found mutated in a human disease was *SELENON* (formerly known as SelN, SEPN1 etc.). For current nomenclature of selenoproteins, see Gladyshev et al. (2016). Mutations in *TXNRD1* have been found in patients with epilepsy (Kudin et al. 2017), and homozygous null mutations in *TXNRD2* lead to a peculiar glucocorticoid deficiency. Null mutations in *GPX4* leads to a perinatal death (FORGE Canada Consortium 2014). Inactivating mutations in deiodinases (DIO1-3) has not yet been discovered, but patients with SECISBP2 mutations have phenotypes associated with deiodinase deficiencies (see below).

Transgenic mouse models have been created for almost all selenoproteins in the last 20 years, and these have helped us understand the individual roles of selenoproteins, albeit mice seem to be more sensitive to loss of selenoproteins than humans.

2.2 *Selenoprotein biosynthesis factors*

Patients with mutations in *SECISBP2* display a growth defect in puberty and, depending on the severity of the deficiency, a range of other phenotypes. They show increased rT3 in line with DIO1 deficiency and impaired hearing in line with DIO2 deficiency.

More recently, a patient with a mutation in the tRNA^{Sec} gene has been identified who displays a phenotype similar to *SECISBP2* deficiency.

SEPSECS is the selenocysteine synthase gene. The enzyme converts Ser-tRNA^{Sec} into Sec-tRNA^{Sec} using selenophosphate. Mutations in SEPSECS lead to a range of neurological phenotypes, from very severe to intellectual disability.

We will present some data on a novel *Sepsesc*-mutant mouse model (Fradejas-Villar, unpubl.).

2.3 *Our new Secisbp2 mutant mouse models*

We have recently created two transgenic mouse models for pathogenic missense mutations in *Secisbp2*. The application of ribosomal profiling has helped us to investigate the role of *Secisbp2* in UGA/Sec codon translation and selenoprotein expression (Zhao et al., in press).

2.4 *Mechanism of selenoprotein mRNA destabilization*

If time allows we will briefly touch on our recent work on the mechanism leading to degradation of selenoprotein mRNAs.

2.5 *The role of tRNA^{Sec} modification*

The modification of tRNASec has received a lot of attention because of its dynamics depending on Se availability and its potential to regulate the hierarchical translation of selenoprotein mRNAs. We have genetically inactivated the tRNA-isopentenyltransferase which adds isopentenyladenosine (i6A) to A37 next to the anticodon of tRNA^{Sec}. We will show the impact of loss of this modification on further tRNA^{Sec} modification and selenoprotein translation in mice.

3 CONCLUSIONS

Selenoproteins are essential for humans. Work on the individual roles of selenoproteins was initially spearheaded by transgenic mouse models, but with the advent of the availability of whole exome sequencing, more and more patients deficient in selenoprotein genes or biosynthesis factors are being discovered. It is now clear that the role of selenoproteins extends beyond the simple idea of "being an antioxidant," and that each selenoprotein has very specific functions. There is much we do not yet know about the functions of many selenoproteins, so the area will remain active and fruitful in the future.

REFERENCES

Gladyshev, V.N., Arnér, E.S., Berry, M.J. et al. 2016. Selenoprotein Gene Nomenclature. *J Biol Chem* 291(46): 24036–40.

FORGE Canada Consortium. 2014. Mutations in the enzyme glutathione peroxidase 4 cause Sedaghatian-type spondylometaphyseal dysplasia. *J Med Genet* 51: 470–474.

Kryukov, G.V., Castellano, S., Novoselov, S.V. et al. 2003. Characterization of mammalian selenoproteomes. *Science* 300: 1439–1443.

Kudin, A.P., Baron, G., Zsurka, G. et al. 2017. Homozygous mutation in TXNRD1 is associated with genetic generalized epilepsy. *Free Radic Biol Med* 106: 270–277.

Schweizer, U. & Fradejas-Villar, N. 2016. Why 21? The significance of selenoproteins for human health revealed by inborn errors of metabolism. *Faseb J* 65(1): 13–16.

Zhao, W.C., Bohleber, S., Schmidt, H. et al. 2019. Consequences of pathogenic Secisbp2 missense mutations probed by ribosome profiling of selenoproteins in vivo. *J Biol Chem* (in press).

Selenium's effect on epidemiology, health, injury, and disease

Selenium Research for Environment and Human Health: Perspectives, Technologies and Advancements – Bañuelos, Lin, Liang & Yin (eds)
© 2020 Taylor and Francis Group, London, ISBN 978-1-138-39014-0

Selenium alleviates oxidative stress in mice fed a serine-deficient diet

X.H. Zhou, Y.H. Liu & Y.L. Yin
Institute of Subtropical Agriculture, The Chinese Academy of Sciences, Changsha, China

1 INTRODUCTION

1.1 *Selenium-dependent antioxidant activity*

Selenium (Se) is an essential trace element in the human body and plays critical roles in a wide variety of physiological processes via selenoproteins. Selenoproteins (glutathione peroxidase, thioredoxin reductase, methionine sulfoxide reductase 1 and endoplasmic reticulum-selenoproteins, etc.) have antioxidant effects and are involved in regulating antioxidant activities. Importantly, glutathione peroxidase is primarily responsible for the antioxidant effects of selenoprotein.

1.2 *Serine in antioxidant activity*

Our previous study demonstrated that serine was important for the glutathione antioxidant system by serving as a substrate of glutathione (Zhou et al. 2017, Zhou et al. 2018). A previous study also showed a synergistic effect of serine with selenocompounds on oxidative stress (Wang et al. 2016).

2 MATERIALS AND METHODS

2.1 *Animal care and experiment design*

C57BL/6J mice (10 weeks old) were randomly assigned into four groups ($n = 8$): (1) mice were fed on a basal diet (CONT); (2) mice were fed on a serine and glycine-deficient diet (SGD); (3) mice were fed on a basal diet supplemented with 4.5 μg of selenium/kg/d from Na_2SeO_3 (SEL); and (4) mice were fed on a serine and glycine-deficient diet supplemented with 4.5 μg of Se/kg/d from Na_2SeO_3 (SGDS). Diet were purchased from Research Diets (New Brunswick, NJ, USA). Treatments were carried out over a period of three months. All the procedures in the present study were approved by the Animal Welfare Committee of the Institute of Subtropical Agriculture, Chinese Academy of Sciences. All procedures were carried out according to the rules established by the committee.

2.2 *Determination of malondialdehyde, glutathione contents, glutathione peroxidase activity, and oxidative products*

Malondialdehyde (MDA), glutathione (GSH) contents, as well as glutathione peroxidase activity in serum and liver were analyzed using commercial kits according to the manufacturer's instructions (Northwest Life Science Specialties, Vancouver, WA, USA). The level of 8-hydroxy-2'-deoxyguanosine (8-OHdG) in liver was measured as previously described. Protein carbonyl was measured by using an ELISA kit (Northwest Life Science Specialties) according to the manufacturer's instructions. Lipids were extracted from the liver using Folch solution, and the level of 8-isoprostane was determined using an ELISA kit (Northwest Life Science Specialties) according to the manufacturer's instructions.

2.3 *qRT-PCR analysis*

RNA was isolated from liver samples using TRIzoL reagent and then reverse-transcribed to cDNA. Real-time PCR was performed as previously described. All the genes were normalized to the housekeeping gene and the relative differences in gene expression among the groups were determined using the comparative Ct value method.

3 RESULTS AND DISCUSSION

3.1 *Selenium maintained MDA level but exerted no effects on GSH level and GSH-Px activity in mice fed a serine-deficient diet*

As shown in Figure 1, serine deficiency caused a significant decrease of GSH level, while an increase of MDA level occurred in both serum and liver. In addition, serine deficiency caused a significant decrease of GSH-Px activity in liver. Selenium significantly decreased MDA level in both serum and liver in mice fed a serine-deficient diet but had no effects on GSH level and GSH-Px activity. These results suggested that Se might alleviate oxidative stress caused by serine deficiency, but not through the glutathione antioxidant system.

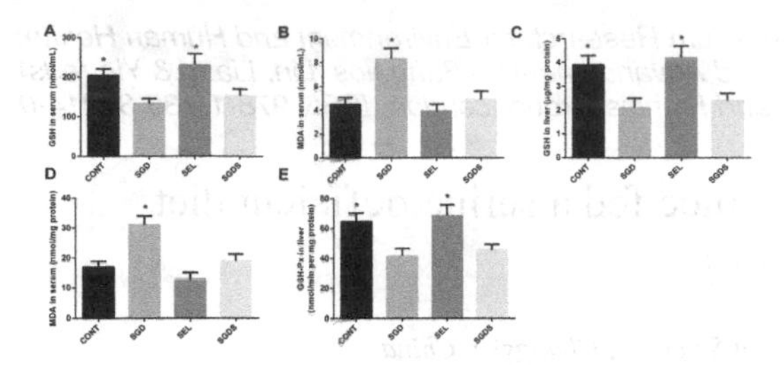

Figure 1. Selenium maintained MDA level but exerted no effects on GSH level and GSH-Px activity in mice fed a serine-deficient diet.

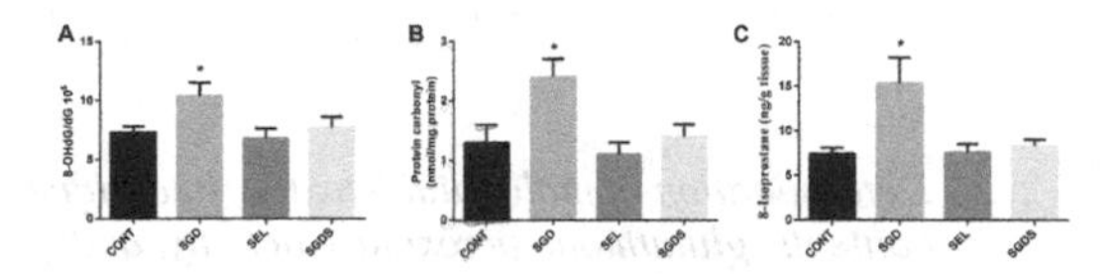

Figure 2. Selenium decreased oxidative damage in mice fed a serine-deficient diet.

3.2 *Selenium decreased oxidative damage in mice fed a serine-deficient diet*

As shown in Figure 2, serine deficiency caused significant increases of 8-OHdG/dG, protein carbonyl and 8-Isoprostane level, while Se significantly decreases these changes. These results suggested that serine deficiency caused lipid, protein and DNA oxidation in liver, while Se protects liver cells from these oxidative damages.

3.3 *Selenium alleviated oxidative stress through thioredoxin antioxidant system in mice fed a serine-deficient diet*

As shown in Figure 3, serine deficiency caused significant decreases of mRNA expression of Gpx (Gpx1 and Gpx2), Txn (Txn1 and Txn2), Txnrd (Txnrd1 and Txnrd2) and Msrb (Msrb1 and Msrb2). However, Se only significantly increased mRNA expression of Txn (Txn1 and Txn2) and Txnrd (Txnrd1 and Txnrd2) but had no effects on mRNA expression of other genes. These results suggested that Se alleviated oxidative stress through thioredoxin antioxidant system in mice fed a serine-deficient diet.

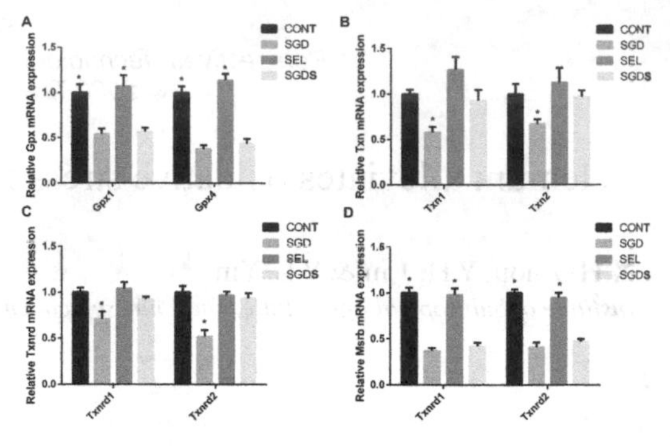

Figure 3. Selenium alleviated oxidative stress through thioredoxin antioxidant system in mice fed a serine-deficient diet.

4 CONCLUSIONS

Others have suggested that Se exerts antioxidant activity primarily through the selenoprotein of glutathione peroxidase (Wang et al. 2016). However, when serine (a major substrate of glutathione) is deficient, the glutathione antioxidant system is impaired. Selenium could not retrieve this antioxidant system, but surprisingly, Se could still alleviate oxidative stress through an alternative antioxidant system, the thioredoxin system.

REFERENCES

Wang, Q., Sun, L.C., Liu, Y.Q., Lu, J.X., Han, F. & Huang, Z.W. 2016. The synergistic effect of serine with selenocompounds on the expression of SelP and GPx in HepG2 cells. *Biol Trace Elem Res* 173: 291–296.

Zhou, X.H., He, L.Q., Wu, C.R., Zhang, Y.M., Wu, X. & Yin Y.L. 2017. Serine alleviates oxidative stress via supporting glutathione synthesis and methionine cycle in mice. *Mol Nutr Food Res* 61(11): 1700262.

Zhou, X.H., He, L.Q., Zuo, S.N. et al. 2018. Serine prevented high-fat diet-induced oxidative stress by activating AMPK and epigenetically modulating the expression of glutathione synthesis-related genes. *BBA-Mol Basis Dis* 1864: 488–498.

Selenium Research for Environment and Human Health: Perspectives, Technologies and Advancements – Bañuelos, Lin, Liang & Yin (eds)
© 2020 Taylor and Francis Group, London, ISBN 978-1-138-39014-0

Selenium-mediated MAPK signaling pathway regulation in endemic osteoarthritis

X.X. Dai, Y. Dai, X.F. Wang, C. Jian, Y.M. Xiong & N. Li
School of Public Health, Xi'an Jiaotong University Health Science Center, Xi'an, Shaanxi, China

1 INTRODUCTION

Kashin–Beck disease (KBD) is a chronic, endemic osteoarthritis (OA) that occurs in limited endemic areas of China. Low dietary levels of selenium (Se) are thought to be the most important biological and environmental factors causing the disease (Yue et al. 2012). Extracellular signal-regulated kinases (ERKs) and C-Jun N-terminal kinase (JNK) are members of the mitogen-activated protein kinase (MAPK) family and are activated by environmental stress (Krens et al. 2006). Selenium has been shown to exhibit a variety of biological functions, including antioxidant functions and maintenance of cellular redox balance, yet low levels of Se can lead to oxidative stress and apoptosis.

2 MATERIALS AND METHODS

2.1 *Study population*

According to national diagnostic criteria of KBD, 110 KBD patients were randomly selected from KBD endemic areas as KBD group and 130 healthy subjects selected from Shaanxi Province served as the control group. Blood samples were drawn from their antecubital vein into tubes containing EDTA for protein extraction or storage at −20°C.

2.2 *Cell culture and tert-butyl hydroperoxide (tBHP) induced injury in human chondrocytes*

The experimental group were divided into 4 subgroups: Basal conditions: C; tBHP (tBHP 300 mmol/L): O; Se treated groups: OS1 and OS2. OS1 and OS2 were supplied with 0.05 μg/mL or 0.1 μg/mL Na_2SeO_3 for 24 h, then treated with 300 mmol/L tBHP for 24 h.

2.3 *Western blotting*

Total protein from blood and chondrocyte samples was extracted. Equivalent amounts of sample protein were separated in 10% SDS-PAGE gel and transferred to nitrocellulose membrane (Millipore, Burlington, USA) using a semi-dry transfer method.

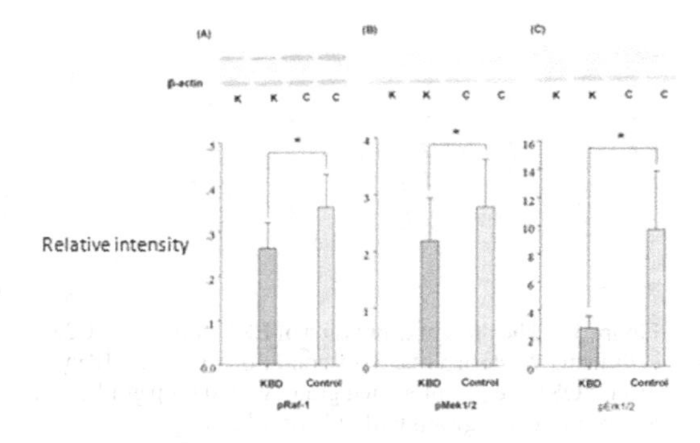

Figure 1. Results of western blots analysis of ERK pathway signal molecule expression in KBD patients (K) and normal controls (C).

3 RESULTS AND DISCUSSION

3.1 *ERK signaling molecules decrease significantly in KBD patients*

The protein expression of pRaf-1, pMek1/2, and pErk1/2 level from whole blood in KBD patients and control group were examined. As shown in Figure 1 the expression level of pRaf-1, pMek1/2, and pErk1/2 decreased significantly in KBD patients compared with controls ($p < 0.01$).

3.2 *Phosphorylated c-Jun N-terminal kinase protein expression was greater in Kashin–Beck disease patients compared with healthy controls*

The protein expression levels of p-JNK in whole blood from KBD patients and healthy controls were detected by Western blot. Western blotting demonstrated that JNK phosphorylation was significantly increased (normalized to α-tubulin) in KBD patients compared with that in the healthy controls ($p < 0.05$, Fig. 2).

3.3 *Effect of Na_2SeO_3 on protein expression of the ERK signal pathway*

To explore the protective effects of Se, we used the tBHP injured chondrocytes to determine the effect of Se on protein expression of the ERK signal pathway.

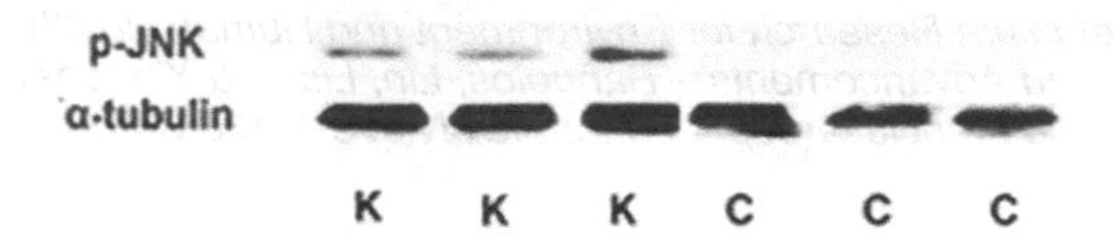

Figure 2. Increased protein expression of phosphorylated c-Jun N-terminal kinase (p-JNK) in whole blood of Kashin–Beck disease (KBD) patients compared with healthy control.

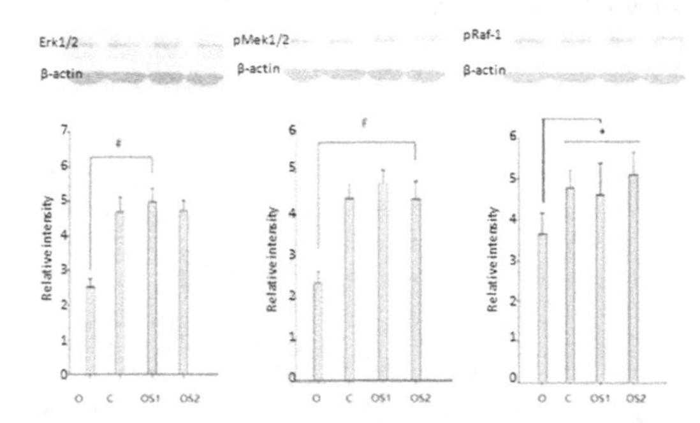

Figure 3. The protein expression of ERK pathway in C28/I2 cells using Western blots. Basal conditions: C; O, tBHP injury group; OS1, Se pre-pretected group with 0.05 μg/mL; OS2, Se pre-protected group with 0.1 μg/mL.

pRafl, pMek1/2 and pErk1/2 Protein expression in chondrocytes were detected by western blot analysis. Chondrocyte protein expression of the ERK signal pathway was significantly decreased by tBHP exposure versus basal conditions ($p < 0.01$), without effect of Se pre-treatment at both concentrations tested. Compared with the tBHP injury group, the expression levels of pRaf-1, pMek1/2 and pErk1/2 in the Se pre-protection group significantly increased ($p < 0.01$, Fig. 3).

3.4 *Effect of Na_2SeO_3 on protein expression of p-JNK*

The JNK pathway plays important roles in the stimulation of apoptotic signaling as well as inflammatory diseases (de Launay et al. 2012). As shown in Figure 4, p-JNK protein levels were significantly increased in tBHP-treated cells. Densitometry analysis showed that p-JNK levels, when normalized to β-actin levels, were significantly increased in the tBHP injury group ($p < 0.01$) compared with the control group. Compared with the tBHP injury group, the expression levels of p-JNK in the Se pre-protection group significantly decreased ($p < 0.01$).

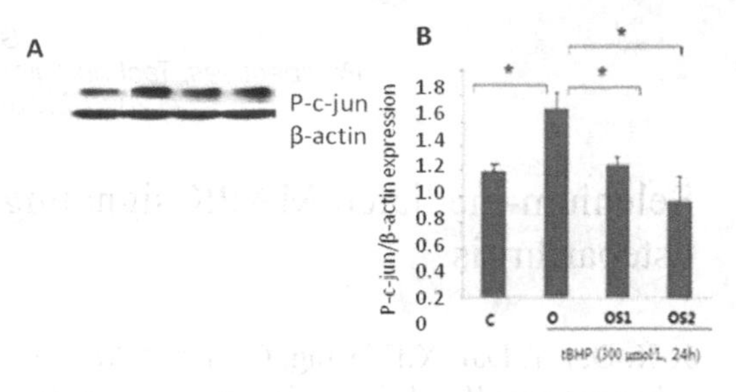

Figure 4. Protein extracts were prepared and analyzed by immunoblotting with antibodies recognizing phosphorylated (p)-c-Jun and β-actin. Signal intensity was then quantified and the results of the densitometric analysis are shown as mean values and standard deviations represented by vertical bars for p-c-Jun expression relative to β-actin (B).

4 CONCLUSIONS

Altered expression levels of p-JNK and ERK signaling molecules were observed in KBD patients. Chondrocyte apoptosis induced by oxidative stress might be mediated via alteration of the JNK and ERK signaling pathway. Selenium exhibited anti-apoptotic effects by down-regulating the p-JNK and stimulates the phosphorylation of the ERK signaling pathway. These findings provide the experimental evidence to elucidate the role of JNK pathway in the pathogenesis of KBD.

ACKNOWLEDGEMENT

This work was supported by National Natural Science Foundation of China Grants (#81673117 and #81573140 to Li).

REFERENCES

de Launay, D., van de Sande, M.G., de Hair, M.J. et al. 2012. Selective involvement of ERK and JNK mitogen-activated protein kinases in early rheumatoid arthritis (1987 ACR criteria compared to 2010 ACR/EULAR criteria): a prospective study aimed at identification of diagnostic and prognostic biomarkers as well as therapeutic targets. *Annals Rheumatic Dis* 71:415–423.

Jirong, Y., Huiyun, P., Zhongzhe, Y. et al. 2012. Sodium selenite for treatment of Kashin-Beck disease in children: a systematic review of randomized controlled trials. *Osteoarthritis Cartilage* 20: 605–613.

Krens, S.F., Spaink, H.P. & Snaar-Jagalska, B.E. 2006. Functions of the MAPK family in vertebrate-development. *FEBS Lett* 580: 4984–4990.

Selenium Research for Environment and Human Health: Perspectives, Technologies and Advancements – Bañuelos, Lin, Liang & Yin (eds)
 ISBN 978-1-138-39014-0

Selenomethionine attenuates D-galactose-induced cognitive deficits by suppressing oxidative stress and neuroinflammation in aging mouse model

J.J. Wang, X.X. Liu, Z.D. Zhang, Q. Zhang & C.Y. Wei*
Institute of Agricultural Quality Standards and Testing Technology, Jilin Academy of Agricultural Sciences, Changchun, China

1 INTRODUCTION

Cognitive deficits are the most common phenotypes in normal aging, including the decline of episodic memory, spatial memory, and attention (Mosher et al. 2016). Others have reported that the major cause of cellular damage during ageing is due to the overproduction of reactive oxygen species (ROS), which are formed via reducing antioxidant defenses (Swomley et al. 2015). Selenium (Se) is a nutritionally essential trace element for human health, possessing diverse pharmacological activities (Rayman et al. 2012). Selenomethionine (SeMet), a major organic form of Se, has greater bioavailability and less toxicity than inorganic Se. Previous studies demonstrated that SeMet played a vital role in the synaptic plasticity and neural protection (Song et al. 2014). In this study, we investigate the protective effects of SeMet supplementation against D-gal-induced memory impairment in mice and explore the potential mechanisms related to the beneficial effects.

2 MATERIALS AND METHODS

2.1 *Experimental design*

All experimental protocols and animal welfare were conducted according to protocols approved by the Institutional Animal Care and Use Committee. Male ICR mice (20–25 g) were obtained from Jiangning Qinglongshan Animal Cultivation Farm and housed in a climate-controlled laboratory with sufficient water and diet. After one-week adaption, mice were randomly allocated to three groups (n = 10): The control group, the D-gal (100 mg/kg) group, and the D-gal + SeMet (10 mg/kg) group. Mice in model group were treated with D-gal by hypodermic injection once daily for 12 weeks. Dosage group were administrated with SeMet intraperitoneally.

2.2 *Morris water maze test*

The test was carried out in a black circular water pool filled with water. Briefly, a black platform was placed 1 cm below the water surface in the center of target quadrant. Mice received training sessions in all four quadrants during visible platform training for 2 days and then subjected to hidden platform training for 3 successive days. The period that mice reached the platform within 90 s was marked as the escape latency. Then on Day 6, a probe trial was performed without platform. The spent time in the target quadrant and the number of platform crossings were measured. All data were recorded using a video-tracking system.

2.3 *Antioxidant assay, cytokine measurement, and Western blot assay*

The activities of antioxidant enzymes superoxide dismutase (SOD), glutathione peroxidase (GSH-Px), catalase (CAT), and the levels of malondialdehyde (MDA) in hippocampus were determined using commercial kits according to the manufacturer's protocol.

The concentrations of TNF-α, IL-1β, and IL-6 in hippocampus were determined by commercially available ELISA kits, according to the manufacturer's instructions (Biolegend, San Diego, USA). The hippocampus tissues were homogenized with ice-cold RIPA lysis buffer and then centrifugated. The proteins were separated by SDS-PAGE, transferred onto PVDF membrane and incubated primary antibodies SIRT1, PPARγ, NF-κB, p-NF-κB, IκBα, p-IκBα, COX-2, iNOS, IL-1β and GAPDH at 4°C overnight. The bands were incubated with a horseradish peroxidase-conjugated secondary antibody for 2 h. The samples were detected with a gel imaging system.

2.4 *Statistical analysis*

All data were presented as means ± SD for at least three independent experiments. The value of $p < 0.05$ was considered to be statistically significant.

3 RESULTS AND DISCUSSION

3.1 *SeMet improved cognitive impairment in aging mice*

D-gal induced behavioral and neurochemical changes can mimic many characters of the natural brain aging process, such as cognitive dysfunction and neurons cell death, which is widely considered as a typical aging model (Rayman 2012). D-gal-treated-alone mice dramatically increased the mean escape latency compared to the control group during trial session, while such an alteration was significantly reversed by SeMet treatment ($p < 0.05$). At the sixth day, the D-gal-treated group displayed a significant decrease in the number of platform crossings ($p < 0.05$) and a decrease in time spent on searching in the target quadrant ($p < 0.01$) compared to the controls, suggesting the remarkable effects of SeMet on D-gal-induced memory and learning deficits.

3.2 *SeMet attenuated oxidative stress in aging mice*

Others have reported that the elevated level of MDA was related to the oxidative damage, and diminished activities of GSH-Px and SOD adversely affected the antioxidant defense in brain, which promoted the process of aging (Kaur et al. 2011). The activities of SOD, GSH-Px, and CAT in hippocampus were significantly lower ($p < 0.01$), and the MDA content was increased in D-gal-treated group compared to the control group ($p < 0.01$). While SeMet treatment partially rescued the reduction of the SOD ($p < 0.01$), GSH-Px ($p < 0.01$) and CAT activities ($p < 0.05$), decreased MDA level ($p < 0.01$), implying that SeMet could alleviate oxidative damage caused by D-gal in aging mice.

3.3 *SeMet inhibited pro-inflammatory cytokines in aging mice*

As documented, inflammation plays a vital role in the development of aging. Pro-inflammatories such as TNF-α, IL-6, and IL-1β in hippocampus were overproduced in D-gal-treated mice in comparison with those in control group ($p < 0.01$). However, mice treated with SeMet showed decreased levels of TNF-α ($p < 0.01$), IL-6 ($p < 0.01$), and IL-1β ($p < 0.05$).

3.4 *SeMet suppressed SIRT1/NF-κB pathway in aging mice*

SIRT1 deacetylase for numerous proteins is involved in several cellular pathways, including stress response and apoptosis, and plays a protective role in neurodegenerative disorders (Godoy et al. 2014). It has been involved in the regulation of signal transduction cascades associated with NF-kB pathways (Fougère et al. 2016). The decreased expression of SIRT1 and PPARγ in D-gal-treated mice is evidenced as described above. SeMet significantly increased the SIRT1 and PPARγ levels ($p < 0.01$), and down-regulated the p-NF-κB ($p < 0.05$), p-IκBα ($p < 0.05$), COX-2 ($p < 0.01$), iNOS ($p < 0.01$), and IL-1β ($p < 0.05$) expressions, indicating the participation of SIRT1/NF-κB signaling pathway and the effectiveness of SeMet in the treatment of D-gal-induced cognitive deficits.

4 CONCLUSIONS

Our findings suggest that SeMet improves cognitive deficits by alleviating oxidative stress and neuroinflammation in aging mice through SIRT1/NF-κB signaling pathway.

ACKNOWLEDGEMENT

Jingjing Wang and Xiaoxiao Liu contributed equally to this work.

REFERENCES

Fougère, A. & Boulanger, E. 2016. Chronic inflammation: Accelerator of biological aging. *J Gerontol (A) Biol* 72: 1218–1225.

Godoy, J.A, Zolezzi, J.M. & Braidy, N. 2014. Role of Sirt1 during the ageing process: relevance to protection of synapses in the brain. *Mol Neurobio* 50(3): 744–756.

Kaur, H. Chauhan, S. & Sandhir, R. 2011. Protective effect of lycopene on oxidative stress and cognitive decline in rotenone induced model of Parkinson's disease. *Neurochem Res* 36: 1435–1443.

Mosher, V., Swain, M.G. & Macqueen, G. 2016. Neuroimaging evidence of hippocampal changes in primary biliary cirrhosis consistent with tissue injury or stress. *J Hepatol* 64(2): S636.

Rayman, M.P. 2012. Selenium and human health. *Lancet* 356(9822): 1256–1268.

Salminen, A. & Huuskonen, J. 2008. Activation of innate immunity system during aging: NF-kB signaling is the molecular culprit of inflamm-aging. *Ageing Res Rev* 7: 83–105.

Song, M. & Zhang, Z. 2014. Selenomethionine ameliorates cognitive decline, reduces tau hyperphosphorylation, and reverses synaptic deficit in the triple transgenic mouse model of Alzheimer's disease. *J Alzheimer's Dis* 41(1): 85–99.

Swomley, A.M. & Butterfield, D.A. 2015. Oxidative stress in Alzheimer disease and mild cognitive impairment: evidence from human data provided by redox proteomics. *Arch Toxicol* 89: 1669–1680.

Selenium Research for Environment and Human Health: Perspectives, Technologies and Advancements – Bañuelos, Lin, Liang & Yin (eds)
© 2020 Taylor and Francis Group, London, ISBN 978-1-138-39014-0

Daily selenium intake from staple food in the Kaschin-Beck disease endemic areas in Tibet

L.S. Yang, H.R. Li & Z. Chen
Key Laboratory of Land Surface Pattern and Simulation, Institute of Geographical Sciences and Natural Resources Research, Chinese Academy of Sciences, Beijing, China
College of Resources and Environment, University of Chinese Academy of Sciences, Beijing, China

H.Q. Gong & M. Guo
The Institute of Endemic Disease Control, Tibet Autonomous Region Center for Disease Control and Prevention, Lhasa, China

1 INTRODUCTION

1.1 *Declining trend of KBD condition in Lhasa*

Previous studies have shown that selenium (Se) deficiency is an important factor in the etiology of Kaschin-Beck disease (KBD) (Tan et al. 1988, Zhang et al. 2011). In recent years, the condition of KBD in Lhasa shows a downward trend and remains at a low level, although a few slight cases can still be detected by X-ray in children (Ci et al. 2014). Based on current observations, we found that a significant positive correlation exists between the declination of KBD condition and the increase of resident's nutritional level of Se in KBD areas in the interior of China, which shows the benefit from the change of diet composition and the input of exogenous staple food with high Se contents (Hou 2000). However, there is insufficient information available on the residents' daily Se intake in the KBD-affected area of Tibet.

1.2 *Study area*

Staple grain is one of the major sources of daily Se intake of rural residents in China. In this study, we conducted a survey on residents' staple food consumption characteristics and assessed daily Se intake from staple food of residents in four KBD-endemic counties of Lhasa, Tibet.

2 MATERIALS AND METHODS

2.1 *Questionnaires and sample collection*

A total of 105 local residents (62 female, 43 male), with a mean age of 50.2 ± 14.2 years (range 18–90 years) from four KBD endemic counties of Lhasa (Linzhou, Mozhugongka, Dazi, and Duilongdeqing) in August 2013 participated in staple food consumption frequency questionnaires survey. Meanwhile, samples of staple grain were collected from the surveyed residents.

2.2 *Determination of selenium in samples*

Selenium concentrations were measured in the samples by hydride generation-atomic fluorescence spectrometry. For quality control, accuracy was guaranteed by using certified reference materials (CRM).

2.3 *Evaluation residents' daily selenium intake from staple food*

$SEIT = \Sigma(A_i \times B_i)$, where $SEIT$ represents the residents' daily Se intake from staple foods; i is the number i type of staple food; A_i means the Se concentration in the number i stable food; and B_i is the amount of the number i stable food consumed daily by residents.

2.4 *Statistical methods*

Data processing and chart production were conducted using SPSS 18.0. *Student's t-test, Mann Whitney U-test* were used for comparison of mean values between two groups. The mean values of different groups were compared by *one-way ANOVA Kruskal-Wallis test.*

3 RESULTS AND DISCUSSION

3.1 *Frequency of staple grain consumption in the investigated residents in KBD area of Lhasa*

72.4% of residents in KBD endemic area consumed local tsampa twice per day. The percentages of inhabitants consumed outsourced rice and flour more than once per day were 48.6% and 66.7%, respectively (Fig. 1).

3.2 *Comparison of consumption frequency of staple grain in different ages and sex groups*

Consumption frequency of outsourced (or not locally grown) rice and flour of youth was significantly higher than that of elderly ($p < 0.05$), whereas local tsampa consumption by youth was lower that of elderly ($p < 0.1$). In addition, there was no obvious

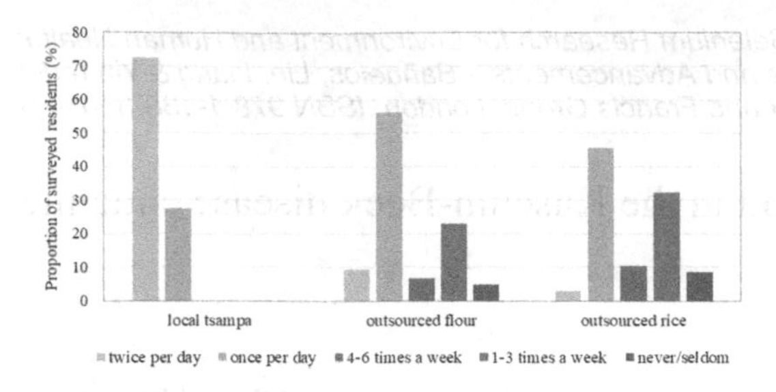

Figure 1. Consumption frequency of staple grain in surveyed residents living in the KBD areas of Lhasa.

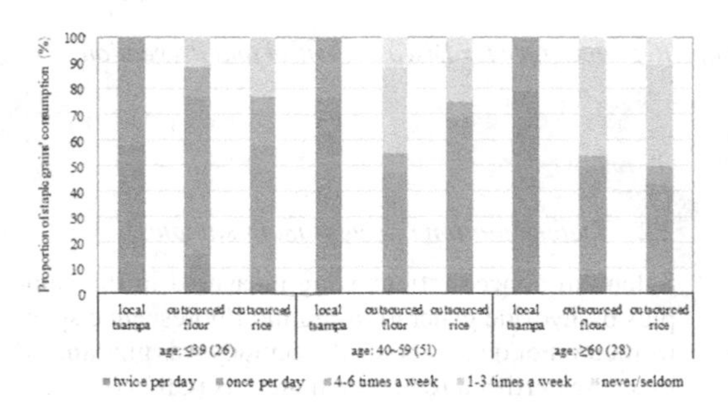

Figure 2. Consumption frequency of staple grain in different age groups in the KBD areas of Lhasa. The number in the bracket is the number of residents surveyed.

difference of staple grain consumption between sex groups (Fig. 2).

3.3 *Selenium content in different staple grain*

The average Se concentrations in self-produced staple grain in Lhasa, such as highland barley (9.07 μg/kg), tsampa (9.63 μg/kg) and wheat (7.46 μg/kg), were much lower than the Se concentration in outsourced rice (36.17 μg/kg) and flour (29.31 μg/kg) from other provinces ($p < 0.01$) (Fig. 3).

3.4 *Daily Se intake from staple grain*

Average daily Se intake of surveyed residents was 8.30 μg from staple foods in KBD endemic area in Lhasa. The daily Se intake from staple grains in

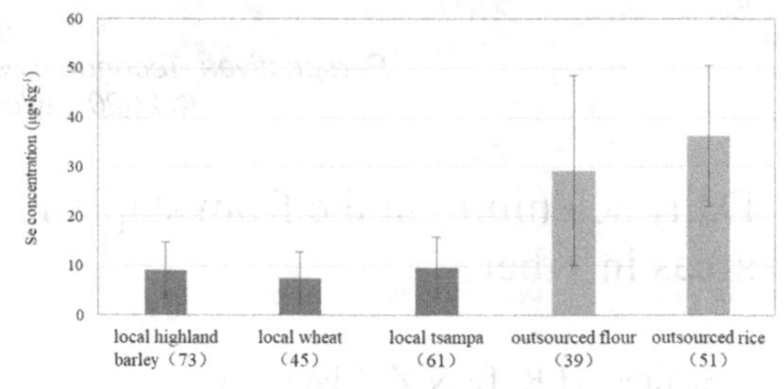

Figure 3. Concentrations of Se in different staple grain consumed in KBD area of Lhasa. The number in the bracket is the number of residents surveyed.

the younger age group (≤39 years old) was higher (9.06 μg) than those of the middle-age group (40–59 years old) (8.48 μg) and the elder age group (≥60 years old) (8.05 μg). 76.1% of daily Se intake was obtained from staple food that were not locally grown.

4 CONCLUSIONS

The increased intake of exogenous rice and flour with high Se contents had a positive significance for enhancing Se intake of local residents in Lhasa KBD endemic area. However, the consumption amount of local low Se staple foods was still relatively high, especially in elder group.

REFERENCES

Ci, Y., Basang, Z.M., Xirao, R.D. et al. 2010. Epidemiological study of Kaschin-Beck disease in Lhasa and Lhoka regions Tibet. *China Journal of Endemiology* 39(5): 529–521. (in Chinese)

Hou, S.F. 2000. Change trend and factors of residential selenium nutrition status in low selenium belt in China. *Geography Research* 19(2): 134–140. (in Chinese)

Tan, J.A., Wang, W.Y., Zhu, Z.Y. et al. 1988. Selenium in environment and Kaschin-Beck disease. *China Journal of Geochemistry* 7(3): 273–280.

Zhang, B.J., Yang, L.S., Wang, W.Y. et al. 2011. Environmental selenium in the Kaschin–Beck disease area, Tibetan Plateau, China. *Environ Geochem Health* 33(5): 495–501.

Selenium Research for Environment and Human Health: Perspectives, Technologies and Advancements – Bañuelos, Lin, Liang & Yin (eds)
© 2020 Taylor and Francis Group, London, ISBN 978-1-138-39014-0

Relation of selenium status to neuro-regeneration after traumatic spinal cord injury

R.A. Heller
Center for Orthopedics, Trauma Surgery and Spinal Cord Injury, University Hospital Heidelberg, Germany
Institute for Experimental Endocrinology, Charité – Universitätsmedizin Berlin, Germany

T. Bock & P. Haubruck
Center for Orthopedics, Trauma Surgery and Spinal Cord Injury, University Hospital Heidelberg, Germany

J. Seelig & L. Schomburg
Institute for Experimental Endocrinology, Charité – Universitätsmedizin Berlin, Germany

P.A. Grützner & B. Biglari
BG Trauma Center, Department of Trauma Surgery and Orthopedics, Ludwigshafen, Germany

A. Moghaddam
Center for Orthopaedics, Trauma Surgery and Sports Medicine, Hospital Aschaffenburg-Alzenau, Aschaffenburg, Germany

1 INTRODUCTION

Traumatic spinal cord injury (TSCI) is a devastating and life-changing event that affects both the local site of injury and the entire body (Kumar et al. 2018). TSCI causes severe medical, psychological, social, and economic challenges for concerned patients, their families, and the health care system.

Despite palliative treatments having improved considerably during recent years, causal therapy options and valid monitoring techniques are either missing or lacking sufficient clinical evidence to implement their routine use. TSCI is characterized by different adjacent phases initially by the mechanical trauma, the spinal damage, and shock (early phase), and finally by a more complex local and systemic inflammatory response during the second phase after injury.

The initial immune response involves a variety of migrating and infiltrating inflammatory cells, such as neutrophils and monocytes, and their immediate activities. The second phase is characterized by a distinct release of cytokines, chemokines, matrix metalloproteinases, and specific growth-factors. The subsequent tissue remodeling necessitates anabolic, proteolytic, and protective pathways acting in concert timely at the site of injury. Also, the release of endogenous signaling substances, such as cortisol, glucagon, and epinephrine is paramount during tissue healing, thereby the dynamics of disease-related factors are traceable in peripheral blood of patients after TSCI.

Next to hormones, certain trace elements are involved in developmental and regenerative processes. The group of selenocysteine (Sec)-containing selenoproteins contributes to protecting cells from oxidative damage, primarily via the families of Se-dependent glutathione peroxidases (GPX) and thioredoxin reductases (TXNRD).

The trace element selenium (Se) is crucial for the biosynthesis of selenoproteins. Both neurodevelopment and the survival of neurons that are subject to stress depend on a regular selenoprotein biosynthesis and sufficient Se supply by selenoprotein P (SELENOP). The expression of neuronal selenoenzymes depends on a sufficiently high Se supply, maintained by SELENOP.

Serum analyses of cytokines, chemokines, growth factors, and trace elements may mirror neuro-regenerative processes during recovery (Moghaddam et al. 2015, Moghaddam et al. 2016a, b, Heller et al. 2017). Therefore, we sought to investigate the correlation between Se and SELENOP concentrations during TSCI and subsequent treatment and neurological remission. The findings of our study are intended to improve clinical risk assessment as a basis for discussing whether a supplemental Se supply should be considered as a meaningful adjuvant treatment option in TSCI or not.

We hypothesize that neuro-regeneration after traumatic spinal cord injury (TSCI) is related to the serum Se status. To investigate this, we have designed a single-center prospective observational study (described below).

2 MATERIALS AND METHODS

Three groups of patients with comparable injuries were studied: vertebral fractures without neurological impairment (n = 10, group C), patients with TSCI showing no remission (n = 9, group G0), and patients with remission developing positive abbreviated injury score (AIS) conversion within 3 months (n = 10, group G1). Serum samples were available from different time points (upon admission; after 4, 9, and 12 h, 1 and 3 days, 1 and 2 weeks, and 1, 2, and 3 months). Serum trace element concentrations were determined by total reflection X-ray fluorescence, SELENOP by an enzyme-linked immunosorbent assay (ELISA), and further parameters by laboratory routine.

3 RESULTS AND DISCUSSION

Serum Se and SELENOP concentrations were higher on admission in the remission group (G1) as compared to G0. During the first week, both parameters remained constant in C and G0, whereas they declined significantly in the remission group (G1).

The highest delta during the first 24h is calculated by subtracting the lowest value of each patient from the corresponding value at admission. The Se concentration differs significantly 24 h after admission (Se 24 h: $p = 0.044$). SELENOP and Zn concentrations showed no significantly different dynamics during the first 24 h in G0 and G1 (results at maximal deviations from admission: SELENOP 24 h: $p = 0.327$; Zn 9 h: $p = 0.236$).

Binary logistic regression analysis, including the delta of Se and SELENOP within the first 24 h indicated an AUC of 90.0% (CI: 67.4%–100.0%) with regards to predicting the outcome after TSCI.

As Se-related effects are exerted mainly through changes in the biosynthesis of selenoproteins, it can be assumed that a sufficiently high Se status at the time of injury contributes to a better expression of antioxidative enzymes, leading to an improved and timely adapted immune response and a more beneficial clinical outcome.

This assumption is supported by the results of our study due to higher Se concentrations in G1 compared to G0 at admission. The observed significant decline of serum Se concentrations in patients with a beneficial neurological remission may be explained by an increased demand of actively regenerating tissue for selenoprotein biosynthesis and consequently might present a surrogate marker of the repair process. Moreover, the absence of changes in Se status after TSCI may inversely indicate an insufficient tissue repair activity, likely associated with a failing regeneration process. If these interpretations of the study data proved reproducible in larger follow-up studies, active Se supplementation after injury might constitute a promising adjuvant treatment option for supporting selenoprotein biosynthesis at the site of injury. Also, such an intervention would increase hepatic SELENOP production as a readily available circulating Se transport form directly available to uptake by the injured tissue for increasing intracellular selenoprotein expression and supporting the regenerative process.

4 CONCLUSIONS

High normal Se levels may contribute to a positive outcome after traumatic spinal cord injury (TSCI). The dynamic changes of serum Se and SELENOP concentrations within the first 24 h of admission were translated into a promising tool for predicting neurological remission with an AUC of 90.0% in the ROC analysis. These results support the importance of trace elements in severe diseases and support strategies considering an adjuvant Se supplementation during therapy. Larger studies and intervention trials are now needed to evaluate whether supplemental Se contributes to recovery and neurological remission after severe trauma.

REFERENCES

Heller, R.A., Raven, T.F., Swing, T. et al. 2017. CCL-2 as a possible early marker for remission after traumatic spinal cord injury. *Spinal Cord* 55(11): 1002–1009.

Kumar, R., Lim, J., Mekary, R.A. et al. 2018. Traumatic spinal injury: Global epidemiology and worldwide volume. *World Neurosurg* 113(0): e345–e63.

Moghaddam, A., Child, C., Bruckner, T., Gerner, H.J., Daniel, V. & Biglari, B. 2015. Posttraumatic inflammation as a key to neuroregeneration after traumatic spinal cord injury. *Int J Mol Sci* 16(4): 7900–7916.

Moghaddam, A., Heller, R.A., Daniel, V. et al. 2017. Exploratory study to suggest the possibility of MMP-8 and MMP-9 serum levels as early markers for remission after traumatic spinal cord injury. *Spinal Cord* 55(1): 8–15.

Moghaddam, A., Sperl, A., Heller, R.A. et al. 2016. Elevated serum insulin-like growth factor 1 levels in patients with neurological remission after traumatic spinal cord injury. *PLoS One* 11(7): e0159764.

Selenium Research for Environment and Human Health: Perspectives, Technologies and Advancements – Bañuelos, Lin, Liang & Yin (eds)
© 2020 Taylor and Francis Group, London, ISBN 978-1-138-39014-0

Anti-inflammatory effect of selenium-enriched *Lycopodiastrum casuarinoides* and *Dendropanax dentiger* on rheumatoid arthritis in rats

G. Chen, L. Zeng, Y.-F. Ba & Y.-P. Zou*
Selenium Technology Innovation Center, College of Life Sciences and Resource Environment, Yichun University, Yichun, China

1 INTRODUCTION

Rheumatoid arthritis (RA) is a multi-systemic inflammatory autoimmune disease involving the surrounding joints, and is one of the main causes of loss of labor and disability in China's population (Zamanpoor 2019).

Plant medicinal resources are widely distributed in Yichun City, Jiangxi Province. Among them, there are two herbs *Lycopodiastrum casuarinoides* and *Dendroanax dentiger* that are effective in treating arthritis. *L. casuarinoides*, a fern in the genus Lycosodiaceae, widely distributes in South China (Liu et al. 2018), while *D. dentiger* is a medicinal plant belonging to the genus Dendropanax in the Araliaceae family (Chien et al. 2014). Both plant species are reported to prevent and treat rheumatoid arthritis (Chien et al. 2014).

There are few pharmacological studies on the treatment of RA using *L. casuarinoides* and *D. dentiger*, and only a few reports showed that their secondary metabolites have anti-inflammatory effects (Chien et al. 2014, Pan et al. 2015). In this study, we separately collected two kinds of herbs (*L. casuarinoides* and *D. dentiger*) that were grown in selenium (Se)-rich and Se-free soils, and then collected water extracts from plant tissues. The two extracts were administered to adjuvant arthritis (AA) rat model to verify the difference in efficacy. The results showed that the Se-enriched Chinese herbal extract has significantly more anti-inflammatory effects than the plant extract from Se-free plant. This result provides a scientific basis for the future development and utilization of Se-enriched herbs.

2 MATERIALS AND METHODS

L. casuarinoides and *D. dentiger* were collected from Mingyue Mountain, Yichun City (MMYC) and Yuhuashan Mountain, Fengcheng City (YMFC), respectively (Fig. 1). Soil samples were also collected near the herbal plants. The total Se content in soil, herbs, and herb extracts were determined by using the atomic fluorescence spectrometry method (Kazi et al. 2014).

The AA rat arthritis model was produced using the method reported in the literature (Asquith et al. 2009). The brief procedure was as follows: The Bacillus Calmette Guerin was inactivated for 1 h in 80°C water bath, thoroughly ground with autoclaved paraffin, and mixed to prepare 10 mg/ml Complete Freund Adjuvant (CFA), from which 0.1 ml was injected subcutaneously into the footpad of a rear paw. The control group was injected with a saline solution.

Figure 1. Morphologies of *L. casuarinoides* (A) and *D. dentiger* (B).

The dosing time was started on Day 7 after modeling. The concentration of the TPT suspension was 2 mg/kg, and the treatment group was treated with Se-enriched *L. casuarinoides* and *D. dentiger* extracts, respectively. The Se-free solution was set as the control group. The dose was 40 g/kg/day once a week, with a continuous gavage for 5 weeks.

Tumor necrosis factor-α (TNF-α) and Interleukin 1 beta (IL-β) were negatively correlated to severity condition of rheumatoid arthritis. The TNF-α and IL-β activities were separately measured according to the ELISA kit instructions.

3 RESULTS AND DISCUSSION

The Se content of soil near *L. casuarinoides* grown in MMYC was 0.68 mg/kg and Se content of stem extract was 0.23 mg/kg. In addition, the Se content of the surrounding soil where *D. dentiger* was grown in YMFC was 0.57 mg/kg, and Se content of stem extract was 0.19 mg/kg. In the selenium-deficient soil, the Se content was only 0.02 mg/kg, and no Se was detected in either of the above two herbs.

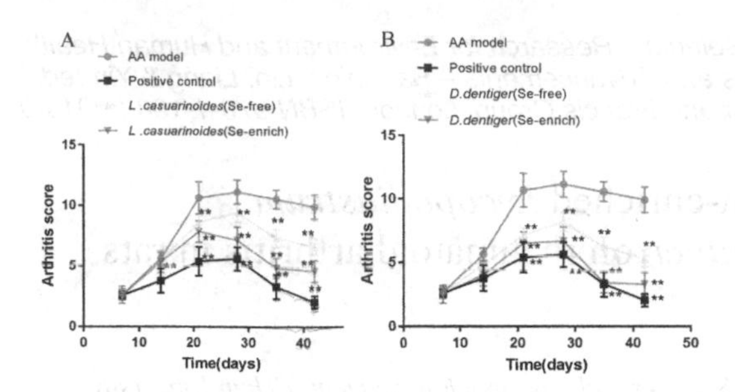

Figure 2. Effects of Se-enriched medicinal plant extracts on arthritis severity of adjuvant arthritis rats. (A): *L. casuarinoides*; (B): *D. dentiger*. **significance at $p < 0.01$, compared to control.

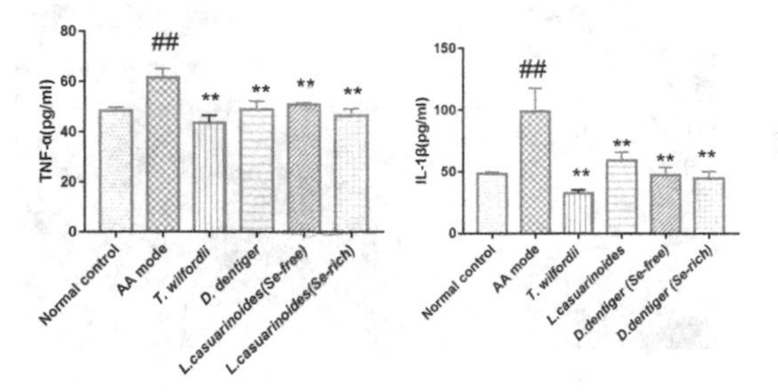

Figure 3. Effects of medicinal plants extracts on serum levels of TNF-α (A) and IL-1β (B). Data shown are mean $\pm$ STD ($n = 6$). ##significance at $p < 0.01$, compared to control; ** significance at $p < 0.01$, compared to control.

The swelling degree of the rear paws of the rats in the selenium-free group of *L. casuarinoides* and *D. dentiger* extract was significantly lower than that of the model group, and the arthritis index was significantly lower ($p < 0.05$ and $p < 0.05$, respectively) (Fig. 2). The Se-enriched groups of *L. casuarinoides* and *D. dentiger* extract was significantly lower than the AA model group, and the arthritis index was significantly lower than that of the AA model group (Fig. 2).

Proinflammatory cytokines serve a central function in the maintenance of chronic inflammation and tissue damage during RA progression. Thus, we evaluated the serum levels of proinflammatory cytokines TNF-α and IL-β by using ELISA. Figure 3A shows that the serum levels of given Se-rich extract of *L. casuarinoides* TNF-α and IL-β (Fig. 3B) and significantly decreased in AA rats. Similarly, the IL-β have also decreased in Figure 3B.

4 CONCLUSIONS

The results showed both Se-enriched herbs of *L. casuarinoides* and *D. dentiger* can effectively alleviate the paw swelling of rat adjuvant-type joints. In addition, the Se-enriched dose group has significantly improved efficacy compared with the Se-free group.

ACKNOWLEDGEMENT

This study was supported by Jiangxi Foreign Science and Technology Cooperation Project (#2015 1BDH80020) and Jiangxi Postdoctoral Fund Project (#JX2016RC40).

REFERENCES

Asquith, D.L., Miller, A.M., McInnes, I.B. & Liew, F.Y. 2009. Animal models of rheumatoid arthritis. *Eur J Immunol* 39: 2040–2044.

Bu, X., Fan, J., Hu, X., Bi, X., Peng, B. & Zhang, D. 2017. Norwegian scabies in a patient treated with Tripterygium glycoside for rheumatoid arthritis. *An Bras Dermatol* 92: 556–558.

Chien, S.C., Tseng, Y.H., Hsu, W.N. et al. 2014. Anti-inflammatory and anti-oxidative activities of polyacetylene from Dendropanax dentiger. *Nat Prod Commun* 9: 1589–1590.

Kazi, T.G., Kolachi, N.F., Afridi, H.I., Brahman, K.D. & Shah, F. 2014. Determination of total selenium in pharmaceutical and herbal supplements by hydride generation and graphite furnace atomic absorption spectrometry. *J AOAC Int* 97: 1696–1700.

Liu, Y., Xu, P.S., Ren, Q. et al. 2018. Lycodine-type alkaloids from *Lycopodiastrum casuarinoides* and their cholinesterase inhibitory activities. *Fitoterapia* 130: 203–209.

Xu, X., Li, Q.J., Xia, S., Wang, M.M. & Ji, W. 2016. Tripterygium glycosides for treating late-onset rheumatoid arthritis: A systematic review and meta-analysis. *Altern Ther Health Med* 22: 32–39.

Zamanpoor, M. 2019. The genetic pathogenesis, diagnosis and therapeutic insight of rheumatoid arthritis. *Clin Genet* 95: 547–557.

Selenium Research for Environment and Human Health: Perspectives, Technologies and Advancements – Bañuelos, Lin, Liang & Yin (eds)
© 2020 Taylor and Francis Group, London, ISBN 978-1-138-39014-0

Study on the mechanism of *GPX3* and selenium in Kashin-Beck disease

B.R. Li, Q. Li, D.D. Zhang, X.N. Yang, R.Q. Zhang, D. Zhang, C. Wang, Z.F. Li, X.L. Yang & Y.M. Xiong*
Institute of Endemic Diseases and Key Laboratory of Trace Elements and Endemic Diseases, National Health Commission of the People's Republic of China
School of Public Health, Xi'an Jiaotong University Health Science Center, Xi'an, Shaanxi, China

1 INTRODUCTION

Kashin-Beck Disease (KBD) is a chronic, degenerative, disabling, and endemic osteoarthropathy with unknown etiology, which is a serious threat to the health of people in endemic areas with KBD. Selenium (Se) deficiency and T-2 toxin were confirmed as important environmental factors of KBD by epidemiological studies. However, excessive apoptosis of chondrocytes in articular and epiphyseal plate cartilages are mainly pathological characteristics of KBD (Guo et al. 2014). The role of oxidative damage and T-2 toxin in KBD and the protective mechanism of Se is still unknown (Chen et al. 2012, Wang et al. 2013). Therefore, this study was conducted to explore the expression of *GPX3* in KBD patients and further clarify the mechanism of *GPX3* in cartilage injuries through establishing the chondrocytes model of oxidative damage and T-2 toxin.

2 MATERIALS AND METHODS

2.1 *Study population*

According to the national diagnostic standard for KBD (WST 207/2010), X-ray examination, and clinical diagnosis, 47 KBD patients were randomly selected as the KBD group, and 31 normal subjects were randomly selected as control group; subjects with any history of KBD and osteoarthrosis were excluded. A peripheral venous blood sample (5 mL) treated with EDTA was collected from KBD patients and control group. Total RNA of blood sample was extracted by Trizol kit and stored at −80°C.

2.2 *Establishment of chondrocytes model by H_2O_2 and T-2 toxin*

The oxidative damage model of human C28/I2 chondrocytes was established by H_2O_2. The experiments were divided into four groups such as control group (C group), supplement Se group (Se group), oxidative damage group (O group), and Se + oxidative damage group (Se + O group).

The T-2 toxin damage model was conducted using T-2 toxin. This experiment included four groups: Control group (C group), supplement Se group (Se group), T-2 toxin group (T group), and Se + T-2 toxin group (Se + T group). The MTT (3-[4,5-dimethylthiazol-2-yl]2,5-diphenyl tetrazolium bromide) colorimetric assay was used to detected chondrocyte activity.

2.3 *Detection the GPX3 mRNA expression level*

Real-time quantitative PCR (RT-qPCR) was performed to detect the expression level of *GPX3* in KBD patients and human C28/I2 chondrocytes with *β-actin* as a reference gene. The primers used for qRT-PCR were F: 5'-TTCACGACATCCGCTGGAA-3' and R: 5'-CATCTTGACGTTGCTGACCGT-3' for *GPX3* as well as F: 5'-GAACGGTGAAGGTGACAGCAG-3' and R: 5'-GTGGACTTGGGAGAGGACT-3' for *β-actin*. The expression level of *GPX3* in chondrocytes model were detected by RT-qPCR similarly. A 12.5-μL reaction mixture contained 6.25 μL SYBR Premix Ex Taq, 0.5 μL of each primer, 1.0 μL cDNA, and 4.25 μL water. Thermal cycler conditions included an initial denaturation step of 94°C for 2min, followed by a three-step PCR program composed of 94°C for 10 s, 60°C for 30 s, and 72°C for 30 s for 40 cycles (Han et al. 2018).

2.4 *Hoechst 33342 staining to detect of chondrocytes apoptosis*

Hoechst 33342 staining solution was added directly to each well and the plates were incubated for 20–30 min at 37°C incubator in the dark. The cells were then washed with PBS and observed using fluorescence microscopy. The apoptotic cells showed bright blue fluorescence and the normal cells showed light blue fluorescence. Cells in each region were photographed and recorded.

3 RESULTS AND DISCUSSION

3.1 *GPX3 mRNA expression in KBD patients*

The expression level of *GPX3* in KBD group was significantly decreased compared with control group ($p < 0.05$). *GPX3* is an important antioxidant enzyme, which plays an important role in maintaining the health of the body (Hughes et al. 2018). The results suggested abnormal expression of *GPX3* may be related to the risk of KBD.

3.2 *The effects of the expression of GPX3 in model of oxidative damage and T-2 toxin*

The chondrocytes model of oxidative damage and T-2 toxin was established to further explore the role of *GPX3* expression in KBD. When cultured chondrocytes were injured with H_2O_2 (600 μmol/L) for 24 hours and T-2 toxin (12 ng/mL) for 72 hours, respectively. Consequently, human C28/I2 chondrocytes' viability were decreased to 40–60% of the C group. In addition, the number of apoptotic cells was markedly increased compared with C group, which suggesting that the cell model was successfully established.

Compared with C group and Se group, the levels of *GPX3* mRNA were significantly decreased and all the differences were statistically significant ($p < 0.05$) in O and T groups. Meanwhile, the number of apoptotic cell was increased ($p < 0.05$). After supplementation with Se, the levels of *GPX3* mRNA in Se + O group and Se + T group were significantly increased ($p < 0.05$). The number of apoptotic cell decreased and, in contrast, the number of normal cell increased.

The results suggested that the abnormal expression of *GPX3* may plays an important role in KBD population. To further explore the mechanism of *GPX3* expression in KBD, the chondrocytes model of oxidative damage and T-2 toxin were established. The results indicated that H_2O_2 and T-2 toxins may decrease the antioxidant capacity of chondrocytes by affecting the expression of *GPX3*, which may lead to oxidative damage and excessive apoptosis of chondrocytes. Selenium may play an anti-apoptosis and anti-toxin effect by regulating the expression of *GPX3* gene and promote its protective effect on KBD.

4 CONCLUSIONS

The decreased expression of *GPX3* may be associated with the risk of KBD. The down-regulation of *GPX3* expression may play an important role in cartilage injury induced by oxidative damage and T-2 toxin. Selenium can up-regulate the expression of *GPX3*, which might exert function in anti-apoptosis and anti-toxin. Therefore, *GPX3* can be used as a biomarker for the diagnosis and prevention of KBD. Importantly, Se may serve important functions in protecting the body from KBD to some extent.

ACKNOWLEDGEMENT

The research is supported by the National Natural Science Foundation of China (No. 81573104 & 81773372).

REFERENCES

Chen, J.H., Xue, S., Li, S. et al. 2012. Oxidant damage in Kashin-Beck disease and a rat Kashin-Beck disease model by employing T-2 toxin treatment under selenium deficient conditions. *J Orthopaedic Res* 30(8): 1229–1237.

Guo, X., Ma, W.J., Zhang, F., Ren, F.L., Qu, C.J. & Lammi, M.J. 2014. Recent advances in the research of an endemic osteochondropathy in China: Kashin-Beck disease. *Osteoarthr Cartilage* 22: 1774–1783.

Han, L.X., Yang, X.L., Sun, W.Y. et al. 2018. The study of GPX3 methylation in patients with Kashin-Beck disease and its mechanism in chondrocyte apoptosis. *Bone* 117: 15–22.

Hughes, D.J., Kunická, T., Schomburg, L., Liška, V., Swan, N. & Souèek, P. 2018. Expression of selenoprotein genes and association with selenium status in colorectal adenoma and colorectal cancer. *Nutrients* 10(11): 1812.

Wang, W., Wei, S., Luo, M. et al. 2013. Oxidative stress and status of antioxidant enzymes in children with Kashin–Beck disease. *Osteoarthritis Cartilage* 21(11): 1781–1789.

Wei, S.X., Shi, B.H., Lyu, A.L., Zhang, F., Zhou, T.T. & Guo, X. 2016. Exploring genome-wide DNA methylation profiles altered in Kashin-Beck disease using Infinium Human Methylation 450 Bead Chips. *Biomed Environ Sci* 29(7): 539–543.

Selenium Research for Environment and Human Health: Perspectives, Technologies and Advancements – Bañuelos, Lin, Liang & Yin (eds)
© 2020 Taylor and Francis Group, London, ISBN 978-1-138-39014-0

The health effects of selenium and selenoprotein in Kashin-Beck disease

Y.M. Xiong*, X.L. Yang, C. Wang, X.N. Yang, Q. Li, X.Y. Wang, Y. Jiang, J.F. Liu, X.L. Du, H. Guo & M.J. Ma

Institute of Endemic Diseases, School of Public Health, Health Science Center, Xi'an Jiaotong University, Xi'an, Shaanxi, China

Key Laboratory of Trace Elements and Endemic Diseases of National Health Commission (Xi'an Jiaotong University), Xi'an, Shaanxi, China

1 INTRODUCTION

Kashin-Beck disease (KBD) is an endemic, disabling and deforming osteoarthropathy and mainly affects children or teenagers in growth and development period (Yamamuro 2001). The mainly pathological changes in KBD are degeneration and necrosis in joint cartilage and epiphyseal plate cartilage (Guo et al. 2014). The disease has been found for over 160 years, but its etiology remains unclear. The epidemiological investigation of environmental risks has shown that selenium (Se) deficiency may contribute to the etiopathogenesis of KBD, and Se supplementation could significantly decrease the incidence of KBD (Mo et al. 1997). Thus, it is considered that Se deficiency is a main environmental factor of KBD, however, the exact molecular mechanism for KBD treatment with Se is still obscure. Screening of KBD susceptibility genes and related functional experiments were conducted to examine the relationship between the molecular mechanism of Se and selenoprotein on chondrocyte oxidative stress, inflammation and apoptosis-signaling pathways. These efforts will further search for new molecular targets for the early diagnosis, warning, prevention, and treatment of KBD.

2 MATERIALS AND METHODS

Gene Ontology (GO) and Kyoto Encyclopedia of Genes and Genomes (KEGG) pathways of selenoprotein family were analyzed by enrichment analysis. The Protein-Protein Interaction (PPI) network was established by STRING 10.5 (https://string-db.org/cgi/input.pl). The selenoprotein gene transcription level and enzymatic activity in whole blood and cartilage samples drawn from KBD and control patients were detected by RT-qPCR and ELISA. Selenoprotein single nucleotide polymorphisms (SNPs) were detected by PCR-RFLP and ARMS-PCR. Methylation of *GPX3* was evaluated by methylation-specific PCR (MSP). The protein levels of inflammation and oxidative-stress-signaling molecules were detected in whole blood and chondrocytes by Western blotting. The chondrocyte oxidative damage model was established using hydrogen peroxide tert-butyl hydroperoxide (tBHP). The oxidative damage effects on apoptosis and oxidative stress and inflammation signaling pathways in chondrocyte were observed in the model to explore for the protective mechanism of Se.

3 RESULTS AND DISCUSSION

3.1 *Bioinformatics analysis of selenoprotein family*

The selenoproteins mainly exhibited biological roles related to redox reactions, such as response to oxidative stress, response to reactive oxygen species, response to lipid hydroperoxide, etc. These roles are mediated mainly by the activation of selenocompound metabolism signal pathways and glutathione metabolism signal pathways.

3.2 *The mRNA levels of selenoprotein*

The mRNA levels of 20 selenoprotein genes were detected in whole blood and 4 in cartilage, mRNA levels of *GPX1*, *GPX3*, *GPX4*, TrxR1, TrxR2, TrxR3, *SEPP*, *DIO2*, and *DIO3* in whole blood decreased in the KBD patients compared to controls, and mRNA levels of *GPX1*, *GPX4,* and *DIO2* in cartilage decreased in KBD patients compared to controls.

3.3 *Genotype analysis*

In this experiment, 11 SNPs from 8 important selenoprotein genes were screened in whole blood. The results showed that *GPX1 Pro198Leu*, *GPX3 rs3792797*, *GPX4* Haplotype (rs713041, rs4807542), *SEPS1* (rs28665122, rs34713741), and *SEP15 rs5859* displayed significant differences in genotypic and allelic frequency between the KBD patients and controls, while *TrxR2* (rs5748469, rs1139793, rs5746841), *SEPP rs7579*, and *DIO2 rs225014* showed no significant differences.

3.4 *Methylation analysis*

The methylation status of *GPX3* and *DIO3* in whole blood was examined using MSP. Results showed that complete methylation of *GPX3* was identified in 16% of KBD patients, but only in 2% of controls ($p < 0.05$). The methylation rate of *DIO3* gene promoter region in whole blood of KBD patients (83.1%) was significantly higher than that of control group (54.5%, $p < 0.05$). These results suggested that the increased methylation levels of *GPX3* and *DIO3* were associated with an increased risk of KBD.

3.5 *The protein expression levels*

The protein expression levels of PI3K/AKt/c-fos, ERK, JNK, Nrf2-ARE, NF-κB, and AP-1 signal molecule in KBD patients were significantly higher than that in controls, while protein expression levels of ERK in KBD patients were decreased compared to controls. This result indicates that signaling pathways related to oxidative stress, inflammation, and apoptosis were disordered in KBD patients. These pathways have similar expressions in the oxidatively damaged chondrocyte model.

3.6 *Genotype subgroup analysis*

GPX enzyme activity decreased in the variant genotype *GPX1 Pro198Leu*, and PI3K/AKt signaling pathway was up-regulated in the variant genotype (AA) individual in *SEPS1-105G > A* (rs28665122). These results suggest that selenoprotein polymorphism had an important role in regulating GPX enzyme activity and PI3K/AKt signaling.

3.7 *Protein levels of PI3K/AKt/c-fos signaling under different methylation status in chondrocytes*

Compared with unmethylated group, protein levels of PI3K/AKt/c-fos signaling were increased in the partial methylated and complete methylated groups (all $p < 0.05$). The protein levels of G$\beta\gamma$, PI3Kp110, and c-fos in partial methylated group were significantly higher than those in complete methylated group, suggesting that the down-regulation of PI3K/AKt/cfos signaling was correlated with a decreased *GPX3* methylation.

3.8 *Protective effect of Na_2SeO_3 on tBHP induced oxidative damage*

In human C28/I2 chondrocyte, tBHP could induce apoptosis and suppress cell survival. The up-regulation of protein levels of c-jun, p-c-jun, MEKK1, p-JNK, AP-1, and PI3K/AKt/c-fos and the down-regulation of Bcl-2 and complete methylation of *GPX3* were observed. Pre-protection with Na_2SeO_3 could ameliorate the cell apoptosis, inhibit the ROS generation, reduce *GPX3* methylation, and regulate the protein levels of these signaling molecules such as c-jun, p-c-jun, MEKK1, p-JNK, AP-1, PI3K/AKt/c-fos, and Bcl-2.

4 CONCLUSIONS

The results indicated that the biological functions of selenoproteins are mainly related to redox reactions. Several SNPs of selenoprotein were associated with the risk of development of KBD, including *GPX1 Pro198Leu*, *GPX4* (rs713041, rs4807542), *SEPS1 G-105A*, and *SEP 15 rs5859*, which might influence inflammation or oxidative stress signal pathways in KBD patients. *GPX3* and *DIO3* hypermethylation were associated with an increased risk of KBD. Furthermore, chondrocyte apoptosis induced by oxidative stress might be mediated via up-regulation of PI3K/AKt/c-fos, JNK, NF-κB, and AP-1 signaling pathways related to inflammation and oxidative stress. Na_2SeO_3 has an effect of anti-apoptosis by down-regulating the signaling pathways and reducing *GPX3* methylation.

ACKNOWLEDGEMENT

This research was supported by National Natural Science Foundation of China (No. 81573104 & 81773372).

REFERENCES

Guo, X., Ma, W.J., Zhang, F., Ren, F.L., Qu, C.J. & Lammi, M.J. 2014. Recent advances in the research of an endemic osteochondropathy in China: Kashin-Beck disease. *Osteoarthritis Cartilage* 22(11): 1774–783.

Mo, D.X., Ding, D.X., Wang, Z.L., Zhang, J.J. & Bai, C. 1997. Twenty-year research on selenium related to Kashin-Back disease. *J Pharm Anal* 1: 79–89.

Yamamuro, T. 2001. Kashin-Beck disease: a historical overview. *Int Orthopaedics* 25(3): 134–137.

Selenium Research for Environment and Human Health: Perspectives, Technologies and Advancements – Bañuelos, Lin, Liang & Yin (eds)
© 2020 Taylor and Francis Group, London, ISBN 978-1-138-39014-0

Biomarkers of selenium and copper status in patients with traumatic spinal cord injury

J. Seelig, R.A. Heller, J. Hackler & L. Schomburg
Institute for Experimental Endocrinology, Charité – Universitätsmedizin Berlin, Berlin, Germany

A. Moghaddam
Department of Trauma and Reconstructive Surgery, Center for Orthopedics, Trauma Surgery and Spinal Cord Injury, Heidelberg University Hospital, Heidelberg, Germany

B. Biglari
BG Trauma Centre Ludwigshafen, Department of Paraplegiology, Ludwigshafen, Germany

1 INTRODUCTION

Traumatic Spinal Cord Injury (TSCI) is damage of the spinal cord resulting in devastating loss of motor and sensory functions. This injury involves complex pathological mechanisms with massive oxidative stress and extensive inflammatory processes, which can bear the risk for permanent paraplegia (Alizadeh et al. 2019). Selenium (Se) is an essential factor for neuronal development, protects from neuron degeneration, and plays a key role in the antioxidative defense. Selenoprotein P (SELENOP) is the transport protein of Se and an essential survival factor for neurons (Pitts et al. 2014). Copper (Cu) serves as an important catalytic cofactor in redox chemistry, e.g., in superoxide dismutase or cytochrome C oxidase. Other copper-containing proteins are relevant for fundamental biological functions, such as lysyl oxidase for the maturation of the extracellular matrix or ceruloplasmin (CP) for the transport of Cu throughout the system. CP accounts for 95% of the Cu content in serum and protects tissue from iron-mediated oxidative damage (Guengerich 2018). Notably, serum Se via SELENOP and Cu via CP are inversely regulated in infection and acute phase response. As TSCI is associated with severe inflammation, we decided to study the potential alterations of the Se and Cu status as potential diagnostic and predictive parameters, as there is a clinical need for informative biomarkers particularly during the first 24 h after injury.

2 MATERIALS AND METHODS

A set of 52 subjects with TSCI were analyzed, 21 of which went into remission (G1), 10 served as controls, and 21 patients developed severe neuronal injury (G0). The G0 group was defined by the absence of positive conversion in the abbreviated injury score (AIS) within three months after injury. The control patients were characterized by vertebral fractures without any neurological impairment. Serum samples were collected from different time points (at admission, and after 4, 9, 12, and 24 h). Sensitive, reliable, and robust sandwich ELISA tests were developed in-house with newly generated monoclonal antibodies for the quantification of SELENOP and CP from minute amounts of sample. Total serum Se and Cu concentrations were measured by total reflection X-ray fluorescence (TXRF) using a validated method. Regression models were calculated using the open-source R statistical environment (Robin et al. 2011).

3 RESULTS AND DISCUSSION

At admission, serum concentrations of Se, Cu, and CP were higher in the remission group (G1) than in the non-remission group (G2). Within the first 24 h Se, SELENOP, Cu, and CP remained unchanged in C and G0. In contrast, concentrations of Se, SELENOP and Cu decreased significantly in G1, whereas CP decreased transiently, reaching a relative minimum 9 h after admission. Similarly, the concentration changes between admission and 24 h were most pronounced in the group of recovering patients (G1). Binary logistic regression analysis including the values at admission of Cu and Se in combination with the 24 h values of Se and CP yielded an area under the curve (AUC) of an impressive 87.7% (CI: 75.1%–100.0%) with regards to predicting the outcome after TSCI.

4 CONCLUSIONS

A significant association between changes in the biomarkers of Se and Cu status with the clinical outcome was identified in this exploratory observational study. The observed alterations in the group of recovering patients may indicate a redistribution of the trace elements in favor of a better anti-inflammatory response and a more successful neurological regeneration.

REFERENCES

Alizadeh, A., Dyck, S.M. & Karimi-Abdolrezaee, S. 2019. Traumatic spinal cord injury: An overview of pathophysiology, models and acute injury mechanisms. *Front Neurol* 10: 282.

Guengerich, F.P. 2018. Introduction to metals in biology 2018: Copper homeostasis and utilization in redox enzymes. *J Biol Chem* 293(13): 4603–4605.

Pitts, M.W., Byrns, C.N., Ogawa-Wong, A.N., Kremer, P. & Berry, M.J. 2014. Selenoproteins in nervous system development and function. *Biol Trace Elem Res* 161(3): 231–245.

Robin, X., Turck, N., Hainard, A. et al. 2011. pROC: an open-source package for R and S+ to analyze and compare ROC curves. *BMC Bioinformatics* 12(1): 77.

Selenium Research for Environment and Human Health: Perspectives, Technologies and Advancements – Bañuelos, Lin, Liang & Yin (eds)
© 2020 Taylor and Francis Group, London, ISBN 978-1-138-39014-0

Genomic analysis for functional roles of thioredoxin reductases and their expressions in osteoarthritis

R.Q. Zhang*
School of Public Health, Shaanxi University of Chinese Medicine, Xianyang, Shaanxi, China
Institute of Endemic Diseases, School of Public Health, Health Science Center, Xi'an Jiaotong University, Xi'an, Shaanxi, China

Q. Li
Institute of Endemic Diseases, School of Public Health, Health Science Center, Xi'an Jiaotong University, Xi'an, Shaanxi, China

Y. Liu
School of Public Health, Shaanxi University of Chinese Medicine, Xianyang, Shaanxi, China

1 INTRODUCTION

Selenium (Se) is an essential trace element for humans, and its physiological functions are mainly realized by selenoproteins (Holmgren 2000, Mustacich & Powis 2000). Thioredoxin reductase (TrxRs) is a family of selenoproteins that is widely expressed in organism cells from humans and consists of three members called TXNRD1, TXNRD2, and TXNRD3. Others have reported that TrxRs mainly exerts biological functions related to anti-oxidative stress, anti-inflammatory, etc. (Koháryová & Kolárová 2008, Neogi & Zhang 2013, Allen & Golightly 2015). However, a comprehensive genomic analysis about *TXNRD1*, *TXNRD2*, and *TXNRD3* has not yet been reported.

Osteoarthritis (OA) is a chronic degenerative steoarthropathy which seriously affects the quality of life of patients. As a disease with unknown etiology and mechanism, OA lacks specific treatment and can only be treated symptomatically. Some studies have suggested that TrxRs may play important roles in the occurrence and development of OA (Bomer et al. 2015, Kamal et al. 2016); however, the relationship between *TXNRD1*, *TXNRD2*, and *TXNRD3* mRNA expressions and OA has not been concluded. Therefore, in this study, an integrative genomic analysis about *TXNRD1*, *TXNRD2*, and *TXNRD3* was performed, and then *TXNRD1*, *TXNRD2*, and *TXNRD3* mRNA levels were analyzed in OA patients based a Chip Array meta-analysis.

2 MATERIALS AND METHODS

2.1 *Integrative genomic analysis*

TXNRD1, *TXNRD2*, and *TXNRD3* were uploaded to STRING 10.5 (www.string-db.org/) software for Gene Oncology (GO) and KEGG enrichment analysis. TXNRD 1, TXNRD2, TXNRD3 proteins were uploaded to STRING software to obtain the information of protein – protein interaction network (PPI). The results were visualized using Cytoscape 3.7.1 (www.cytoscape.org) for visualization. To explore the effects of environmental chemicals on TrxRs, an environmental chemical-TrxR interaction network was established by Network Analyst (www.networkanalyst.ca/) and Comparative Toxicogenomics Database (www.ctdbase.org/).

2.2 *TXNRD1, TXNRD2, and TXNRD3 mRNA levels in OA patients*

To investigate the *TXNRD1*, *TXNRD2*, and *TXNRD3* mRNA levels in OA patients, seven eligible datasets (GSE191, GSE32317, GSE4103, GSE46750, GSE48556, GSE55235, and GSE55457) were retrieved from the GEO database (www.ncbi.nlm.nih.gov/gds/) according to an inclusion and exclusion criteria. The raw expression data of *TXNRD1, TXNRD2, and TXNRD3* were extracted from the seven datasets and were logarithmically converted to obtain stable data. Means and standard deviations were then calculated for meta-analysis by Review Manager 5.3 (The Cochrane Collaboration, Oxford, UK). The standard mean differences (SMD) between OA patients and controls were used to evaluate the mRNA changes in OA. Datasets meeting the criteria for homogeneity ($I^2 < 50\%$ or $p > 0.05$) were analyzed with a fixed effects model. If they met the criteria for heterogeneity ($I^2 > 50\%$ or $p < 0.05$), they were analyzed using a random effects model.

3 RESULTS AND DISCUSSION

3.1 *Integrative genomic analysis*

The results of GO enrichment analysis showed that the main biological roles of *TXNRD1*, *TXNRD2*, and *TXNRD3* were antioxidant activity, cell homeostasis, cell oxidant detoxification, coenzyme binding, etc. The results of KEGG enrichment analysis showed that TrxRs were mainly involved in selenocompound metabolism, cysteine and methionine metabolism, NOD-like receptor signaling pathways, etc. The PPI network suggested that TXNRD1 interacted with eight other proteins (TXNRD2, SEPHS1, TXN2, etc.). TXNRD1 was the most important core protein, followed by TXNRD2 and TXN, which interacted with seven other proteins, respectively.

Based on the CTD database, we observed that *TXNRD1* was affected by 16 environmental chemicals, such as sodium selenite, auranofin, Se, and acrolein, while *TXNRD2* were affected by auranofin and mercury chloride. Both of them interacted with auranofin and mercuric chloride.

3.2 *TXNRD1, TXNRD2, and TXNRD3 mRNA levels in OA patients*

Results of meta-analysis showed that there was no significant difference in the mRNA levels of *TXNRD1, TXNRD2*, and *TXNRD3* between OA patients and normal controls. *TXNRD1*: $SMD_{OA\text{-}control} = 0.11$ $(-0.42, 0.65)$ $(P = 0.68)$, *TXNRD2*: $SMD_{OA\text{-}control} = -0.21$ $(-0.49, 0.006)$ $(P = 0.13)$, *TXNRD3*: $SMD_{OA\text{-}control} = 0.64$ $(-0.12, 1.39)$ $(P = 0.10)$. Thioredoxin reductases (TrxRs) are homodimeric flavin enzymes containing selenocysteine, and TrxRs catalyze the NADPH-dependent reduction of oxidized thioredoxins. Increasing evidence indicates that TrxRs play important roles in regulating cellular oxidative stress. TrxRs can interact with various downstream proteins to regulate certain signaling pathways, thus affecting the oxidative stress process. Over expression or dysfunction of TrxRs can lead to various diseases. For example, TrxRs is overexpressed in multiple tumors, which can maintain the growth of cancer cells and affect the tumor phenotypes. Others have reported that as Se proteins, the decrease of TrxRs expressions can lead to the increase of free radicals in cancer tissues, suggesting that TrxRs plays an important role in maintaining the balance of antioxidant systems. These observations suggested that TrxRs might be closely related to the occurrence and development of various diseases (Zhang et al. 2019).

However, a genomic analysis about TXNRD1, TXNRD2, and TXNRD3 is necessary to conduct to help understand the function roles of TrxRs comprehensively. In this study, results from GO enrichment analysis showed that TXNRD1, TXNRD2, and TXNRD3 mainly exert biological functions such as antioxidant activity, cell homeostasis, cell oxidant detoxification, and coenzyme binding, and are mainly involved in selenocompound metabolism, cysteine and methionine metabolism, NOD-like receptor signaling pathways, etc. These results are similar to those from previous studies. In the PPI network of TrxRs, TXNRD1 was the core protein, followed by TXNRD2 and TXN, which showed that TXNRD1 and TXNRD2 were key members in their family. TXN acts as a homodimer and is involved in many redox reactions. The encoded protein is active in the reversible S-nitrosylation of cysteines in certain proteins, which is part of the response to intracellular nitric oxide (Zhang et al. 2019, Kamal et al. 2016).

Osteoarthritis is the most common joint disorder worldwide. Experiments showed that reactive oxygen species (ROS) play important role in the pathogenesis of OA. Evidence shows major selenoproteins may be important modifiers of the joint to inflammatory responses. Evidence is presented that the local environment in the rheumatic joint contributes to increased TRX production (Bomer et al. 2015). However, the results of meta-analysis showed that there was no significant difference in the mRNA levels of *TXNRD1, TXNRD2*, and *TXNRD3* between OA patients and normal controls, which may be related to a small sample size of the datasets. These results need to be further validated in larger population in the future.

4 CONCLUSIONS

The biological roles of TrxRs are mainly to regulate redox balance, cell growth and apoptosis, etc. As the core protein of the PPI, TXNRD1 may be affected by Se and selenium-containing compounds, leading to the occurrence of certain diseases. There was no significant difference in the mRNA levels of TXNRD1, TXNRD2, and TXNRD3 between OA patients and normal controls, which need to be further validated in larger population in the future.

ACKNOWLEDGEMENT

This study was funded by Research Project from Health Commission of Shaanxi Province (No. 2018A017).

REFERENCES

Allen, K.D. & Golightly, Y.M. 2015. State of the evidence. *Curr Opin Rheumatol* 27(3): 276–283.

Bomer, N., den Hollander, W., Ramos, Y.F. et al. 2015. Underlying molecular mechanisms of DIO2 susceptibility in symptomatic osteoarthritis. *Ann Rheum Dis* 74(8): 1571–9.

Holmgren, A. 2000. Redox regulation by thioredoxin and thioredoxin reductase. *Biofactors* 11(1/2): 63–64.

Kamal, A.M., El-Hefny, N.H., Hegab, H.M. et al. 2016. Expression of thioredoxin-1 (TXN) and its relation with oxidative DNA damage and treatment outcome in adult AML and ALL: A comparative study. *Hematology* 21(10): 567–575.

Koháryová, M. & Kolárová, M. 2008. Oxidative stress and thioredoxin system. *Gen Physiol Biophys* 27(2): 7184.

Mustacich, D. & Powis, G. 2000. Thioredoxin reductase. *Biochem J* 346: 1–8.

Neogi, T. & Zhang, Y. 2013. Epidemiology of osteoarthritis. *Rheum Dis Clin North Am* 39(1): 1–19.

Zhang, J., Zhang, B., Li, X. et al. 2019. Small molecule inhibitors of mammalian thioredoxin reductase as potential anticancer agents: An update. *Med Res Rev* 39(1): 5–39.

Selenium Research for Environment and Human Health: Perspectives, Technologies and Advancements – Bañuelos, Lin, Liang & Yin (eds)
© 2020 Taylor and Francis Group, London, ISBN 978-1-138-39014-0

Commonly used drugs affect hepatic selenium metabolism

J. Hackler, K. Renko, S. Diezel, Q. Sun, W.B. Minich & L. Schomburg
Institut für Experimentelle Endokrinologie, Charité-Universitätsmedizin Berlin, Berlin, Germany

1 INTRODUCTION

The life expectancy is constantly increasing in developed countries, partly due to improved medical care. On average, every senior in Germany > 60 years of age is taking three or more medications per day, increasing risk for developing side effects due to polypharmacy (Moßhammer et al. 2016). Little is known about interfering effects of drugs on selenium (Se) metabolism and selenoprotein expression. Certain examples have been identified, e.g. statins affecting tRNA-Sec maturation (Moustafa et al. 2001) and aminoglycosides altering selenoprotein P (SELENOP) expression in HepG2 cells (Renko et al. 2017), causing a replacement of selenocyseine by other amino acids.

We hypothesize that commonly used and novel drugs, alone and in combination, impair regular Se homeostasis by interfering with selenoprotein biosynthesis and thereby contribute to side effects and multi-morbidity in seniors. By identifying strongly interfering drugs and drug combinations disrupting hepatic Se metabolism, we will generate a valuable data resource, and highlight problematic drugs and drug combinations. This new information will contribute to a better counselling for diseased elderly with respect to medication choice, nutritional support, and control of Se status.

2 MATERIALS AND METHODS

A stably transfected HEK293 cell line harboring a selenium-dependent reporter construct is used for the screening of a set of FDA-approved drugs (Martitz et al. 2016). The reporter construct consists of full-length Firefly luciferase (Fluc) und Renilla luciferase (RLuc) open reading frames that are separated by an in frame UGA codon and carries an additional selenoprotein-specific SECIS-element of GPX4 further downstream.

20.000 cells/well are plated in pre-coated (2.5 µg/mL poly–L-lysine) 96-well plates and starved in DMEM/ F12+2.5%(v/v) FCS (SM) for 24 h. Medium is removed and replaced by SM + 5 nmol/L Na_2SeO_3 and containing drugs at a final concentration of 10 µmol/L and 0.05% (v/v) DMSO. The medium is removed after 48 h, and cells are lysed with 40 µL lysis buffer for 10 min at RT. 20 µL lysate are transferred to white 96-well plates and RLuc activity is determined using Beetle-Juice detection systems (PJK, Kleinblittersdorf, Germany). RLuc activity is measured as relative light units (RLU) using a microplate reader (Berthold Technologies, Bad Wildbad, Germany). Control wells are treated with 5 and 10 nmol/L of Na_2SeO_3 ± DMSO and with the positive read-through modulator G418 (50 µg/mL).

3 RESULTS AND DISCUSSION

From a total of 1953 drugs approved by the US Food and Drug Administration (FDA), 16 plates with 1280 drugs were each tested, including controls, for calculating the UGA read-through factors as an indicator of interference with regular selenoprotein translation. Twenty-eight out of 1280 (2.2%) and 52 out of 1280 (4.1%) compounds were identified as positive and negative modulators, respectively. These results are currently in more detailed analyses with respect to concentration dependence, selenoprotein-specific effects, and selenium-dependent activity. We expect that these results will help in identifying critical drugs in medical use that negatively affect the Se status, and most importantly provide insights into dangerous combinations of drugs used by the elderly.

4 CONCLUSIONS

The dual reporter system was suitable for the identification of modulating compounds. Obviously, a number of selenium-relevant drugs and drug combinations can be identified by this high-throughput screen that needs to be evaluated for health-relevant effects in seniors, to avoid unexpected severe side effects and ameliorating multi-morbidity.

ACKNOWLEDGEMENT

Conflict of Interests and Funding: LS holds shares in selenOmed GmbH, a company involved in selenium status assessment and supplementation. JH, RK and QS have nothing to declare. Research is funded by

Charité Medical School Berlin and the Deutsche Forschungsgemeinschaft (DFG Research Unit 2558 TraceAge, Scho 849/6-1).

REFERENCES

Martitz, J. 2016. Factors impacting the aminoglycoside-induced UGA stop codon readthrough in selenoprotein translation. *J Trace Elem Med Biol* 37: 104–110.

Moßhammer, D. 2016. Polypharmacy-an upward trend with unpredictable effects. *Dtsch Arztebl Int* 113: 627–633.

Moustafa, M.E. 2001. Selective inhibition of selenocysteine tRNA maturation and selenoprotein synthesis in transgenic mice expressing isopentenyladenosine-deficient selenocysteine tRNA. *Mol Cell Biol* 21: 3840–52.

Renko, K. 2017. Aminoglycoside-driven biosynthesis of selenium-deficient selenoprotein P. *Sci Rep* 7: 4391.

Selenium Research for Environment and Human Health: Perspectives, Technologies and Advancements – Bañuelos, Lin, Liang & Yin (eds)
© 2020 Taylor and Francis Group, London, ISBN 978-1-138-39014-0

Autoimmunity to selenoprotein P in thyroid patients

Q. Sun, S. Mehl, C.L. Görlich, J. Hackler, W.B. Minich, K. Renko & L. Schomburg
Institut für Experimentelle Endokrinologie, Charité-Universitätsmedizin Berlin, Berlin, Germany

1 INTRODUCTION

Autoimmunity is characterized by an impaired self-tolerance and is the major cause for the two common autoimmune thyroid diseases (AITD), Hashimoto's thyroiditis (HT), and Graves' disease (GD) (Taylor et al. 2018). In AITD, the autoantibodies (aAb) to thyroglobulin (Tg), thyroperoxidase (TPO), and TSH receptor (TSH-R), respectively, are characteristic diagnostic markers. The thyroid gland is rich in the trace element selenium (Se), which may modulate the endocrine-immune interface and affect hydrogen peroxide metabolism, inflammation, and aAb generation (Duntas et al. 2015, Wu et al. 2015). Selenium transport is mediated by selenoprotein P (SELENOP) (Burk et al. 2015), and low Se supply is known to increase HT risk (Schomburg et al. 2011). In this study we hypothesize that autoimmunity to SELENOP may be associated with AITD.

2 MATERIALS AND METHODS

2.1 *Immunoluminometric assay for detection of aAb to SELENOP*

An immunoluminometric assay for detection of aAb to SELENOP (SELENOP-aAb) was established and used to analyze serum samples from a cohort of thyroid patients (n = 320) and healthy subjects (n = 400). The immunoluminometric assay is based on the binding of autoantibodies to SELENOP fused to a reporter, secreted embryonic alkaline phosphatase (SEAP), followed by a precipitation of this antibody-antigen-reporter complex by protein A. After immunoprecipitation, the SEAP activity was measured as a luminescence signal, which is proportional to the SELENOP-aAb titer in the sample.

2.2 *Analysis of Se-status*

Three biomarkers of Se-status were analyzed, i.e. serum Se concentrations by total reflection X-ray fluorescence (TXRF), serum SELENOP by ELISA, and serum glutathione peroxidase (GPX3)-activity by an enzymatic test. The enzymatic test measures the decreasing absorbance (ΔE) at 340 nm due to NADPH consumption using a spectrophotometer.

Statistical analyses were performed using GraphPad Prism (version 7). Data were tested for normal distribution and compared using unpaired t-test.

2.3 *Isolation of immunoglobulins (IgG) from serum samples*

Total IgG was isolated of three SELENOP-aAb positive and six SELENOP-aAb negative sera by precipitation with Protein A. Serum samples were incubated with protein A slurry in PBS overnight at 4°C and centrifuged the next day. Supernatants were discarded, and pellets were washed six-times with PBS. Precipitated IgG was eluted with 25 mM citric acid (pH 2.0) and was neutralized by addition of 1 M HEPES (pH 8.0). The Se concentration in the protein A precipitates was analysed by TXRF.

3 RESULTS AND DISCUSSION

3.1 *Prevalence of SELENOP-aAb in control and in thyroid patients*

Prevalence of SELENOP-aAb in thyroid patients was higher than in controls (5.3% vs. 0.5%), and higher in AITD as compared to other thyroid diseases. A higher prevalence of SELENOP-aAb was found in female patients compared to males (5.8% vs. 2.2%).

3.2 *Se status in SELENOP-aAb positive patients*

SELENOP-aAb positive patients displayed significantly higher serum Se and SELENOP concentrations in comparison to SELENOP-aAb negative patients (Se: 89.8 ± 20.7 μg/L vs. 75.8 ± 17.5 μg/L [$p = 0.0016$]; SELENOP: 5.2 ± 1.6 mg/L vs. 4.4 ± 1.5 mg/L [$p = 0.0289$]). GPX3 activities were not different between SELE-NOP-aAb positive and negative samples (256.5 ± 26.8 U/L vs. 253.7 ± 50.6 U/L).

3.3 *Verification of the specificity of SELENOP-aAb*

Immunoglobulins isolated from SELENOP-aAb positive samples precipitated measurable amounts of Se in contrast to control immunoglobulins from SELENOP-aAb negative samples. Non-labeled recombinant SELENOP showed a competitive binding of SELENOP-aAb in SELENOP-aAb positive samples but not in the negative serum samples. These results verified the binding specificity of the measured aAb to SELENOP.

4 CONCLUSIONS

These results indicate that a considerable fraction of thyroid patients develops SELENOP-aAb. Whether these autoantibodies increase disease risk or develop during course of disease is unknown. The presence of SELENOP-aAb affects Se and SELENOP status and may thus be of pathophysiological relevance in AITD. This hypothesis needs to be tested in larger and prospective clinical studies.

REFERENCES

Burk, R.F. & Hill, K.E. 2015. Regulation of selenium metabolism and transport. *Annu Rev Nutr* 35: 109–134.

Duntas, L.H. 2015. The role of iodine and selenium in autoimmune thyroiditis. *Horm Metab Res* 47(10): 721–726.

Schomburg, L. 2011. Selenium, selenoproteins and the thyroid gland: interactions in health and disease. *Nat Rev Endocrinol* 8(3): 160–171.

Taylor, P.N., Albrecht, D., Scholz, A. et al. 2018. Global epidemiology of hyperthyroidism and hypothyroidism. *Nat Rev Endocrinol* 14(5):301–316.

Wu, Q., Rayman, M.P., Lv, H. et al. 2015. Low population selenium status is associated with increased prevalence of thyroid disease. *J Clin Endocrinol Metab* 100(11): 4037–4047.

Selenium Research for Environment and Human Health: Perspectives, Technologies and Advancements – Bañuelos, Lin, Liang & Yin (eds)
© 2020 Taylor and Francis Group, London, ISBN 978-1-138-39014-0

Impact of dietary diversity on Se intake adequacy in Kenya

P.B. Ngigi, G. Hanley-Cook, C. Lachat & G. Du Laing
Ghent University, Ghent, Belgium

P.W. Masinde
Meru University of Science & Technology, Meru, Kenya

1 INTRODUCTION

Selenium (Se) deficiency in humans has been reported particularly in geographical regions characterized by low soil Se concentrations and over-reliance on a narrow range of staple foods produced on these soils, as is the case for subsistence farming households in sub-Saharan Africa. In addition, the food preferences of various social groups, food preparation methods, and changes in eating habits also affect dietary Se intake. Furthermore, agricultural and environmental factors, such as land management practices and climatic conditions, also affect Se concentration in foods through altered availability of soil Se for plant uptake (Ngigi et al. 2019).

High variability of Se concentrations in foodstuffs highlights the importance of inter and intra food group consumption to attain daily Se requirements. However, dietary diversity remains a challenge in food insecure countries and regions characterized by a high dependence on subsistence farming based on few staple cereal, legume, and tuber crops. In addition, agricultural biodiversity, which comprises a variety of cultivated plants and animals, indigenous foods, and other products gathered by rural populations within traditional subsistence systems, has been adversely affected by the degradation of ecosystems and climate change. As a result, the contribution of plant and animal biodiversity to food and nutrition security has drastically diminished in the last decades.

In this study, we assessed to which extent the food biodiversity and the dietary diversity affect Se intake adequacy in Kenya. Therefore, the diet composition and dietary Se intake were compared between two Kenyan regions, the Central Highlands around Mount Kenya, and the Lake Basin region near Lake Victoria.

2 MATERIALS AND METHODS

The study targeted children aged 6–59 months and women of reproductive age aged 19–39 years. In total, 155 children and 131 women participated in the Central Highlands, while 29 children and 24 women participated in Lake Basin. Average intake of foodstuffs was estimated through a single 24-hour recall (24 h). To assess the Se concentration in the foodstuffs consumed, plant and animal foods were sampled from the households' food basket and analyzed for Se concentration. Average dietary Se intake was computed based on estimated average amount of food intake from the 24 h (taking into account the effect of cooking) and the actual food Se concentrations.

Dietary Species Richness (DSR) was calculated as an indicator of food biodiversity. Based upon the count of the number of distinct food species consumed by an individual per day. Moreover, the Nutrient Adequacy Ratio (NAR) of Se was calculated as the ratio of an individual's Se intake divided by the EAR for the subject's sex and age bracket.

3 RESULTS AND DISCUSSION

3.1 *Dietary diversity*

Maize was found to be a predominant staple food in both regions, while green banana is a common food for children. Beans and potatoes are largely consumed in the Central Highlands, while cassava is largely consumed in the Lake Basin. In general, no significant differences in the average daily amounts (g/d) of foods consumed were found between the two regions among children under 5 ($p > 0.05$). However, for women, a significant difference was found in the average amount of beans ($p = 0.044$), wheat ($p = 0.040$), and kale ($p = 0.016$) consumed between the two regions. Women in Central Highlands consumed more beans, while their counterparts in Lake Basin consumed more wheat products and kales. Milk is the only animal source food that is largely consumed in both regions. Intake of other animal source foods is significantly lower ($p < 0.05$) in Central Highlands (3% of children and 6% of women consumed beef), than in Lake Basin (48% of children and 64% of women consume fish). Another key difference is a number of species that were only consumed in Central Highlands. Among them, important dietary Se sources include pigeon peas, green grams, cow pea, and amaranth leaves.

The study finds no significant difference in food biodiversity and dietary diversity. In both regions, 7 to 8 food species dominate the diets of both women and children, with the women having consumed only 4 food groups and the children 4 to 5 food groups.

3.2 *Selenium concentrations in foodstuffs*

No significant differences in Se concentration were observed for most foods except for milk ($p < 0.001$) and eggs ($p = 0.016$) having a higher Se concentration in Central Highlands, and maize having a lower concentration in that region ($p < 0.05$). Selenium concentration in the other commonly consumed foods does not differ significantly between the two regions.

3.3 *Dietary Se intake*

Significant differences in dietary Se intake from wheat products were found for children ($p = 0.036$) and from milk for women ($p = 0.002$). Besides the wheat products from Lake Basin having a higher Se concentration, the children from this region also consumed more wheat products than in Central Highlands. On the other hand, women in Central Highlands consumed more milk compared to their counterparts in Lake Basin. There were no significant differences in Se intake from the other foods between the two regions ($p < 0.05$). The main dietary Se sources in Central Highlands are milk (3.49 μg/d) and maize (2.82 μg/d) for children, and maize (8.05 μg/d), milk (3.97 μg/d), and beans (3.74 μg/d) for women. In Lake Basin, the main dietary Se sources for both women and children are fish, maize, and wheat products at 28.33, 8.86, and 4.04 μg/d for women, and 15.92, 2.98, and 2.52 μg/d for children, respectively.

Notably, fish intake contributes a large portion of daily dietary Se in Lake Basin: 45 % in children and 44% in women. On the other hand, in Central Highlands, beef intake contributes 14% and 24% of the daily dietary Se intake for children and women, respectively. Remarkably, the foods that were only consumed in Central Highlands and not in Lake Basin contribute on average up to 20% and 19% of the daily dietary Se intake for children and women, respectively.

In summary, fish intake contributes a larger proportion of the daily dietary Se intake in Lake Basin, while plant-based diets are predominant in Central Highlands.

3.4 *Association between dietary diversity and Se intake*

The associations between DSR and dietary Se intake and NAR are positive and significant for both women

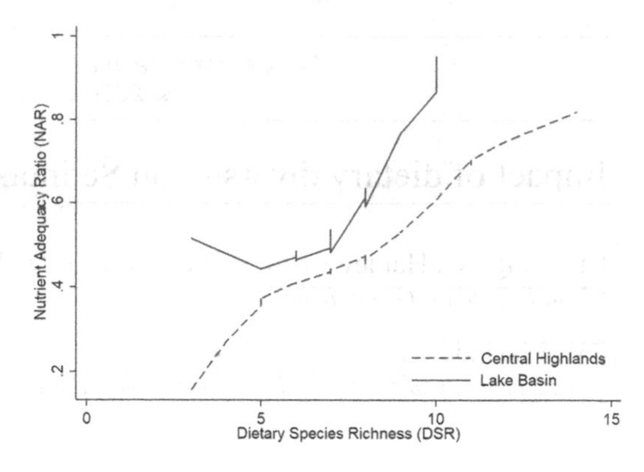

Figure 1. Association of DSR with NAR of Se for 131 mothers in Central Highlands and 24 mothers in Lake Basin, Kenya.

and children in Central Highlands. For women, a one-unit increase in DSR is associated with 2 μg/p/d increment in dietary Se intake ($p < 0.001$) and 5-percentage-points increase in NAR (Se) (Fig. 1). For children, every one-unit increase in DSR is associated with an increase in dietary Se intake of 1 μg/p/d and a 5-percentage-points increase in NAR (Se).

4 CONCLUSIONS

In conclusion, diets in both regions are characterized by a low food biodiversity and dietary diversity, particularly among women. Coupled with the low-selenium concentration in local foods, the resulting inadequate dietary Se intakes result in a high risk for dietary Se deficiency. Since food biodiversity is positively and significantly associated to dietary Se intake, adequacy, and Se status, dietary diversification has intrinsic potential to improve dietary Se intake to adequate intake levels in Kenya.

REFERENCE

Ngigi, P.B., Lachat, C., Masinde, P.F. & Du Laing, G. 2019. Agronomic biofortification of maize and beans in Kenya through selenium fertilization. *Environ Geochem Health* doi: 10.1007/s10653-019-00309-3.

Selenium Research for Environment and Human Health: Perspectives, Technologies and Advancements – Bañuelos, Lin, Liang & Yin (eds)
© 2020 Taylor and Francis Group, London, ISBN 978-1-138-39014-0

Effects of selenium compounds on oxidative stress and apoptosis in HepG-2 cells

Y. Qin, H. Xu, H. Liang, F. Shen, Y. Wei & M. Gu*
Cultivation Base of Guangxi Key Laboratory for Agro-Environment and Agro-Products Safety, College of Agriculture, Guangxi University, Nanning, China

1 INTRODUCTION

Selenium (Se) is an essential trace element that is composed a variety of important enzymes. The relationship between low-serum Se levels to a higher cancer risk has attracted considerable attention in the area of cancer prevention (Fernandes et al. 2015). Adverse or beneficial health effects of Se are strongly dependent on the Se species (Marschall et al. 2016). However, how these Se compounds potentially exhibit anti-cancer properties are not fully understood.

The aim of the present study was to investigate the effects of six different Se species, including selenite, selenate, selenomethionine (SeMet), L-selenocystine (L-Secys2), Se-methylselenocysteine (SeMeSecys), Se nanoparticles (SeNPs) on cell viability, oxidative stress, and apoptosis in HepG-2 cells.

2 MATERIALS AND METHODS

2.1 *Cell culture and chemical treatments*

HepG-2 cells were exposed to selenite (5–100 μmol/L), selenate (25–1000 μmol/L), L-Secys2 (5~100 μmol/L), SeMet (25–1000 μmol/L), SeMeSecys (25–1000 μmol/L), and SeNPs (5–100 μM μmol/L) for 24 h.

2.2 *Cell viability, superoxide dismutase (SOD) activity, reactive oxygen species (ROS), and cell apoptosis determination*

The cell viability was determined with the 3-(4,5-dimethylthiazol-2-yl)-2,5-diphenyl (MTT) assay. The SOD activity was performed according to the reagent kit manufacturer's instructions (Solarbio, Beijing, China). Intracellular ROS levels were determined by DCFH-DA assay, and the cell apoptosis was measured by Annexin V-FITC/PI assay. Fluorescence intensity was then monitored using an Accuri C6 flow cytometry (BD Biosciences, USA).

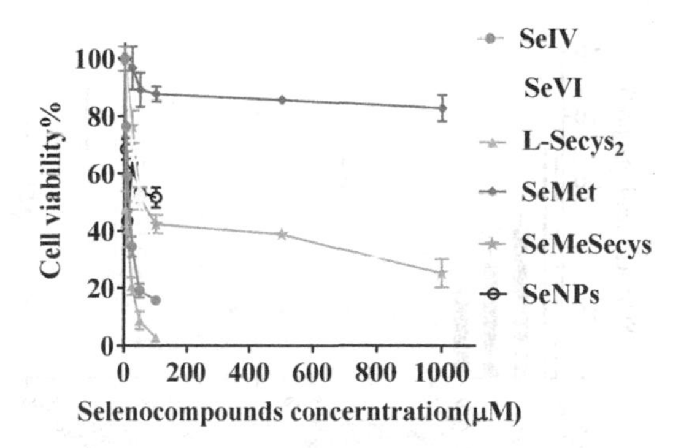

Figure 1. Effect of Se compounds on HepG-2 cell viability.

2.3 *Statistical analysis*

Analysis of variance was carried out using Tukey post-hoc test with SPSS 24.0 software. The level of significance was 0.05.

3 RESULTS AND DISCUSSION

3.1 *Effect of Se compounds on cell viability*

Results showed that Se inhibited the growth of HepG-2 human cells in a dose-dependent manner (Fig. 1). Compared with the same dose of Se compounds, SeMet exhibited the lowest cytotoxic response, while L-Secys2 exhibited the highest cytotoxic response. The different cytotoxic response of Se compounds might be related to their intracellular Se uptake and metabolite production (Marschall et al. 2016).

3.2 *Effect of Se compounds on ROS and SOD levels*

All six Se compounds increased intracellular ROS levels (Fig. 2). Among them, selenite, L-Secys2, and SeNPs generated larger amount of intracellular ROS than other compounds. Meanwhile, selenite and L-Secys2 (above 1 μmol/L) decreased antioxidant enzyme SOD activity, while other Se compounds decreased SOD activity at high concentrations

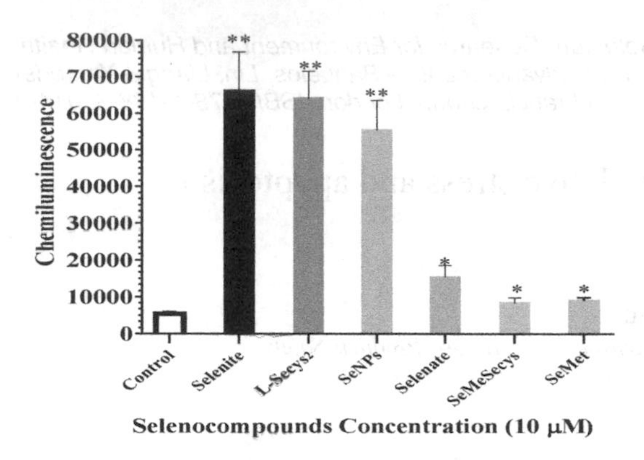

Figure 2. Effects of Se compounds on ROS level. * Significance at $p < 0.05$; ** Significance at $p < 0.01$.

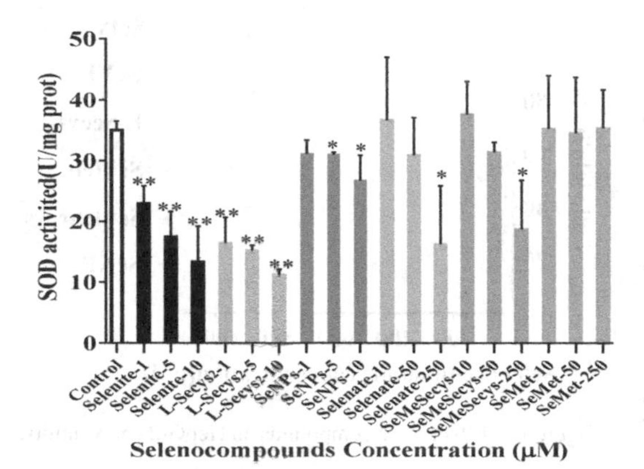

Figure 3. Effects of selenium compounds on SOD activity. * significance at $p < 0.05$; ** significance at $p < 0.01$.

(Fig. 3). The metabolites of selenite, L-Secys2, and SeNPs induced oxidative stress that contributed to the cytotoxicity in cancer cells (Menon et al. 2018).

3.3 *Effect of Se compounds on cell apoptosis*

Selenium compounds induced apoptosis in HepG-2 cells in a dose-dependent manner (Fig. 4). Compared with the same dose of Se compounds, the group treated with selenite and L-Secys2 showed higher apoptosis rates compared to other Se compounds. The results indicate that Se can inhibit cancer cells proliferation at higher concentrations and even result in apoptosis (Takahashi et al. 2017).

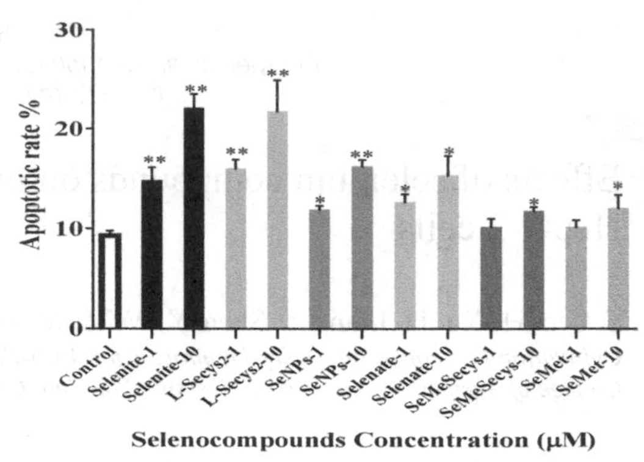

Figure 4. Effects of Se compounds on apoptosis of HepG2 cells. Two-dimension scatter plots depicting the distribution of cells positively stained for Annexin V, propidium iodide (PI). * Significance at $p < 0.05$; ** Significance at $p < 0.01$.

4 CONCLUSIONS

These results suggest that higher concentrations of Se compounds can increase oxidant stress, inhibit cell growth, and induce apoptosis in HepG-2 cells. Notably, L-Secys2 and selenite can be useful in therapeutic effects.

ACKNOWLEDGEMENT

This research work was financially supported by the Innovation Driven Special Foundation of Guangxi (No.AA17202038), Natural Science Foundation of Guangxi (No. 2016GXNSFAA380308)

REFERENCES

Ferndes, A.P. & Gandin, V. 2015. Selenium compounds as therapeutic agents in cancer. *Biochimica Biophysica Acta* 1850: 1642–1660.

Marschall, T.A., Bornhorst, J., Kuehnel, D. & Schwerdtle, T. 2016. Differing cytotoxicity and bioavailability of selenite, methylselenocysteine, selenomethionine, selenosugar 1 and trimethylselenonium ion and their underlying metabolic transformations in human cells. *Mol Nutr Food Res* 60: 2622–2632.

Menon, S., KS, D.S., R, S., S, R. & S, V.K. 2018. Selenium nanoparticles: A potent chemotherapeutic agent and an elucidation of its mechanism. *Colloids Surf B Biointerfaces* 170: 280–292.

Takahashi, K. Suzuki, N. & Ogra, Y. 2017. Bioavailability comparison of nine bioselenocompounds in vitro and in vivo. *Int J Mol Sci* 18(3): 506.

Selenium Research for Environment and Human Health: Perspectives, Technologies and Advancements – Bañuelos, Lin, Liang & Yin (eds)
© 2020 Taylor and Francis Group, London, ISBN 978-1-138-39014-0

Innovating animal models of aging: Selenium, frailty index, and mortality in pet dogs with exceptional longevity

D.J. Waters, E. Chiang, C. Suckow & A. Maras
Center for Exceptional Longevity Studies, Gerald P. Murphy Cancer Foundation, West Lafayette, IN, USA

A.-C. Kruger, Q. Sun & L. Schomburg
Institut für Experimentelle Endokrinologie, Charité-Universitätsmedizin Berlin, Berlin, Germany

1 INTRODUCTION

Aging is characterized by the accumulation of macromolecular damage, which leads to the accumulation of functional deficits. People are living longer lives, but persons with the same chronological age display considerable heterogeneity in their accumulation of deficits. Frailty index (FI) operationalizes frailty as the proportion of health deficits present in each individual, providing vital insights into the aging process and its consequences in terms of mortality risk and healthy life expectancy. In older adults, low selenium (Se) status has been associated with frailty and decreased muscle strength (Beck et al. 2007, Lauretani et al. 2007), but, to date, no studies have examined whether Se status significantly influences the lethality of frailty. The objective of this research was to advance our understanding of the importance of Se for frailty and the aging process and to minimize adverse health consequences associated with increased life expectancy by using a novel animal model of highly successful human aging.

2 METHODS

To achieve the research objective, we launched the first systematic scientific study of the oldest-living pet dogs in North America, gathering detailed data on exceptionally long-lived dogs that are physiologically equivalent to human centenarians. Frailty index was constructed assessing accumulation of 34 deficits using information from personal interviews with dog owners, validated through veterinary examination.

Selenium concentrations were determined by total reflection X-ray spectroscopy (TXRF) from plasma samples. Cox proportional hazard was used to determine relationships between plasma Se level, functional deficit accumulation, and mortality risk. Analysis included data collected from 122 canine centenarians (81 females, 41 males) who had a median evaluation-to-death interval of 7.0 months (range, 0.5–35.0 months).

3 RESULTS AND DISCUSSION

Overall, an increase in functional deficit accumulation in canine centenarians was associated with increased mortality risk. Age-adjusted hazard ratio (HR) for mortality per 0.01-unit FI increase was 1.04 (1.01–1.06), which closely mimics the HR of 1.05 (1.04–1.05) reported in humans (Rockwood et al. 2017). In females, longer duration of endogenous ovary exposure buffered the adverse impact of deficit accumulation on mortality. Among females with highest ovary exposure, the relationship between plasma Se concentration and mortality was U-shaped (Waters & Chiang 2018), revealing an optimal range for plasma Se equivalent to 0.77–0.88 mg/kg toenail Se. This range is similar to the optimal range of Se status for reducing human cancer risk determined by dose-response meta-analysis (0.85–0.94 mg/kg) (Hurst et al. 2012). In our study population, this range of Se status was not associated with lower functional deficit accumulation or mitigation of the mortality consequences of frailty.

4 CONCLUSIONS

This is the first report describing the relationship between plasma Se level, functional deficit accumulation, and mortality risk in pet dogs with exceptional longevity, a model of highly successful aging. Future research using the dog model will focus on the relationship between selenoproteins, such as plasma GPX, deficit accumulation, and life expectancy.

REFERENCES

Beck, J., Ferrucci, L., Sun, K. et al. 2007. Low serum selenium concentrations are associated with poor grip strength among older women living in the community. *Biofactors* 29(1): 37–44.

Hurst, R., Hooper, L., Norat, T. et al. 2012. Selenium and prostate cancer: systematic review and meta-analysis. *American J Clin Nutr* 96(1): 111–122.

Lauretani, F., Semba, R.D., Bandinelli, S., Ray, A.L., Guralnik, J.M. & Ferrucci, L. 2007. Association of low plasma selenium concentrations with poor muscle strength in older community-dwelling adults: The InCHIANTI Study. *Am J Clinic Nutr* 86(2): 347–352.

Rockwood, K., Blodgett, J.M., Theou, O. et al. 2017. A frailty index based on deficit accumulation quantifies mortality risk in humans and in mice. *Sci Rep* 7: 43068.

Waters, D.J. & Chiang, E.C. 2018. Five threads: how U-shaped thinking weaves together dogs, men, selenium, and prostate cancer risk. *Free Radic Biol Med* 127(1): 36–45.

Selenium Research for Environment and Human Health: Perspectives, Technologies and Advancements – Bañuelos, Lin, Liang & Yin (eds)
© 2020 Taylor and Francis Group, London, ISBN 978-1-138-39014-0

Selenium content in maize in high oesophageal cancer prevalence counties in Kenya

S.B. Otieno*
Department of Community Health, Nairobi Outreach Centre, Great Lakes University, Westlands, Nairobi, Kenya

T.S. Jayne & M. Muyanga
Michigan State University, East Lansing, Michigan, USA

1 INTRODUCTION

Selenium (Se) is required for many essential enzymes, and provides protection against various diseases (Pickering et al. 2011). Several studies have shown a significant inverse relationship between cancer prevalence and Se levels in diet (the unholy-cross) (Glattre et al. 1989, Dreher et al. 1996). Serum Se levels have been observed to be lower in variety of cancer cases compared to controls (Ujie et al. 1998, Helzsouer et al. 1989). Several intervention studies have demonstrated the link between Se and incidence of various cancers (Yu et al. 1997). The role of Se in cancers is considered to be due to its effect in cell cycle and maintenance of homoeostasis (Huawei 2009, Ip et al. 2001). Studies have shown the up-regulation effect of selenite and SeMet on cell-cycle-related genes (Zeng et al. 2009). A cytokine, Cyclin d4 (cdk4) that controls cell cycle progression responds to mitogenic stimuli of G1 phase of cell cycle by phosphorylation of tumor suppressor protein pRb (Gille et al. 1999). Furthermore, others have suggested that Se anticancer properties be due to enhanced expression of humoral genes (A2M) and tumor suppressor gene-IGFBP3, HHIP (Huawei. 2009). L-Methionine-gamma-lyase in tumor cells converts SeMet into methylselenol, which activates caspase cascade and apoptosis in cancer cells (Miki et al. 2001). Both selenite (inorganic Se) and organic Se induce anti-cancer effects through different pathways (Huawei 2009). The evidence available shows that the inorganic Se induces anti-cancer effects (genotoxicity) by inducing DNA single strand breaks, which affects cell cycle at S-phase, while organic Se in turn exerts its genotoxicity effects at G1 phase of cell cycle.

Most of the oesophageal cancer cases are in western region counties of Uasin Gishu, Trans Nzoia, and Kakamega (called the esophageal cancer corridor) (Odera et al. 2017). Currently 11% of all new cancer cases in Kenya are esophageal cancer, of which 90% are squamous cell carcinoma (Cheng et al. 2015). Western Kenya, including Eldoret Town (MTRH) in Uasin Gishu County, reports the highest incidences in both male and females (Ahmed et al. 1969) with ratio of male to female being 1.5:1 suggesting a common risk factor (Odera et al. 2017).

Selenium-rich food crops and food supplements are used to counteract low dietary uptake of S (Lavu et al. 2011). *Zea mays* is consumed by 96% of Kenyans and is therefore cheap and an inexpensive way to provide Se to large number of population (Otieno et al. 2015). This study was conducted to determine Se levels in *Zea mays* grains in the Maize belt in Kenya, a region of high esophageal cancer prevalence. Kenya has one of the highest esophageal cancer incidences of 17.6 per 100,000 (Schaafsma et al. 2015, Patel et al. 2013).

2 MATERIALS AND METHODS

Two hundred and fifty grams of *Zea mays* grain samples were collected at the time of interview and placed into separate Ziploc bags from the study sites. The samples were grouped into batches. They were transported in cool boxes at 4°C to KEBS laboratory and analyzed for Se by Atomic Absorption Spectroscopy (AAS) (Perkin Elmer A Analyst 300, Germany). About 0.25 g maize were weighed and placed in a graduated cylinder. They were then mixed with hydrogen peroxide, followed by digestion in nitric acid (0.75 ml) and hydrochloric acid (2.25 ml). The contents were thoroughly mixed with test tube shaker. The mixture was heated up to 80°C for one hour on aluminum heated block and allowed to cool. Then 11.5 ml of distilled water were added, mixed thoroughly, and allowed to settle. A portion of the solution was centrifuged for analysis by AAS. Standards were prepared from TITRISOL Stock solutions with same amount of acids.

3 RESULTS AND DISCUSSION

The mean value of Se in the *Zea mays* grains varied from 1.82 ± 0.76 mg/kg in Batch 4 (Kakamega) to 2.11±0.86 mcg/kg in Batch 3 (Trans Nzoia) (Table 1). The overall mean was 1.94 mg/kg. No significant difference was observed in Se content between

Table 1. Mean concentrations of Se in *Zea maize* grains.

	Se Concentration (mg/kg)
Batch 1 (Uasin Gishu)	1.88 ± 0.98 (n = 19)
Batch 2 (Uasin Gishu)	1.90 ± 1.07 (n = 11)
Batch 3 (Trans Nzoia)	2.11 ± 0.86 (n = 19)
Batch 4 (Kakamega)	1.82 ± 0.76 (n = 19)
Batch 5 (Kisii)	1.97 ± 0.99 (n = 29)

the maize batches $p > 0.05$ (Table 1). The mean Se levels in all batches were slightly higher than those reported in other countries (Chilima et al. 2011) but were lower than the one earlier study conducted in Kenya (Otieno et al. 2014). Selenium intake by *Zea mays* roots depends on several soil physical and chemical characteristics (Zang et al. 2011). If the soils in study site were deficient of S, then the levels of Se increased in grains. Our data show that there are 70–90 µg Se in maize grains. Processing of maize leads to loss of about 10–20% of Se (almost 40 µg). These losses translate to an average consumption of only 30–50 µg Se per day. While this value meets average nutritional value (USDA 2009), consumption of supra-nutritional level 100 mcg to 200 mcg per day are needed to inhibit genetic damage and cancer development in humans (Whanger et al. 2004).

4 CONCLUSIONS

We conclude that, although *Zea mays* grains have the Required Daily Allowance (RDA) level of 70–90 µg Se, the Se content is not sufficient to inhibit cancer development and genotoxicity in human subjects.

REFERENCES

Ahmed, N. & Cook, P. 1969. The incidence of cancer of the oesophagus in Western Kenya. *Br J Cancer* 23(2): 302–312.

Cheng, M.L., Zhang, L., Borok, U. et al. 2015 Selenium as an essential micronutrient: Roles in cell cycle and apoptosis. *Cancer Epidemiol* 39(2): 143–149.

Chilima, A.D.C., Young, S.D., Black, C.R., Meachim, M.C., Lammel, J. & Broadly, M.R. 2011. Agronomic biofortification of maize with selenium in Malawi. In G.S. Banuelos, Z.Q. Lin, X.B. Yin & D. Ning (eds), *Selenium Global Perspectives on Human, Animals and Environment*: 79–80. Hefei: University of Science and Technology of China Press.

Dreher, I. & Kohrle, J. 2013. The role of selenium and selenoprotein in thyroid tissue. *Thyroid* 6: 145.

Glattre, E., Thomassen, Y. & Thoresen, S.O. 1989. Prediagnostic serum selenium in a case control study of Thyroid Cancer. *Int J Epidemiol* 18: 4549.

Gille, H. & Downward, J. 1999. Multiple ras effector pathways contribute to GI cell cycle progression. *J Biol Chem* 274:22033–22040

Ip, C. & Ganther, H. 1988. Efficacy of trimethylselenonium versus selenite in cancer chemoprevention and its modulation by arsenite. *Carcinogenesis* 9: 1481–1484.

Zeng, H. 2009. Selenium as an essential micronutrient: Roles in cell cycle and apoptosis. *Molecules* 14(3): 1263–1278.

Patrick, L. 2009. Selenium biochemistry and cancer: A review of the literature. *Altem Med Rev* 9(3): 239–258.

Miki, K., Xu, M., Gupta, A. et al. 2001. Methioninase cancer gene therapy with selenomethionine as gene as suicide prodrag substrate. *Cancer Res* 61: 6805–6810.

Otieno, S.B., Were, F., Kabiru, E.W. & Waza, K. 2014. Study of selenium content of foods in a high HIV prevalence Community; A case of Pala Bondo District Kenya. In G.S. Banuelos, Z.Q. Lin, X.B. Yin & D. Ning (eds), *Selenium Global Perspectives on Human, Animals and Environment*: 62-65. Hefei, China: University of Science and Technology of China Press.

Otieno, S.B., Jayne T.S. & Muyanga, M. 2015. The effects of soil chemical characteristics on accumulation of native selenium by *Zea mays* grain in maize belt in Kenya. *Int J Biol Biomol, Agric Food Biotechnol Eng* 9(11): 1031–1035.

Odera, O.J., Odera, E., Githonga, J. et al. 2017. Esophageal cancer in Kenya. *Am J Dis* 4(3): 23–33.

Schaafsma, T., Wakefield, J., Hanish, R. et al. 2015. Africa's oesophageal cancer corridor: Geographic variations in incidence correlates with certain micronutrients deficiencies. *Plos One* 10: e0140107.

Ujie, S., Itoh, Y. & Kikuchi, H. 1998. Serum selenium contents and the risk of cancer. *J Cancer Res* 25: 1891–1897.

Whanger, P.D. 2004. Selenium and its relationship with cancer: an update. *Br J Nutr* 91: 11–28.

Zang, L.H., Yu, F.Y., Li, Y.J. & Miao, Y.F. 2011. Effect of soil pH on physiological characteristic of selenite uptake by maize roots. In G.S. Banuelos, Z.Q. Lin, X.B. Yin & D. Ning (eds), *Selenium Global Perspectives on Human, Animals and Environment*: 87–88. Hefei, China: University of science and technology of China Press.

Zeng, H. & Botnen, J.H. 2007. Selenium is critical for cancer-signaling gene expression but not cell proliferation in human colon Caco-2 cells. *Br J Cancer* 24(3): 250–254.

Selenium Research for Environment and Human Health: Perspectives, Technologies and Advancements – Bañuelos, Lin, Liang & Yin (eds)

Biomarkers of selenium status

L. Schomburg
Institute for Experimental Endocrinology, Charité-Medical School Berlin, Germany

1 INTRODUCTION

1.1 *The selenium status*

The term "selenium status" is not defined clearly, yet everyone in the field intuitively understand the concept, i.e. reflecting the nutritional and physiological state of selenium (Se) supply of a given subject and describing a potential deficient, adequate, or surplus situation. The major determinants of the Se status relate to nutritional intake, metabolism, and excretion.

1.2 *Nutritional intake assessments*

There are several analytic methods to monitor the nutritional intake of Se, differing by accuracy and precision, degree of efforts, time and resources needed, and dependence on learned assumptions and extrapolations. Two examples are given below to highlight this issue. Food-frequency questionnaires can nicely cover all dietary items consumed with little extra efforts, costs, and technical challenges, especially when combined with modern IT-based tools (Conrad et al. 2018). However, the extracted information relies heavily on having correct knowledge of the Se content of the different nutritional items, which is difficult and varies strongly for Se analysis. At the other end of the scale are techniques with duplicate diet analyses involving an analytical technique, where identical portions to the consumed ones are prepared, stored, and analyzed later (Isaksson 1993). Duplicate diets can provide very precise information, and thus constitute the gold standard of dietary analysis. However, these analyses are difficult to conduct with large groups of participants, because they are time consuming and expensive, and they require an enormous degree of discipline, organization, and effort. Yet, in the case of Se, this type of study can accurately consider the different Se contents of similar dietary items, determine the actual Se intake, and provide solid insights and the relevant data with excellent precision.

1.3 *Selenium metabolism*

The different nutritional forms of Se share some specific and some common metabolic fates. The contribution of the microbiota and the gastrointestinal uptake rates in relation to the actual Se status are not yet understood well. In human subjects, the liver is the major organ to receive nutritional Se and systematically channels it into the circulation and to other tissues. Several excellent reviews cover the next steps finally leading to two Sec-loaded tRNA isoforms that can both be used during translation of the different selenoproteins (Hatfield et al. 2002). Interestingly, this step is precisely controlled at several molecular levels, including gene transcription rates, mRNA stability, translation initiation, progression, and termination in combination with additional factors affecting the pattern of selenoproteins being expressed dynamically. The balance between stress-related and housekeeping selenoproteins is affected by external stimuli and modifies the ratio of biosynthesis of liver-resident selenoproteins versus secreted selenoprotein P (SELENOP). These reactions are at the center of Se status control, display some age-, Se status-, and sex-specific differences, and are modified by medication and disease, e.g. severe liver disease always negatively impacts the SELENOP biosynthesis rate and systemic Se transport.

1.4 *Excretion of Se*

The excretion of Se proceeds via a number of routes, i.e., via air and skin, urine, and feces. An unpleasant smell is an early sign of toxic Se intake, brought about by the exhalation of dimethyl-selenide and other volatile selenocompounds via the lungs and breath, but likely also via the skin. Similarly, tri-methyl-selenonium, a charged and hydrophilic selenocompound originating again mainly from liver, can be built and secreted. Selenosugars constitute the excreted selenocompounds via the urine under physiological conditions (Kobayashi et al. 2002). Finally, a constant loss of cells in combination with leftovers from the nutrition and microbial conversions contribute to Se content of feces. A detailed characterization and quantification of excretory selenocompounds should thus enable a reliable assessment of Se status, yet collecting and analyzing the respective biospecimens is both difficult and unpleasant, and it probably needs to be conducted for at least 24 h.

2 MATERIALS AND METHODS

2.1 *Biomarkers of Se status*

Potential biomarkers of Se status are discussed in view of their potential advantages and limitations, and

PubMed was screened accordingly at the end of June 2019. Search criteria included the term "Se status" in combination with "clinical study" and "biomarker." Respective hits were assessed for quality and relevance. It was not our intent to report on all related available literature.

3 RESULTS AND DISCUSSION

3.1 *Theoretical considerations*

There are several requirements that need to be met for a suitable Se status biomarker. Firstly, Se intake, metabolism, and excretion should be reflected. Secondly, it needs to be accessible, i.e. from matrices that are typically used in the clinics. Thirdly, analytical methods that can be standardized are needed. Fourthly, the parameter should not depend on biomaterial that is unstable or extensively manipulated.

3.2 *Potential candidates*

The following potential biomarkers have been used in some clinical studies: Total Se in full blood, serum, or plasma; glutathione peroxidase (GPX) activity determined in blood cells, serum or plasma; and SELENOP concentrations in serum or plasma. Similarly, hair and nail samples, buccal cells, and Se or selenocompounds in urine samples have been used.

3.3 *Preferences in clinical studies*

The search terms "selenium status AND clinical study AND biomarker" yielded 75 PubMed-listed hits. In combination with "food frequency", the number was reduced to twenty, combining "total concentration" limited the number to six. The term "buccal" appeared in one hit, "GPX" was mentioned in four, "urine" in nine and "hair" or "nail" in three hits, respectively. In comparison, "selenoprotein" was reduced to thirteen hits, with two studies mentioning "selenium-binding protein." Conducting these searches without the term "status" and looking for "selenium AND clinical study AND biomarker," the number of total hits directly amounted to 233, which would provide a more solid and complex database for such an overview.

4 CONCLUSIONS

Different biomarkers are used for approaching the Se status of human subjects in clinical studies. The dietary intakes, as well as total Se and selenoproteins in serum or plasma are most often determined in clinical studies. Urine or buccal cells are rarely analyzed, and hair and nail samples carry the risk for artefacts. Food frequency analyses are valid methods but they need to be combined with a thorough analysis of the Se contents of the major dietary components. When it comes to clinical trials, total Se concentrations and/or selenoproteins are the readouts of choice. The former can be standardized with reference material. Enzymatic activities of the GPX isoenzymes (GPX1 and GPX3, respectively) can be reliably measured, but reference standards are missing, and the activities depend on the pre-analytic history of the samples. The Se transport protein SELENOP can best be determined by immunological techniques, and a standard reference material (NIST SRM 1950, metabolites in frozen human plasma) has been used for standardization (Hybsier et al. 2017). SELENOP may constitute the most suitable biomarker in humans (Hurst et al. 2010). Similar analyses and validated assays are also needed for experimental studies with rodents, cattle, swine, and chickens as the most relevant agricultural species for human nutrition.

REFERENCES

Conrad, J., Koch, S.A.J. & Nöthlings, U. 2018. New approaches in assessing food intake in epidemiology. *Curr Opin Clin Nutr Metab Care* 21(5): 343–351.

Hatfield, D.L. & Gladyshev, V.N. 2002. How selenium has altered our understanding of the genetic code. *Mol Cell Biol* 22(11): 3565–3576.

Hurst, R., Armah, C.N., Dainty, J.R. et al. 2010. Establishing optimal selenium status: results of a randomized, double-blind, placebo-controlled trial. *Am J Clin Nutr* 91(4): 923–931.

Hybsier, S., Schulz, T., Wu, Z. et al. 2017. Sex-specific and inter-individual differences in biomarkers of selenium status identified by a calibrated ELISA for selenoprotein P. *Redox Biol* 11: 403–414.

Isaksson, B. 1993. A critical evaluation of the duplicate-portion technique in dietary surveys. *Eur J Clin Nutr* 47(7): 457–460.

Kobayashi, Y., Ogra, Y., Ishiwata, K., Takayama, H., Aimi, N. & Suzuki, K.T. 2002. Selenosugars are key and urinary metabolites for selenium excretion within the required to low-toxic range. *Proc Natl Acad Sci USA* 99(25): 15932–15936.

Selenium Research for Environment and Human Health: Perspectives, Technologies and Advancements – Bañuelos, Lin, Liang & Yin (eds)
© 2020 Taylor and Francis Group, London, ISBN 978-1-138-39014-0

A new function of glutathione peroxidase-1 in regulating transcription of regenerating islet-derived protein-2 in pancreatic islets of mice

J.-W. Yun, Z.P. Zhao, X. Yan, M.Z. Vatamaniuk & X.G. Lei*
Department of Animal Science, Cornell University, Ithaca, NY, USA

1 INTRODUCTION

Glutathione peroxidase-1 (GPX1) is the first identified selenoprotein and selenoperoxidase in mammals. We have reported that overexpression of GPX1 in mice induces type-2-like diabetes phenotypes, including hyperinsulinemia (McClung et al. 2004, Wang et al. 2008). In searching for the underlying mechanism for these phenotypes, we have discovered a diminished expression of regenerating islet-derived protein 2 (REG2) in pancreatic islets of the GPX1 overexpressing (OE) mice (Yun et al. 2019). This study was performed to test if and how GPX1 suppressed Reg2 expression via scavenging reactive oxygen species (ROS).

2 MATERIALS AND METHODS

Wild-type (WT, C57B1xC3H) and OE mice (male, 2-months of age, n = 7) were treated with ROS-generating diquat (24 mg/kg body weight) and streptozotocin (150 mg/kg body weight) for 48 h. Their islets (n = 70) were treated with H_2O_2 (0, 25, and 50 μM), ROS-scavenging ebselen (0, 50, and 100 μM) and N-acetylcysteine (NAC; 0, 3, and 6 mM) for 6–12 h. Responses of Reg2 mRNA and/or protein levels in the pancreas, islets, and islet-culture media to these treatments were determined. Thereafter, 13 transcriptional factors (TFs) with putative binding sites in the Reg2 proximate promoter were identified, and their mRNA and protein levels were analyzed in the OE and WT islets treated with ebselen, NAC, and H_2O_2. Chromatin immunoprecipitation of OE and WT islets and subsequential deletions of the binding sites in bTC-3 cells were performed to reveal the binding of TFs to the Reg2 promoter and the inhibition of the Reg2 promoter activation by ebselen using luciferase reporter assays. In the end, RNA interference of c-jun in single islet cells was performed to confirm the overall findings.

3 RESULTS AND DISCUSSION

Pancreatic and islet REG2 protein production and (or) secretion were positively correlated ($p < 0.05$) with the intracellular ROS status and inhibited by the ROS-scavengers or antioxidants. Among the 13 TFs with putative binding sites in the Reg2 proximate promoter, only activator protein-1 (AP-1) and albumin D box-binding protein (DBP) mRNA and protein levels were elevated ($p < 0.05$) in the OE pancreatic islets compared with the WT islets. Their mRNA abundances in the cultured islets were elevated ($p < 0.05$) by ebselen and NAC, but decreased ($p < 0.05$) by H_2O_2. These responses were contrary to those of Reg2 expression. The two TFs were bound to the Reg2 promoter at the location of -168 to 0 base pair (bp). Deleting the AP-1 (−143/−137 and −60/−57 bp) and/or DBP (−35/−29 bp) binding domains in the Reg2 promoter attenuated and/or abolished the inhibition of Reg2 promoter activation by ebselen. Suppressing the c-jun mRNA expression in the OE islets by c-jun siRNA elevated the Reg2 mRNA expression.

4 CONCLUSIONS

The down-regulation of Reg2 expression in the GPX1-overproducing pancreatic islets was mediated by a transcriptional inhibition of the gene via two ROS responsive transcription factors: AP-1 and DBP. Our findings reveal GPX1 as a novel regulator of Reg2 expression, and linking these two previously-unrelated proteins will have broad biomedical implications.

REFERENCES

McClung, J.P., Roneker, C.A., Lisk, D.J., Langlais, P., Liu, F. & Lei, X.G. 2004. Development of insulin resistance and obesity in mice overexpressing cellular glutathione peroxidase. *Proc Natl Acad Sci USA* 101: 8852–8857.

Wang, X., Vatamaniuk, M.Z., Wang, S., Roneker, C.A., Simmons, R.A. & Lei, X.G. 2008. Molecular mechanisms for hyperinsulinemia induced by the overexpression of Se-glutathione peroxidase-1 in mice. *Diabetologia* 51: 1515–24.

Yun, W., Zhao, Z.P., Vatamaniuk, M.Z. & Lei, X.G. 2019. Glutathione peroxidase-1 inhibits transcription of regenerating islet-derived protein-2 in pancreatic islets. *Free Radic Biol Med* 134: 385–393.

Analytical methodology

Selenium Research for Environment and Human Health: Perspectives, Technologies and Advancements – Bañuelos, Lin, Liang & Yin (eds)
© 2020 Taylor and Francis Group, London, ISBN 978-1-138-39014-0

Determination of selenium species in black bean protein powder by HPLC and ICP-MS

L. Meng, Y.L. Xu, B. He, M. Qi, Z.R. Xia & D.J. Tang*
Ankang R&D Center for Se-enriched Products, Ankang, China
Key Laboratory of Se-enriched Products Development and Quality Control, Ministry of Agriculture and Rural Affairs, Ankang, China

1 INTRODUCTION

Selenium (Se) is an essential trace element for humans and has many significant physiological functions, even though the range between the essentiality and toxicity of Se is very narrow (Jagtap et al. 2016). The toxicity, bioavailability, environmental transport mechanism, and chemoprotective activities of Se are highly related to its chemical form and oxidation state (Wang et al. 2016). Inorganic Se can be toxic at high concentrations, while the bioavailability of selenomethionine (SeMet) is higher than that of organic Se in living organisms (Hu et al. 2017). Therefore, it is of great significance to conduct Se speciation analyses.

High-performance Liquid Chromatography (HPLC) coupled to Inductively-coupled Plasma Mass Spectrometry (HPLC-ICP/MS) is presently the most common means for Se species analysis, which has the advantages of high efficiency separation, relatively low detection limit, wide linear range, and high precision analysis (Sele et al. 2018, Pyrzynska et al. 2019). The primary goal of this work is to develop a convenient operation, rapid, efficient, and sensitive Se speciation method for black bean protein powder using HPLC-ICP/MS.

2 MATERIALS AND METHODS

2.1 *Materials and reagents*

Black Bean Protein Powder was provided by Ankang local enterprise. The regents including diammonium phosphate, formic acid, and hydrochloric acid used in this study were of analytical pure (Tianli, Tianjin, China). The HPLC grade methanol was purchased from Sinopharm (Shanghai, China). The proteinase XIV was supplied by Yuanye (Shanghai, China). The standard substances selenocystine ($SeCys_2$), selenite (SeO_3^{2-}), and selenomethionine (SeMet) were purchased from Sigma-Aldrich (St. Louis, MO, USA), and selenate (SeO_4^{2-}) solution was obtained from the National Institute of Metrology (Beijing, China).

2.2 *Sample preparation*

We weighed 0.2 g of the pulverized sample and placed it in a centrifuge tube, added 4 mg/mL protease XIV solution, vortexed for 1 min, and sonicated for 3 h at 37°C in water bath (ultrasonic cleaner, Hechuang, Kunshan, China). After standing for 10 min, the sample solution was then centrifuged for 10 min at 4°C with 9000 rpm (high-speed refrigerated centrifuge, Cence, Changsha, China) and filtered the supernatant through a 0.22 μm membrane for testing. $SeCys_2$, Se(IV), SeMet, and Se(VI) were separated by HPLC (Agilent 1260, Santa Clara, USA), and the characterization of four Se species was examined by ICP/MS (Agilent 7900, Santa Clara, USA).

3 RESULTS AND DISCUSSION

3.1 *Chromatogram of mixed standard solution of Se species*

Using 40 mmol/L $(NH_4)_2HPO_4$ (pH 5.0) and 60 mmol/L $(NH_4)_2HPO_4$ (pH 6.0) as mobile phase, $SeCys_2$, Se(IV), SeMet, and Se(VI) as four standards were respectively separated in gradient elution. The peaks were clearly identified at respective peak times. The chromatogram of 10 μg/L mixed standard solution of Se species was shown in Figure 1.

3.2 *Determination of actual sample*

Selenium was determined in black bean protein powder according to the established method, and the chromatogram is shown in Figure 2. The main chemical forms of Se were identified as SeMet, Se(IV), and SeCys2, along with two unknown Se species.

3.3 *Recoveries of four Se species*

At three mass concentration levels, the recovery rate of spiked $SeCys_2$ was low because $SeCys_2$ was unstable and easily decomposed. The recoveries of the other

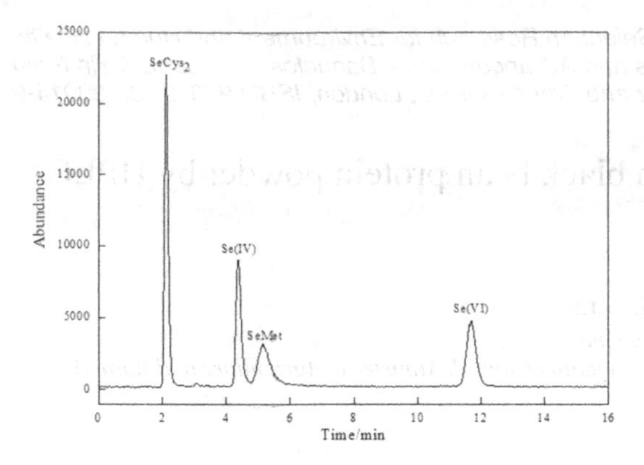

Figure 1. The chromatogram of 10 μg/L mixed standard solution of Se species.

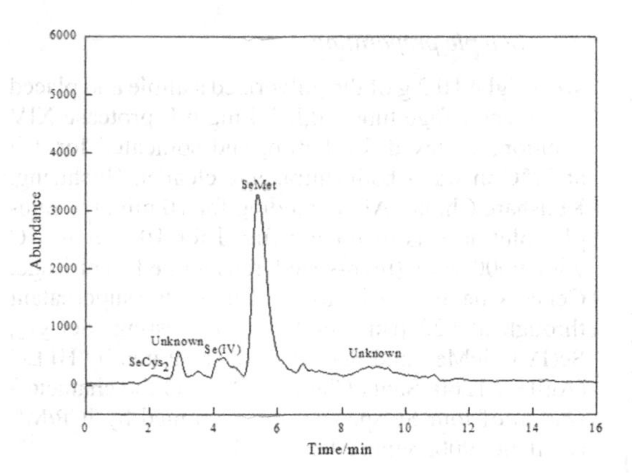

Figure 2. The chromatogram of Se species in black bean protein powder.

Table 1. Recoveries of four Se species.

Se species	Original Se Concentration (μg/L)	Spiked Se (μg/L)	Observed Se (μg/L)	Recovery (%)
$SeCys_2$	0.94	0.5	1.35	82.0
		1	1.65	71.0
		2	2.33	69.5
Se(IV)	2.59	1	3.45	86.0
		2	4.85	113.0
		5	7.64	101.0
SeMet	51.54	25	74.32	91.1
		50	104.11	105.1
		100	170.71	119.2
Se(VI)	0.09	0.5	0.53	88.0
		1	1.21	112.0
		2	2.17	104.0

three Se species were between 80% and 120%, indicating that this method is suitable for this analysis. The results are shown in Table 1.

4 CONCLUSIONS

A method has been developed for the determination of $SeCys_2$, Se (IV), SeMet, and Se(VI) in black bean protein powder by HPLC-ICP/MS. Four Se species were successfully separated. Using protease XIV as extractant, Hamilton PRP-100 anion-exchange column as analytical column, 40 mmol/L $(NH_4)_2HPO_4$ (pH 5.0) and 60 mmol/L $(NH_4)_2HPO_4$ (pH 6.0) as mobile phase, the RSD were less than 3%, and the recoveries were between 69.5% and 120.0%. Hence, the results were satisfactory. The experimental results show that the described method for the analysis of Se species has the advantages of convenient operation, rapidity, high efficiency, and sensitivity.

REFERENCES

Jagtap, R. & Maher, W. 2016. Determination of selenium species in biota with an emphasis on animal tissues by HPLC-ICP-MS. *Microchem J* 124: 422–529.

Wang, X.J., Wu, L., Gao, J.Q. et al. 2016. Magnetic effervescent tablet-assisted ionic liquid dispersive liquid–liquid micro-extraction of selenium for speciation in foods and beverages. *Food Addit Contam Part A* 33(7): 1190–1199.

Hu, T., Liu, L.P., Chen, S.Z. et al. 2017. Determination of selenium species in *Cordyceps militaris* by High-performance Liquid Chromatography coupled to Hydride Generation Atomic Fluorescence Spectrometry. *Anal Lett* 51(14): 2316–2330.

Sele, V., Ørnsrud, R., Sloth, J.J. et al. 2018. Selenium and selenium species in feeds and muscle tissue of Atlantic salmon. *J Trace Elem Med Biol* 47: 124–133.

Pyrzynska, K. & Sentkowska, A. 2019. Liquid chromatographic analysis of selenium species in plant materials. *TrAC Trends in Analy Chem* 111: 128–138.

Selenium Research for Environment and Human Health: Perspectives, Technologies and Advancements – Bañuelos, Lin, Liang & Yin (eds)
© 2020 Taylor and Francis Group, London, ISBN 978-1-138-39014-0

Selenium speciation in cereals by ultrasonic-assisted enzyme extraction and HPLC-ICP-MS

S.Z. Chen, L.P. Liu, Y. Liu & T.H. Zhou
Beijing Center for Diseases Prevention and Control, Beijing, China

1 INTRODUCTION

Selenium (Se) is one of the essential trace elements in human body, with a very narrow concentration range from sufficient to deficient or toxic (Olmedo et al. 2013, Sun et al. 2013). Selenium deficiency can lead to anemia, coronary heart disease, Kashin-beck disease, diabetes, and more than 40 other diseases (Rayman 2000). High levels of Se can, however, also cause chronic toxic symptoms, such as cirrhosis, liver cancer, tooth, hair and nail loss, eye irritation, and paralysis (Bem 1981, Najafi et al. 2012).

Cereals are commonly consumed by humans. In recent years, selenium-enriched cereals have become more and more popular with people, as well as selenium-enriched corn, selenium-enriched rice, selenium-enriched wheat, etc. The study of Se species in selenium-enriched cereals has become a popular research topic, but the analytical methods used for detecting Se species in cereals are less reported.

High performance liquid chromatography-inductively-coupled plasma mass spectrometry (LC-ICP-MS) has been widely used for its high separation efficiency, high sensitivity, wide linear range, good selectivity, and simple operation. In this study, five Se species in cereals were studied by LC-ICP-MS. Separation conditions, detection modes, and extraction methods were optimized.

2 MATERIALS AND METHODS

2.1 *Instruments*

The detection of Se was performed with an Agilent 7700x ICP-MS (Agilent, Tokyo, Japan) equipped with a micromist nebulizer (100 μL, natural aspiration) and a Scott spray chamber (2°C). The high energy helium mode (HEHe) of ICP-MS was used to reduce polyatomic interference. Selenium species were separated by using a Hamilton PRP-X100 (250 mm × 4.6 mm, 5 μm) anion-exchange column (Hamilton, Reno, NV, USA) and a HPLC 1260 system (Agilent, Karlsruhe, Germany) with 100-μL injection loop. Extraction of Se species was performed with a digital control ultrasonic cleaner.

2.2 *Separation and detection conditions*

The mobile phase consisted of A: 40 mmol/L $(NH_4)_2HPO_4$ (pH = 5.0) and B: 60 mmol/L $(NH_4)_2HPO_4$ (pH = 6.0). The flow rate is 1 mL/min. Selenium species can achieve baseline separation in 15 minutes.

RF power: 1550 W, sampling depth: 5 ~ 8 mm, carrier gas flow rate: 1.05 L/min, cooling air flow rate: 14 L/min; Helium collision gas flow rate: 4.8 mL/min, dwell time: 0.3 s, m/z: 78.

2.3 *Pretreatment method*

A portion of dried sample powder (0.1–0.3 g) was accurately weighed in a 15-mL polypropylene centrifuge tube, and 5 mL ultrapure water and 20 mg protease XIV were added. The centrifuge tubes were closed and placed into the ultrasonic water bath system. The temperature of ultrasonic water bath was maintained at 37°C for 3 h. The extract was centrifuged at 9000 rpm for 10 min. The supernatants were filtered through a 0.45 μm Millipore syringe filter and stored at 4°C before analysis. Reagent blank was prepared in the same way.

3 RESULTS AND DISCUSSION

3.1 *Linear range and detection limit*

Under the optimal experimental conditions, the correlation coefficients (r) were greater than 0.999 in the range of 0.5 to 200.0 μg/L, indicating that the linear relationships of $SeCys_2$, MeSeCys, Se(IV), SeMet, and Se(VI) were good.

If the sampling amount was 0.1 g, the volume of extracting solution was 5 mL. The detection limits of $SeCys_2$, MeSeCys, Se(IV), SeMet, and Se(VI) were 2.5, 5.0, 2.5, 10.0, and 5.0 μg/kg, respectively.

3.2 *Accuracy and repeatability*

Precision and spike recovery experiments were performed according to the established sample pretreatment method. The results showed that the recoveries of $SeCys_2$, MeSeCys, Se(IV) SeMet, and Se(VI) were 60.5–65.0%, 119.3–120.8%, 95.0–105.3%, 86.7–89.1%, and 94.6–97.0%, respectively.

Table 1. Concentrations (mg/kg) of different Se species in cereals (n = 3).

Sample	$SeCys_2$	MeSecys	Se(IV)	SeMet	Se(VI)
Rice 1	0.015	ND*	0.005	0.198	0.009
Rice 2	0.015	ND	0.007	0.358	0.008
Rice 3	0.003	ND	0.003	0.021	0.004
Wheat	0.856	0.044	0.151	10.19	0.598
Corn 1	0.105	0.035	0.014	3.528	0.032
Corn 2	0.452	0.039	0.017	10.250	0.105

* ND: not detectable.

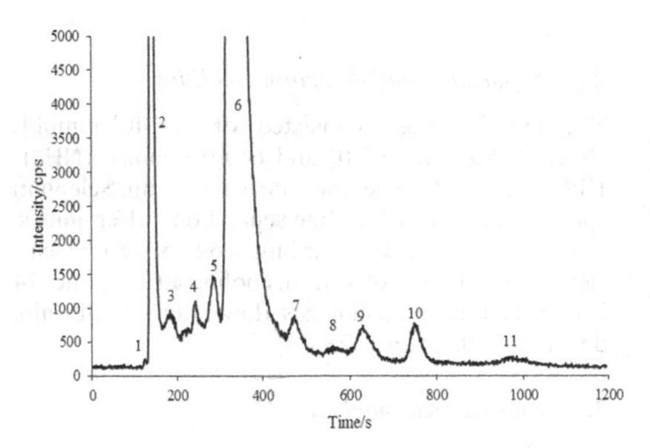

Figure 1. Selenium species chromatogram of a corn sample (Corn 1). Peaks 1 = unknown, 2 = SeCys2, 3 = MeSeCys, 4 = unknown, 5 = Se(IV), 6 = SeMet, 7 = unknown, 8 = unknown, 9 = unknown, 10 = Se (VI), 11 = unknown.

The method was applied to determine SeMet in wheat flour certified reference material (ERM®-BC210a, the standard value of SeMet is 11.18 ± 1.06 mg/kg). The determined concentration of SeMet was 10.52 ± 0.28 mg/kg, which was within the certified value range.

3.3 *Selenium species in cereal samples*

Selenium species in cereal samples were analyzed by optimized pretreatment and detection methods. The results showed that $SeCys_2$, MeSeCys, Se(IV), SeMet, and Se(VI) were detected in wheat and maize samples, and $SeCys_2$, Se(IV), SeMet, and Se(VI) were detected in rice samples except MeSeCys (Table 1). The extraction efficiency was from 62.4% to 103.3%. SeMet accounted for 54.2 to 86.4% of total Se content, which indicated that selenomethionine was the main Se species in cereals. Figure 1 shows that eleven Se compounds were detected in a corn sample (Corn 1) after ultrasound-assisted enzymatic extraction. Due to the lack of corresponding reference materials, qualitative and quantitative analysis of unknown Se compounds was not carried out.

4 CONCLUSIONS

Selenium species in cereals were studied by ultrasonic-assisted enzyme extraction combined with HPLC-ICP-MS. The results showed that the main Se species in cereals was selenomethionine. A small amount of selenocystine, selenite, selenate, and some unknown Se compounds were also detected. The accuracy of the results can be guaranteed by detecting selenomethionine within 5 hours after extraction.

REFERENCES

Bem, E.M. 1981. Determination of selenium in the environment and in biological material. *Environ Health Perspect* 37(37): 183–200.

Najafi, N. M., Tavakoli, H., Abdollahzadeh, Y., & Alizadeh, R. 2012. Comparison of ultrasound-assisted emulsification and dispersive liquid-liquid microextraction methods for the speciation of inorganic selenium in environmental water samples using low density extraction solvents. *Analytica Chimica Acta* 714(3): 82–88.

Olmedo, P., Hernández, A.F., Pla, A., Femia, P., Navas-Acien, A. & Gil, F. 2013. Determination of essential elements (copper, manganese, selenium and zinc) in fish and shellfish samples. Risk and nutritional assessment and mercury–selenium balance. *Food Chem Toxicol* 62: 299–307.

Rayman, M.P. 2000. The importance of selenium to human health. *Lancet* 356(9225): 233–241.

Sun, M., Liu, G. & Wu, Q. 2013. Speciation of organic and inorganic selenium in selenium-enriched rice by graphite furnace atomic absorption spectrometry after cloud point extraction. *Food Chem* 141(1): 66–71.

Selenium Research for Environment and Human Health: Perspectives, Technologies and Advancements – Bañuelos, Lin, Liang & Yin (eds)

Analytical methods involve speciation analysis and elemental mapping to describe processes in biogeochemistry: A review

J. Feldmann & E.M. Krupp
TESLA-Trace Element Speciation Laboratory, University of Aberdeen, Aberdeen, Scotland, UK

1 INTRODUCTION

Selenium (Se) in geological samples can occur as selenide, selenite, and selenate with oxidation stages of: −2, +4, and +6, and Se often substitutes sulfur (S) in minerals. There is, however, limited information describing the occurrence as elemental Se (Se^o). In biological samples, Se may occur additionally in a number of biological molecules, such as the two amino acids selenocysteine (SeCys) or selenomethione (SeMet), which can be incorporated into proteins. While SeCys is an essential amino acid for most animals and forms selenoproteins such as glutathione peroxidase (GPx) or selenoprotein P (SelP). SeMet is only replacing cysteine (Cys) in Se-containing proteins and is not essential. Other environmental processes generate volatile Se species, such as dimethylselenide and dimethyldiselenide. More detail about the plethora of Se speciation is described elsewhere (Wallschlager & Feldmann 2010). Each group of Se species needs a bespoken methodology. In this mini-review, we are lining out the different types of analytical methods used to describe processes in three different case studies of Se.

2 CASE STUDIES WITH DIFFERENT ANALYTICAL APPROACHES

Selenium is not considered to be essential for plants, but there is strong initiative amongst nutritionists and soil scientist to enhance the Se concentration in foodstuff (e.g. cereals) to enhance the Se consumed by human populations exposed to low levels of Se. Hence, studies which can identify how Se is taken up by plants and translocated in plants are pertinent. Earlier studies look at total Se levels throughout the plant, but to get an insight into the molecular processes, Se speciation methodologies are needed. Selenium in the form of selenate or selenite is often used in fertilizers, and it has been shown that the uptake of Se into the upper parts of the plants is often inefficient. In plants, Se speciation would involve the detection of Se biotransformations into selenide, which is then incorporated into glutathione or phytochelatins. The analysis of any Se-phytochelatin complexes involves a soft extraction followed by reverse-phase liquid chromatography coupled to ICPMS as selenium-specific detector and simultaneously to electrospray mass spectrometry as (RP-HPLC-ICPMS/ESI-MS). This technique has been recently demonstrated, and some of the Se biomolecules were detected (Bluemlein et al. 2009, Aborode et al. 2016). Whether their complexation in the plant roots prevents the translocation of Se remains to be studied.

The use of HPLC-ICPMS makes the use of mass balance approaches possible, since all Se species can be quantified with regards to their Se content, even if not all Se species are available as calibrants. This is possible because the ICPMS can be used as a selenium-specific detector in the species-independent calibration mode as described elsewhere (Amayo et al. 2011). The quantification of Se as eluted species (Sum of Se_{org}) can be compared to total Se (Se_{Total}) in the sample, which makes the identification of unaccounted Se ($Se_{unknown}$) species visible.

The mass balance often identifies losses of Se due to the formation stable elemental Se (Se^o) in the soil. Se^o is thermodynamically stable under slightly reduced acidic conditions, as indicated in the Se-O-H-Pourbaix diagram (Brookins 1988). Hence, a bespoke extraction method using sulfite to form selenosulfate ($SeSO_3^{2-}$) is necessary to make Se^o amenable to chromatographic separation, as shown elsewhere (Aborode et al. 2015). In *T. alata* plants $Se_{unknown}$ was not significantly different from the amount of Se^o. EXAFS at the synchrotron confirmed the occurrence of elemental Se in the plant roots (Aborode et al. 2015).

Pilot whales are top predators in the marine food chain and accumulate enormous concentration of Se in their tissues such as liver, kidney, and muscle (Gajdosechova et al. 2015). Concentrations up to 500 mg/kg can be detected. These concentrations seem to reach toxic levels. Speciation analysis is necessary to identify the Se speciation in the tissues to see which proportion of Se is biological active. When extraction efficiencies using enzymatic extractions and other soft extractions are used, only a small part (<20%) can often be extracted, as shown in the pilot whale tissues (Gajdosechova et al. 2016). Speciation methods involving proteolytic digestions, which are used to determine SeCys and SeMet, need to be employed to identify whether the Se status as essential Se is guaranteed. In this regard, the subsequent chromatographic analysis separates selenite and selenate as well as SeCys and SeMet so that essential from non-essential elemental species can be separated. Often anion-exchange liquid chromatography coupled to ICPMS has been used (Gajdosechova et al. 2016). A problem using this approach is that SeCys is not stable enough and derivatization reactions is necessary using iodoacetamide to establish a stable complex with only SeCys (Bierla et al. 2018).

Using XANES/EXAFS revealed that most of the Se was bound to Hg to form inert insoluble HgSe in the liver of the older pilot whales.

Is this detoxification taken place in every cell? To analyze for this spatial resolution analysis of Se and Hg is necessary. HgSe was not homogeneously distributed, which could be identified by LA-ICPMS with a resolution of 10 μm. Using synchrotron x-ray fluorescence mapping with 0.8-μm resolution revealed that HgSe were clustered in nanoparticles (Gajdosechova et al. 2016).

How are these particles formed and where? The dynamic of these HgSe nanoparticles can be revealed by using isotope ratio analysis. Selenium should show different isotope ratios of the stable isotopes at different trophic levels, and it can be expected that the different tissues have unique Se isotope ratios. Therefore, isotope ratio analysis using a multi-collector (MC)-ICPMS is necessary to determine the isotope ratio with high precision to identify isotope fractionation of heavy elements such as Se. MC-ICPMS analysis should reveal if the organic Se species have the same isotope level than HgSe particles. This has so far not been done using an enzymatic extraction, but instead measuring the Se isotope ratio Hg isotope fractionation was performed (Bolea-Fernandez et al. 2019). The study revealed that the HgSe nanoparticles are formed in the same tissue as they were found. Hence, the Hg/Se nanoparticles are not transported from one organ to the other inside the whale.

Whether the size and the number of Hg/Se particles are different in the different age groups could be analyzed by using single-particle ICPMS (*sp*ICPMS). Quadrupole ICPMS can only measure one isotope in *sp*ICPMS mode. This analysis was done for Se and Hg in sequential measurement of an enzymatic extract of the tissue liver and brain (Gajdosechova et al. 2016). When this technique is used and more than 1000 particles are measured, then the number of particles /g tissue and the maximum size and size distribution can be determined. The analysis of the liver and brain extract in the pilot whale group delivered an interesting result: The number of Hg/Se nanoparticles increased with age in the pilot whale liver and brain. Additionally, the maximum size of the particles was larger for older than younger pilot whales (Gajdosechova et al. 2016). Whether the formation of Hg/Se nanoparticles is a successful Hg detoxification method of the pilot whales or the whale is getting selenium-deficient cannot directly be answered. For this answer, it would be necessary to determine the individual essential selenoproteins.

In summary the different case studies show that it is necessary to use a whole array of analytical techniques to determine how Se is distributed, and in which molecular species.

REFERENCES

Aborode, F.A., Raab, A., Foster, S. et al. 2015. Selenopeptides and elemental selenium in *Thunbergia alata* after exposure to selenite: a quantification method for elemental selenium. *Metallomics* 7:1056–1066.

Aborode, F.A., Raab, A., Voigt, M., Malta-Costa, L., Krupp, E.M. & Feldmann, J. 2016. The importance of glutathione and phytochelatins on the selenite and arsenate detoxification in *Arabidopsis thaliana*. *J Environ Sci* 49: 150–161.

Amayo, K.O., Pettursdottir, A., Newcombe, C. et al. 2011. Identification and quantification of arsenolipids using reversed-phase HPLC coupled simultaneously to high-resolution ICPMS and high-resolution electrospray MS without species-specific standards. *Anal Chem* 83: 3589–3595.

Bolea-Fernandez, E., Rua-Ibarz, A., Krupp, E.M, Feldmann, J. & Vanhaecke, F. 2019. High precision isotopic analysis sheds new light on mercury metabolism in long-finned pilot whales (*Globicephala melas*). *Sci Rep* 9: 7262.

Bierla, K., Lobinski, R. & Szpunar, J. 2018. Determination of proteinaceous selenocysteine in selenized yeast. *Int J Mol Sci* 19:543.

Bluemlein, K., Klimm, E., Raab, A. & Feldmann, J. 2009. Selenite enhances arsenate toxicity in *Thunbergia alata. Environ Chem* 6:486–494.

Brookins, D.G. 1988. *Eh-pH Diagrams for Geochemistry.* Berlin: Springer

Gajdosechova, Z., Lawan, M.M., Urgast, D.S. et al. 2016. In vivo formation of natural HgSe nanoparticles in the liver and brain of pilot whales. *Sci Rep* 6: 34361.

Rondan, F.S., Henn, A.S., Mello, P.A. et al. 2019. Determination of Se and Te in coal at ultra-trace level by ICP-MS after microwave-induced combustion *J Anal At Spectrom* 34: 998–1004.

Wallschläger, D. & Feldmann, J. 2010. Formation, occurrence significance and analysis of organoselenium and organotellurium compounds in the environment, In A. Sigel, H. Sigel & R.K.O. Sigel (eds), *Metals in Life Sciences (Met. Ions Life Sci), Organometallics in Environment and Toxicology* Nr **7:** 319–364. Hoboken, NJ: Wiley.

Selenium nanoparticles

Selenium Research for Environment and Human Health: Perspectives, Technologies and Advancements – Bañuelos, Lin, Liang & Yin (eds)
© 2020 Taylor and Francis Group, London, ISBN 978-1-138-39014-0

Biogenic selenium nanoparticles as co-catalysts for enhancing photocatalytic activity of Se-ZnO nanocomposites

A. Vaishnav, R. Prakash & B. Pal
School of Chemistry and Biochemistry, Thapar Institute of Engineering and Technology, Patiala, India

N. Joshi
Department of Biotechnology, Thapar Institute of Engineering and Technology, Patiala, India

N. Tejo Prakash
School of Energy and Environment, Thapar Institute of Engineering and Technology, Patiala, India

1 INTRODUCTION

This study deals with the biosynthesis of selenium (Se) nanoparticles and their impregnation as co-catalyst on ZnO to prepare Se-ZnO nano-catalyst for studying its photocatlytic activity. Selenium nanoparticles was isolated from Se-tolerant bacteria using trypton soya broth supplemented with Na_2SeO_4. The prepared Se-ZnO phtocatalyst was characterized for its physicochemical properties using UV-DRS, DLS, TEM, SEM-EDS and XRD. The potential of biosynthesized SeZnO was tested against the photocatalytic degradation of recalcitrant compound 4-chloroguaiacol and real pharmaceutical industry effluent under sunlight irradiation (Fig. 1). Various concentrations of Se were doped on ZnO semiconductor, and 0.5wt% was found to be the most optimum. The effect of photocatalyst on different concentrations of model compound was also evaluated. Kinetic study of degradation reaction was performed and it was observed to follow first-order reaction. Evolution of 88% of CO_2 under solar radiations further confirms the complete mineralization of the compound.

2 MATERIALS AND METHODS

2.1 *Isolation of Se-tolerant bacterial strains*

Selenium tolerant bacterial strains were isolated from soil samples collected from Se-affected agricultural land geographically located at 31°13′N, 76°21′E, in the Jainpur village of Nawanshahar district, Punjab, India. Soil samples were inoculated into flasks containing sterile Trypton Soya Broth (TSB) and incubated for 24 h. Bacterial suspension was spread aseptically over petri dish (90 mm) containing Trypton Soya Agar (TSA) medium supplemented with Se (as sodium selenite). Bacteria having selenate ion reducing capability formed visually distinct red or orange-red colonies due to enzymatic reduction of selenate ion into red color, amorphous, elemental SeNPs (Ahluwalia et al. 2016).

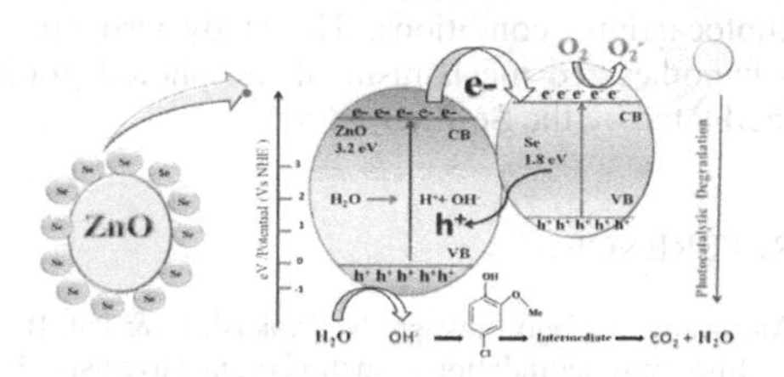

Figure 1. Scheme elucidating the proposed mechanism of photocatalytic reaction.

2.2 *Biosynthesis and isolation of extracellular SeNPs*

For the screening of bacterial strain that produces extracellular SeNPs, culture broth was cell-free supernatants showing orange-red color colonies, which confirmed the presence of extracellular SeNPs. For isolation of extracellular Se NPs, the supernatant containing nanoparticles was further centrifuged at 10,000 rpm for 10 mins. The obtained Se NPs pellet was further washed and stored as suspension in distilled water at 4°C.

2.3 *Synthesis of Se-impregnated ZnO nanocomposites (Se-ZnO-NCs)*

Commercially available ZnO was mixed with distilled water under continuous stirring. Different volumes of SeNPs suspension were added dropwise to the ZnO solution and kept for overnight stirring. The Se-ZnO nanocomposites (NCs) were recovered through centrifugation and further washed three times with distilled water. The nanocomposite was dried and crushed into fine powder. The powder obtained was further sintered in muffle furnace at 200°C. Different wt% Se-doped ZnO photocatalysts were prepared by varying the amount (100 μl to 400 μl) of SeNPs.

2.4 *Photo degradation of 4-chloroguaiacol (4-CG)*

4-chloroguiacol treated with ZnO and Se-ZnO was used to study the enhanced degradation activity of Se-impregrenated ZnO photocatalyst. The ZnO and different wt % Se doped photocatalyst was weighed

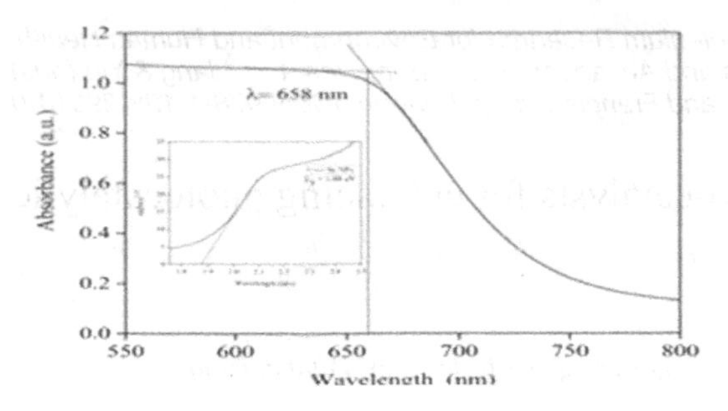

Figure 2. UV visible spectrum of SeNPs corresponding to particle size.

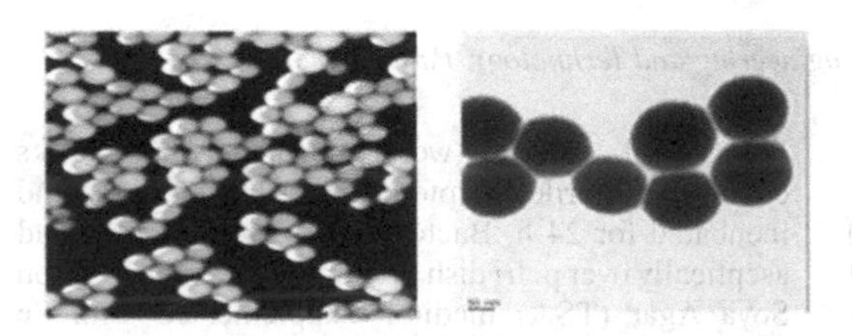

Figure 3. SEM (left) and TEM (right) images of SeNPs.

and suspended in 4-choroguaiacol solution. The photocatalytic activity was carried out in test tubes containing 20 μg/ml 4-CG solution (from the stock solution of 1 mg/ml) and kept under solar radiation with intensity of 791 ± 20 Wm^2, 39 ± 2°C for 15 min with continuous stirring. At different times, samples were taken. The supernatant was filtered, and absorbance of the supernatant was measured by scanning the sample between the range 200–700 nm in the UV-Visible spectrophotometer.

2.5 *Characterization of nanomaterials*

Diverse characterization techniques viz., UV-visible spectroscopy, DLS scattering, x-ray diffraction, SEM, and HR-TEM were used to characterize the nanoparticles as well as the degradation profile of 4-CG.

3 RESULTS AND DISCUSSION

The UV-Vis spectrometric analysis, shown in Figure 2, was done by scanning the SeNPs suspension in the range of 200–1100 nm. The characteristic absorbance peak was observed at 658 nm which corresponds to particle size of 160 ± 10.2 nm.

Spherical shaped SeNPs were observed with uniform distribution. The SeNPs were in size range of 130-160 nm. Highly monodisperse nature of the particles was seen. The following STEM and HR-TEM images (Fig. 3), supported by EDX of the prepared SeNPs, confirm the presence of Se in the solution.

On sintering biogenic SeNPs with ZnO, a decrease in the Energy gap (E_g) (2.8 eV) was observed for Se-ZnO NCs compared to pure ZnO. It was observed that as the amount of Se doping over ZnO was increased from 0.5 to 2 wt%, the Energy gap further decreased to 2.72 eV. The crystalline structure of synthesized photocatalyst was confirmed using X-ray diffractometer. The XRD result showed the successful synthesis of Se-ZnO NCs through the sintering process. The crystalline structure of synthesized Se-ZnO NC photocatalyst was confirmed using X-ray diffractometer. The strong diffraction peak appearing at 30.30° (101) corresponded to the hexagonal selenium phase. Hence, the XRD result showed the successful synthesis of Se-ZnO NCs (Fig. 4).

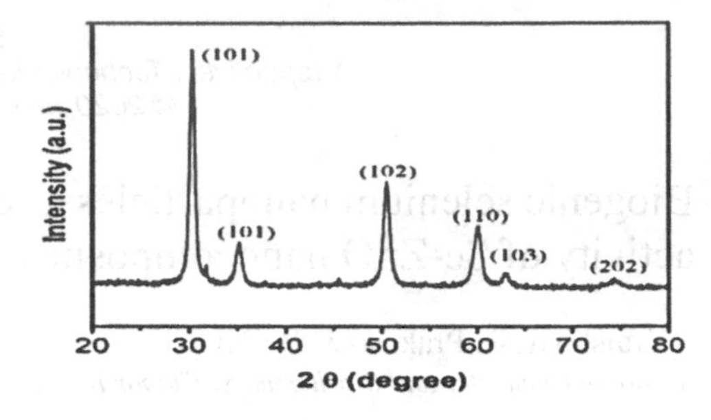

Figure 4. XRD spectrum of Se-ZnO composite.

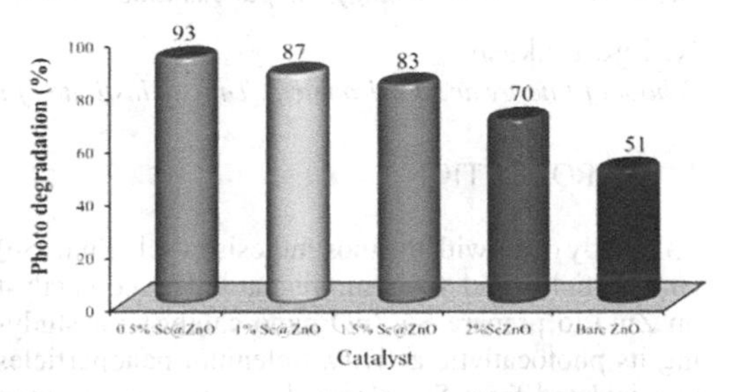

Figure 5. Percent degradation of 4-CG at varying of Se in ZnO.

The absorption spectrum of 4-chloroguiacol of 20 μg/ml photo degradation by Se-ZnO indicated a 93% degradation 4-chloroguiacol with 0.5%Se-ZnO after 15 min sunlight irradiation (Fig. 5).

4 CONCLUSIONS

A green and eco-friendly biological approach was used to synthesize SeNPs by bio-reduction of selenite using *Bacillus sp*. Application of synthesized NCs in photocatalytic degradation of 4-chloroguiacol demonstrated significant levels of degradation under solar photocatalytic conditions. The study also presented a hypothesized mechanism of the photodegradation facilitated by the Se-ZnO-NCs.

REFERENCE

Ahluwalia, S., Tejo Prakash, N., Prakash R. & Pal, B. 2016. Improved degradation of methyl orange dye using bio co-catalyst Se nanoparticles impregnated ZnS photocatalyst under UV radiation. *Chem Eng J* 306: 1041–1048.

Selenium Research for Environment and Human Health:
Perspectives, Technologies and Advancements – Bañuelos, Lin, Liang & Yin (eds)

Insights into the surface chemistry of BioSeNPs produced by *Bacillus lichenformis*

Y.Q. Yuan* & R.L. Qiu
School of Environmental Science and Engineering, Sun Yat-sen University, Guangzhou, China
Guangdong Provincial Key Laboratory of Environmental Pollution and Remediation Technology, Guangzhou, China

J.M. Zhu & C.Q. Liu
State Key Laboratory of Environmental Geochemistry, Institute of Geochemistry, Chinese Academy of Sciences, Guiyang, China

1 INTRODUCTION

The two main Se oxyanions (SeO_3^{2-} and SeO_4^{2-}) in the environment have received considerable public concern, due to the bioavailability and potential to be toxic to biological systems. Strategies are needed to remediate or immobilize the oxyanion forms of Se. Microorganisms play an important role in Se transformations from soluble toxic forms to insoluble non-toxic Se(0), which is considered a promising technology for Se remediation (Nancharaiah et al. 2016). Furthermore, others have confirmed that bioreduction of Se oxyanions usually formed elemental Se nanoparticles (~50–500 nm; SeNPs) (Jain et al. 2015).

However, the biogenic SeNPs (BioSeNPs) are in stable colloidal spherical forms, quite different from their allotrope with crystalline structures (Lenz et al. 2011). Jain et al. (2015) have reported that produced BioSeNPs are usually coated with an organic layer, e.g. extracellular polymeric substances (EPS). To understand how organic molecules regulated BioSeNPs formation, it is critical to understand what biogenic Se(0) is. Thus, we investigated the properties of the extracellular SeNPs produced by a model of Se(IV)-reducing bacterium *Bacillus licheniformis* in an effort to understand how these properties facilitate formation of BioSeNPs.

2 MATERIALS AND METHODS

2.1 *Bacillus lichenformis cultures*

Bacillus lichenformis strain SeRB-1 was used in this study (Yuan et al. 2014). For routine experiments, the isolate was grown in LB medium at 37°C with a shaking speed of 180 rpm (Sukun shaker, SKY −200B, Shanghai). Primary cultures from cryo-stocks were transferred twice into fresh LB medium for about ~24 h to ensure cultures at exponential phase, and then they were used in the subsequent analyses (Yuan et al. 2015).

2.2 *BioSeNPs production and purification*

The experiments were carried out in 100-ml glass serum bottles with 80 ml LB medium. Ten percent of exponential growth cell culture was added to the medium, and incubated at 37°C with constant shaking. Sodium selenite (10 mM) was supplied as electron acceptors. Meanwhile, EPS was extracted by sonication and centrifugation (Desmond et al. 2018) with a modification. This method was used to produce SeNPs with the same method for intact cells when extracellular reduction was a vital detoxification route (Jain et al. 2015). Samples were collected from growing cultures at about 24 h and centrifuged at 10,600 × g for 10 min at 4°C to separate the pellets and supernatant. The pellet was purified by rinsing in 1M Tris-HCl buffer (pH 7.2) three times before characterization studies.

2.3 *Selenium nanoparticles characterization*

Pellets used for microscopy analysis were treated by the methods described by Yuan et al. (2014) with a Tecnai G20 TWIN (FEI, USA) transmission electron microscope (TEM), attached with an Oxford-INCAX EDS analysis system (Oxford Instruments Analytical, UK) at 200 kV. Selected-area electron diffraction (SAED) pattern was performed in the diffraction mode. The functional groups of proteins and carbohydrates on the SeNPs were characterized by fourier transform infrared spectrometer (FT-IR) (IS10, USA) at 4000–400 cm^{-1} in the transmittance mode, with a spectral resolution of 4 cm^{-1} and an average of 32 scans (Jain et al. 2015).

3 RESULTS AND DISCUSSION

3.1 *TEM-EDS analysis of BioSeNPs*

Similarly, TEM images showed 10 to 250 nm granules formed both extracellularly (mainly) and intracellularly, and the nanoparticles were attached with EPS (Fig. 1a) (Jain et al. 2015). EDS analysis confirmed Se was the main element of the nanoparticles. The

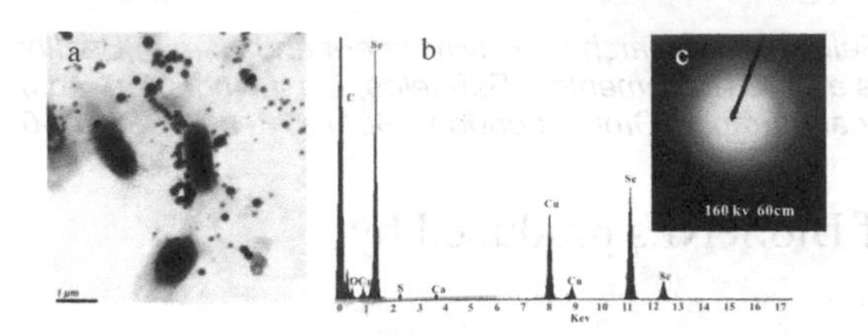

Figure 1. Electron micrographs and EDS analysis of the pellets at 24 h. (a) TEM, (b) EDS, (c) SAED (inset).

presence of C, O, S, and Ca might be attributed to the EPS coating of the BioSeNPs (Fig. 1b). SAED analysis showed the extracellular BioSeNPs were mainlyamorphous (Fig. 1c), probably indicating the transient precursor phase of crystalline Se(0) (Jin et al. 2018).

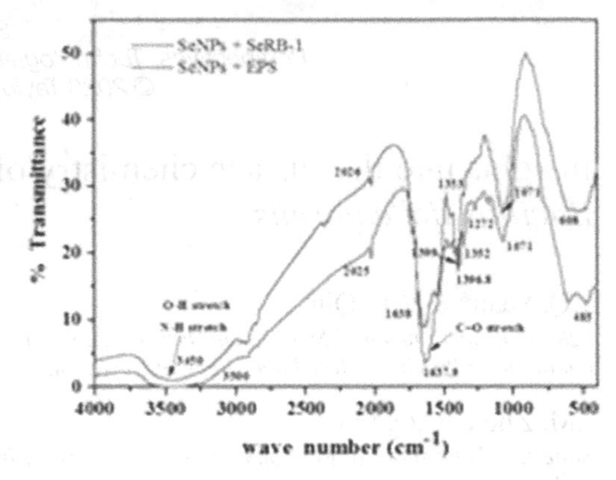

Figure 2. FT-IR spectra of the BioSeNPs (pink purple) and SeNPs produced by EPS (green) at 24 h.

3.2 *FT-IR analysis of BioSeNPs*

The analysis showed that the FT-IR spectra of SeNPs produced by intact cells of SeRB-1 and EPS were similar. Both SeNPs had -OH and -NH stretching vibrations of amine and carboxylic group at around 3450 cm^{-1} (Jain et al. 2015). Strong feature of C=O stretching vibration of proteins (amide I) was observed at 1638 cm^{-1} (Xu et al. 2011). The weak presence of C-N stretching and N-H bending vibrations of proteins (amide III) were also observed at around 1250 cm^{-1} (Wang et al. 2012), whereas the feature at 1399 $cm^{-1}$is probably indicating -OH bending in carboxylic acid or -COO- symmetric stretching (Xu et al. 2011). Another strong peak at 1071 cm^{-1}might be attributed to C-O-C and C-H stretching from the carbohydrate groups (Xu et al. 2011, Jain et al. 2015). Another broad feature at around 700 - 600 cm^{-1} might correspond to CH out of aromatic ring or CH_2 bending of polysaccharides, proteins and lipids (Xu et al. 2011). The small spectral deviations of the two SeNPs might be due to more functional groups from intact cells (e.g. P=O at 1240 cm^{-1}; CH_2 vibrations at 485 cm^{-1}) (Jain et al. 2015, Wang et al. 2012). FT-IR spectroscopy confirmed the presence of proteins (amide I, 1638 cm^{-1}; amide III, 1350–1250 cm^{-1}) and carbohydrates (1071 cm^{-1}), and possible lipids (~3450 cm^{-1} and 700–600 cm^{-1}) present on the BioSeNPs.

4 CONCLUSIONS

Extracellular biomineralization of BioSeNPs is one of the significant detoxification mechanisms of toxic Se oxyanions. The presence of functional groups of carbohydrates and proteins on EPS produced SeNPs further verified extracellular Se(0) biomineralization mechanism, and thus further confirmed the layer on BioSeNPs might be from EPS (Jain et al. 2015). The surface organic layers might make the spherical BioSeNPs stable in the colloidal suspension against crystallization. Further investigations on how the extracellular biomacromolecular regulate mineral precursor's assembly and mineral growth and structure is necessary, which will help further elucidate phase transformation pathways concerning biomineralization.

ACKNOWLEDGEMENT

This work was supported by the Natural Science Foundation of China (41703073) and the Fundamental Research Funds for Central Universities (18lgpy44).

REFERENCES

Desmond, P., Best, J.P., Morgenroth, E. et al. 2018. Linking composition of extracellular polymeric substances (EPS) to the physical structure and hydraulic resistance of membrane biofilms. *Water Res* 132: 211–221.

Jain, R., Jordan, N., Weiss, N. et al. 2015. Extracellular polymeric substances govern the surface charge of biogenic elemental selenium nanoparticles. *Environ Sci Tech* 49: 1713–1720.

Jin, W.J., Jiang, S.Q., Pan, H.H. et al. 2018. Amorphous phase mediated crystal-lization: Fundamentals of biomineralization. *Crystals* 8(1): 48.

Lenz, M., Kolvenbach, B., Gygax, B. et al. 2011. Shedding light on selenium bio-mineralization: Proteins associated with bionano-minerals. *Appl Environ Microbiol* 77: 4676–4680.

Nancharaiah, Y.V., Mohan, S.V. & Lens, P.N.L. 2016. Biological and bioelectro-chemical recovery of critical and scarce metals. *Trends Biotechnol* 34: 137–155.

Wang, L.L., Wang, L.F. & Ren, X.M. 2012. pH dependence of structure and surface properties of microbial EPS. *Environ Sci Tech* 46: 737–744.

Xu, C., Zhang, S.J., Chuang, C.Y. et al. 2011. Chemical composition and relative hydro-phobicity of microbial exopolymeric substances (EPS) isolated by anion exchange chromatography and their actinide-binding affinities. *Marine Chem* 126: 27–36.

Yuan, Y.Q., Zhu, J.M., Liu, C.Q. et al. 2014. Three high-reducing selenite tolerance bacteria from Se-laden carbonaceous mudstone. *Earth Sci Front* 21: 331–341.

Yuan, Y.Q., Zhu, J.M., Liu, C.Q. et al. 2015. Biomineralization of Se nanoshpere by *Bacillus Licheniformis*. *J Earth Sci* 26: 246–250.

Selenium Research for Environment and Human Health: Perspectives, Technologies and Advancements – Bañuelos, Lin, Liang & Yin (eds)
 ISBN 978-1-138-39014-0

Poria cocos polysaccharide decorated selenium nanoparticles attenuate colitis by suppressing hyper inflammation

H. Yang, Y.F. Yang, L.Q. Duan, Q.J. Ling & Z. Huang*
College of Life Science and Technology, Jinan University, Guangzhou, Guangdong, China

1 INTRODUCTION

Selenium (Se) is an essential trace element for mammals. Inadequate Se can aggravate the development of inflammatory bowel diseases (IBD). Selenium nanoparticles (SeNPs) are characterized by low toxicity and anti-inflammatory properties with functional modifications. In this study, we attached SeNPs with natural polysaccharides from *Poria cocos* polysaccharide (Pachyman, PYN), which has an anti-inflammatory effect. Herein, we explored PYN-SeNPs potential beneficial effects on IBD by the mice model subjected to acute colitis.

2 MATERIALS AND METHODS

2.1 *Materials and chemicals*

Poria cocos polysaccharide (PYN) was purchased from Elicityl (Grenoble, France). Antibodies included anti-COX-2 and anti-TNF-α were obtained from Cell Signaling Technology. All chemicals were of analytical grade and purchased from Sigma.

2.2 *Preparation and characterization of PYN-SeNPs*

We first attached SeNPs with PYN to synthesize PYN-SeNPs. We confirmed the particle size and dispersity by dynamic light scattering (DLS) and transmission electron microscopy (TEM). Cell uptake of PYN-SeNPs and SeNPs during the time course was compared and evidenced by images taken by fluorescence microscope.

2.3 *Effect of PYN-SeNPs on the DSS-treated mice*

Acute colitis mice were treated with/without PYN-SeNPs and the body weights were measured daily. After 11 days, mice were humanely euthanized, and colon lengths were measured. The disease activity index of the colitis was evaluated. Meanwhile, Hematoxylin-Eosin (H&E) staining of colon tissues was conducted to show the effect of PYN-SeNPs on colonic tissue injury in colitis. In addition, immunohistochemistry and immunofluorescence were conducted to show the effect of PYN-SeNPs on inflammatory markers in Dextran Sodium Sulfate (DSS)-induced colitis.

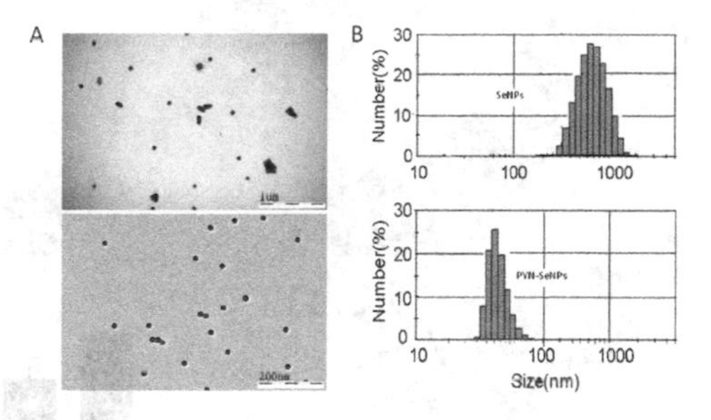

Figure 1. Preparation and characterization of PYN-SeNPs.

3 RESULTS AND DISCUSSION

3.1 *Characterization of PYN-SeNPs*

As shown in Figure 1a, the particle-size distribution of PYN-SeNPs was 35–80 nm, which is smaller compared to 305–900 nm of SeNPs. The dispersity of PYN-SeNPs is also better than SeNPs itself (Fig. 1b). The cellular uptake of PYN-SeNPs in Bone Marrow Derived Macrophage (BMDM) was in a time-dependent manner, and BMDM macrophages uptake much more PYN–SeNPs than PYN.

3.2 *Effects of PYN-SeNPs on the development of DSS induced colitis in mice*

As shown in Figure 2A, the disease activity index (DAI) was significantly lowered in the PYN-SeNPs group compare to DSS group, and Figures 2b, c show that PYN-SeNPs could reduce the effect of DSS on the shortening of colon length in mice.

3.3 *Effect of PYN-SeNPs on the colonic damage in DSS treated mice*

The effect of PYN-SeNPs on colitis was detected by H&E staining. Compared to NC group, the DSS group shows severe histological damage: Inflammatory response was observed, epithelial cells were

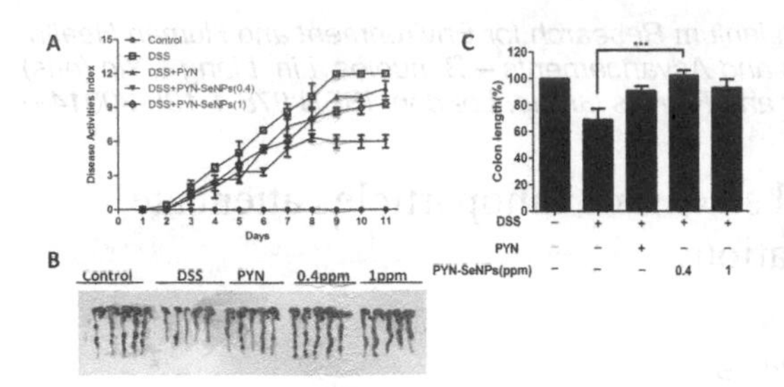

Figure 2. PYN-SeNPs relieve the clinical symptoms of DSS-induced mouse colitis.

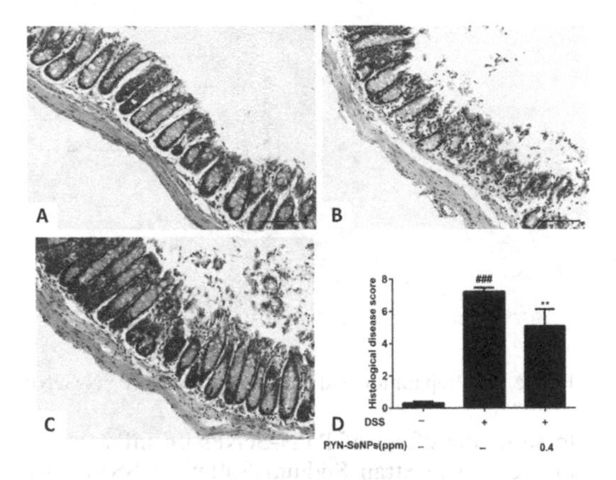

Figure 3. Effects of PYN-SeNPs on colonic tissue injury in colitis.

disrupted, muscle layer became thicker, goblet cells were lost, and severe inflammatory cell infiltration occurred. However, the group-obtained PYN-SeNPs show a reduced degree of histological damage and a reduced inflammatory cell infiltration, along with a milder loss of epithelial cells and less expansion of the lesion area in the colon tissue (Fig 3a-c). Consistent with the pathological changes in colons, PYN-SeNPs treatment significantly reduced the total histological score of the colitis colon tissues (Fig. 3d).

3.4 *PYN-SeNPs affect the expression of inflammatory markers in DSS–treated mice*

As shown in Figure 4a, the amount of TNF-α in the DSS group was significantly increased compared with the NC group (Fig. 4b). COX-2 (Fig. 5) consistently shows the same trend. The quantifications are shown in Figures 4d and 5d.

4 CONCLUSIONS

PYN-SeNPs supplementation may offer therapeutic potential to reduce the symptoms of acute colitis through its anti-inflammatory actions.

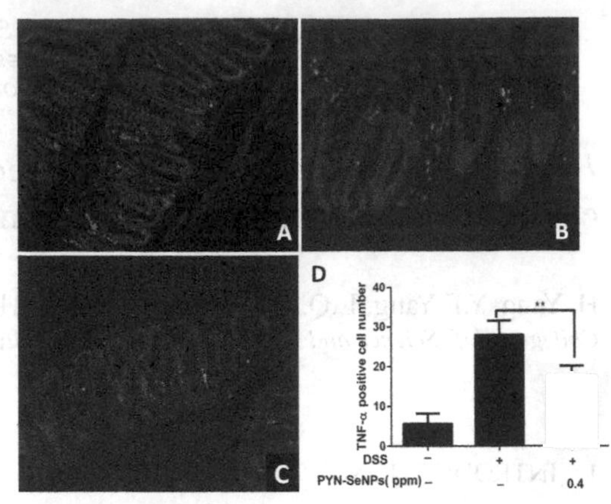

Figure 4. Effects of PYN-SeNPs on inflammatory marker TNF-α in DSS-induced colitis.

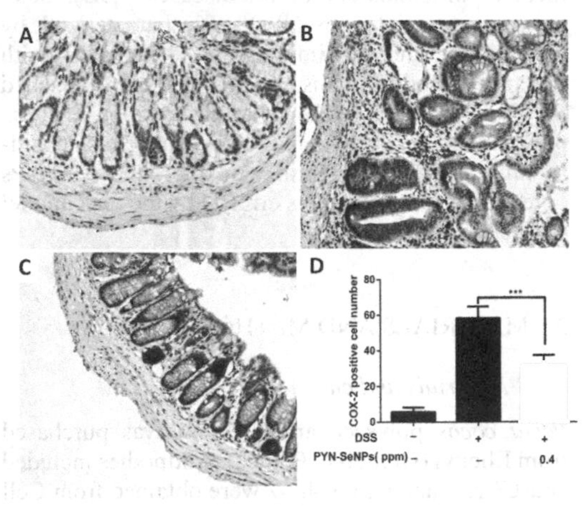

Figure 5. Effects of PYN-SeNPs on inflammatory marker COX-2 in DSS-induced colitis.

REFERENCES

Huang, Z., Rose, A.H. & Hoffmann, P.R. 2012. The role of selenium in inflammation and immunity: from molecular mechanisms to therapeutic opportunities. *Antioxid Redox Signal* 16(7): 705–743.

Huang, Z., Rose, A.H., Hoffmann, F.W. et al. 2013. Calpastatin prevents NF-κB-mediated hyperactivation of macrophages and attenuates colitis. *J Immunol* 191(7): 3778–88.

Zhu, C., Zhang, S., Song, C. et al. 2017. Selenium nanoparticles decorated with Ulva lactuca polysaccharide potentially attenuate Colitis by inhibiting NF-κB mediated hyper inflammation. *J Nanobiotechnol* 15(1): 20.

Selenium Research for Environment and Human Health: Perspectives, Technologies and Advancements – Bañuelos, Lin, Liang & Yin (eds)
© 2020 Taylor and Francis Group, London, ISBN 978-1-138-39014-0

Biotransformation and volatilization of nanoscale elemental selenium

J. Wang, F.B. Joseph & R. Uppala
Department of Environmental Sciences, Southern Illinois University-Edwardsville, Edwardsville, Illinois, USA

Z.-Q. Lin*
Department of Environmental Sciences and Department of Biological Sciences, Southern Illinois University-Edwardsville, Edwardsville, Illinois, USA

1 INTRODUCTION

With the rapid development of nanotechnology, nanoscale elemental selenium (Se) particles (SeNPs) have been applied for various uses, especially in biomedicine, electronics, catalysis, and food supplement production. Thus, there is increasing concern for SeNPs as emerging contaminants released into the environment. Due to their unique physical and chemical properties, Se nanoparticles (<100 nm in diameter) may pose potential adverse impacts on the environment. However, little is known about the chemical behaviors of SeNP contaminants in the environment. In particular, the biological processes that control the transport and fate of SeNPs in the soil-plant system have not been well elucidated. Some major research questions need to be answered experimentally. Will SeNPs in soil be bioavailable to soil microbes and plants? Will SeNPs be biologically transformed into other Se compounds after their release into the soil-plant system? We speculated that SeNPs could be significantly biotransformed in the soil-plant system by microorganisms such as fungi or soil bacteria that associate with certain plant species. This presentation will report some of research findings obtained from a series of case studies conducted in recent years.

2 CHEMICAL CHANGES OF SELENIUM NANOPARTICLES IN THE SOIL ENVIRONMENT

To determine the chemical changes of SeNPs in the soil environment, laboratory experiments were conducted using chemically synthesized elemental SeNPs with a diameter of 17 to 69 nm. We have evaluated the stability and partitioning of elemental SeNPs among different fractions in the soils containing different levels of soil organic matter. The fractionation and speciation analyses of Se in soil treated with different levels of elemental SeNPs indicated that less Se from SeNPs was found in the exchangeable fraction with increasing soil incubation time. Water soluble and exchangeable Se concentrations tended to change to more stable organic matter (OM)-bound and residual Se during the soil incubation period. Thus, increasing interaction time of SeNPs with soil components, environmental impacts on Se from SeNPs might potentially decrease in the soil environment. Soil OM content has a significant effect on bioavailability of SeNPs, likely through adsorption, complexation, and aggregation. Elemental SeNPs showed different chemical behaviors compared to selenate (SeO_4) in soil. With low soil OM content, concentrations of water soluble Se in the SeNP-treated soil were significantly ($p < 0.05$) lower than that in SeO_4-treated soil but were higher than those in SeO_4-treated soil containing high OM content.

Selenium speciation analysis using HPLC-ICP/MS and XANES suggested that, in soils treated with SeNPs, nanoscale elemental Se was transformed into other chemical forms, e.g. selenite and selenate, during the soil incubation process.

3 BIOTRANSFORMATION OF NANOSCALE ELEMENTAL SELENIUM TO VOLATILE SE BY SOIL BACTERIA

High levels of Se volatilization were previously observed in fields growing *Stanleya pinnata* (Freeman & Bañuelos 2011). Under laboratory conditions, rates of Se volatilization from a soil-*S. pinnata* system treated with different levels of SeNPs were determined compared to selenate and selenite. In addition, the cumulative Se mass volatilized from the soil-root compartment (88.6 ± 10.6 μg/pot) was significantly ($p < 0.05$) higher than from the shoots (0.2 ± 0.1 μg/pot) during the same experimental time period. These results demonstrated that soil microbes play an important role in the volatilization of SeNPs in the soil-*S. pinnata* system. The microbial volatilization of SeNPs by *Pseudomonas fuscovaginae* that was previously isolated from the rhizosphere soil of *S. pinnata* was dependent on Se concentration in the substrate. In comparing different chemical forms of Se in the substrate, the Se volatilization rate from SeNPs by *P. fuscovaginae* was lower than from selenate but higher than from bulk elemental Se.

Salicornia bigelovii and *Polypogon monspeliensis* have been previously identified as good Se volatilization plant species (Lin & Terry, 2002, Lin et al. 2001). *Stanleya pinnata*-associated *P. fuscovaginae* also exhibited higher rates of Se volatilization from SeNPs than *S. bigelovii*-associated soil bacterium *Corynebacterium propinquum* but was similar with *Cellulomonas cellasea* isolated from the rhizosphere soil of *P. monspeliensis*.

Selenium speciation analysis by XANES showed that *P. fuscovaginae* was able to biotransform SeNPs into organic forms of Se, including ~67.5% of methylselenocysteine (MeSeCys), and ~15.5% selenocysteine (SeCys), along with small portions of selenate and selenite (~8% each) in *P. fuscovaginae* cultural solution, while the dominant chemical form of Se with the bacterial cells was elemental Se. In conclusion, our results demonstrated that SeNPs in the soil-plant system have partially become bioavailable and further biotransformed into volatile Se compounds.

4 BIOTANFORMATION AND VOLATILIZATION OF NANOSCALE ELEMENTAL SELENIUM BY MACROFUNGI

Effects of different chemical forms of Se, including SeNPs on bioaccumulation and volatilization of Se by mycelium tissues of different fungal species have been examined, including Reishi (*Ganoderma lucidum*), Shiitake (*Lentinula edodes*), Lion's mane (*Hericium erinaceus*), Oyster Pearl (*Pleurotus ostreatus*), and Oyster Blue (*Pleurotus columbinus*). Our results showed that significant amounts of Se accumulated in mycelium tissues but varied significantly ($p < 0.05$) among those species. Among different Se chemical species, concentrations of Se in Oyster Blue mycelia were 155.99 ± 30.47 mg/kg when the growth substrate was treated with SeNPs of 10 mg Se/L, compared to the Se concentration of 315 ± 19.4 mg/kg when the substrate was treated with Na_2SeO_4. Concentrations of Se accumulated in mycelium tissues of Reishi and Oyster Blue increased with increasing level of Se in growth media, from 1, 2.5, 5, to 10 mg/L in the form of selenate (SeO_4) or SeNPs. In addition, SeNPs can also be biotransformed and volatilized by fungal mycelia. Oyster Blue volatilized significant amounts of Se at a rate of 6449 ng/flask/day when the substrate was treated with sodium selenate but at a much lower volatilization rate of 6.5 ng/flask/day when the substrate was treated with SeNPs.

REFERENCES

Freeman, J.L. & Bañuelos, G.S. 2011. Selection of salt and boron tolerant selenium hyperaccumulator Stanleya pinnata genotypes and characterization of Se phytoremediation from agricultural drainage sediment. *Environ Sci Technol* 45: 9703–9710.

Lin, Z.-Q. & Terry, N. 2003. Selenium removal by constructed wetlands: quantitative importance of biological volatilization in the treatment of selenium-laden agricultural drainage water. *Environ Sci Technol* 37(3): 606–615.

Lin, Z.-Q., Cervinka, V., Pickering, I.J., Zayed, A. & Terry, N. 2002. Managing selenium-contaminated agricultural drainage water by the integrated on-farm drainage management system: Role of selenium volatilization. *Water Res* 36: 3150–3160.

Author index

9781032570617